BROMELIADS

For Home, Garden and Greenhouse

BROMELIADS

For Home, Garden and Greenhouse

Werner Rauh

Institute of Systematic Botany
University of Heidelberg, Germany

with

Herbert Lehmann, Julien Marnier-Lapostolle
and Richard Oeser

Edited by
Peter Temple, RIBA, PPInst RA, FRSA
British Bromeliad Society

Translated by
Peter Temple and Harvey L. Kendall

BLANDFORD PRESS

Poole   Dorset

First published in the U.K. by Blandford Press Ltd.,
Link House, West Street,
Poole, Dorset, BH15 1LL

English language edition copyright ©1979 Blandford Press Ltd.

Rauh, Werner
  Bromeliads for home, garden and greenhouse.
  1. Bromeliaceae
  I. Title
  635.9′34′22   SB413.B7

ISBN 0 7137 0845 X

Originally published in Germany as *Bromelien*
World copyright ©1970 (Vol. 1) and 1973 (Vol. 2) Verlag Eugen Ulmer, Stuttgart

Filmset by Keyspools Ltd., Golborne, Lancashire
Printed in Great Britain by Butler & Tanner Ltd, Frome and London

Contents

In recognition of his great service to bromeliad research, this book is dedicated with grateful respect to L. B. Smith, Senior Botanist at the Smithsonian Institution, Washington, U.S.A. His close co-operation has united us in our efforts.

Preface

Hardly any other tropical plants have had such a difficult history before finding their way into our homes and in gaining acceptance as house or hobbyists' plants. Not until window greenhouses and conservatories became popular, and not until the interest in the construction of epiphyte 'trees', terrariums or inexpensive miniature greenhouses did the interest in bromeliads really begin. With their rigid, firm, almost sword-shaped, beautifully marked leaves and their bizarre forms, the bromeliads were really predestined for modern style bright, spacious dwellings. With the colourful splendour of their inflorescences, which often last for months, they bring into homes that exotic 'breath of tropical air'.

However, the selection offered to us by flower shops and nurseries is still relatively small in comparison with the actual extent of the family. Of the nearly 2,000 species known today, at least 500 make good house plants. Yet of these, relatively few species, plus a number of hybrids, comprise the available assortment of those commercial nurseries who specialize in the cultivation of bromeliads. Only someone who knows and has travelled in places which make up the native habitats of bromeliads can comprehend the enormous extent, multiplicity and beauty of these plants. If nurseries and florists would import large amounts of bromeliads, especially the grey-leafed tillandsias, the authors translators and editor of this book are convinced that the enthusiasm for bromeliads would experience an upsurge similar to that which cacti have enjoyed in recent decades. Indeed, the peculiar lifestyle of bromeliads means that they require much less care than most plants of tropical origin. They do not even need a pot! A piece of wood to which the plants can be attached and a piece of wire to hang them up is often all that is needed.

The amateur and professional grower's lack of knowledge about this remarkable plant family, whose value as house plants has long been recognized, may be traced back to the fact that, in contrast to orchids, cacti and other house plants, there are few books to help with any selection or details of culture and care. The authors therefore gladly accepted the suggestion to produce a book which would fill this gap. The work was originally produced in German as two volumes. The first volume dealt with the subfamily *Tillandsioideae*; within this group there are some 400 species of tillandsias, which are especially notable since they are not only the most interesting to the serious collector, but also offer many miniature forms for enthusiasts who posses only limited space for cultivation. The second volume contained the remaining subfamilies, the *Pitcairnioideae* and the *Bromelioideae*. The two volumes are now presented here as a single unit.

May the present work contribute to the expansion and the deepening of the love of bromeliads. If it introduces new members into the already existing circle of bromeliad fanciers, then it will have fulfilled its purpose.

Werner Rauh and Collaborators
Heidelberg, 1969

Author's Note

Whilst this English language edition was in the final stages of being prepared for press, some changes in nomenclature and my further researches indicated that one or two species were incorrectly described and identified in the original volumes. I thus take the opportunity of adding this brief note of correction and ask that readers consult the following paragraphs when using the book.

Plate 5 shows *T. kalmbacheri* Matuda and not *T. intumescens* as stated in the caption and in the main description of the plant (page 141).
The references to *T. festucoides*, *T. linearis* and their appropriate illustrations should be disregarded on page 5.
Ill. 168 is a photograph of *Vriesea incurva* (Griseb.) R. W. Read and not *V. patula* as stated in the caption; the reference to Ill. 168 on page 244 is thus incorrect.
Ill. 137 is a photograph of *Tillandsia spiculosa* Griseb. var. *ustulata* (Reitz.) L. B. Smith and not *T. triticea* as stated in the caption; the references to Ill. 137 on pages 175 and 178 are thus incorrect.
Also, please note that: *T. ignesiae* is incorrectly spelt on page 139 in the fifth line of its description; the correct synonym of *T. ionantha* is *T. scopus* on page 142; that *T. werdermannii* is affiliated to *T. latifolia* (not latipolia) on page 191.

According to some recent botanical opinion, *Tillandsia ebracteata* (see pages 12, 13 and 121) is now being reclassified as *T. multiflora*. Ills 19 and 20 would thus now show *T. multiflora* Benth. var. *multiflora* according to this renaming.
Additionally, according to recent publications by L. B. Smith, *Tillandsia decomposita* should now be *T. duratii* var. *saxatilis* (Hassler) L. B. Smith wherever it occurs in this book. In addition, L. B. Smith states that *T. plumosa* (Plate 23) and *T. atroviridipetala* (Plate 22) are the same species, being merely synonyms. I for myself am of the opinion that these *are* two different species.

Werner Rauh
September, 1978

Editor's Foreword

This work does not attempt to compete with any of the standard scientific and purely botanical works, whose merits are of the highest and which form the backbone of our current knowledge of the bromeliads. The object in producing this English language edition has been rather to make good a deficiency in the available literature, which I consider has been much felt by the bromeliad enthusiast.

The book is intended for use by persons having no previous knowledge of botany. The illustrations and text will enable them to identify and learn about those bromeliads they may possess—or which they may acquire in the course of their travels and through their purchases.

Whatever the botanical merits of the scientific publications, and there are many, the previous scientific knowledge required precludes their use by the beginner or by the amateur—and even by many a grower and florist.

This is another reason why the English translation of Werner Rauh's volumes of bromeliads has been undertaken; to present bromeliads in such a way that they may be identified by the unlearned eye. So far as lay within my power, the descriptive text is written in normal everyday language, using only such technical terms as seem to be indispensable for the sake of accuracy and whose meaning can be explained within the book.

As a practical grower and collector of bromeliads over many years, it has been my pleasure and honour to translate and to edit this work, and in undertaking this I have had the delicate task of producing the work as complete as possible without tampering with Professor Rauh's text. I have been assisted greatly in the work of translation by Professor Harvey L. Lendall of the California State University at Long Beach, U.S.A.

I have not amplified or modified the descriptions of the plants which constituted the original German text.

The colour plates, black and white photographs and line illustrations of the original German volumes have been retained and all appear in the present English language edition.

The book is exciting because it covers the entire field, acclimatizing, seed sowing, growing and propagation—and the illustrations are clear and beautifully produced. This book will be of great interest to the general reader for the vivid accounts of the collection of many of these plants, as well as being a mine of information to the amateur and professional alike.

Peter Temple
London, 1978

Introduction

Plant lovers, of course, know the pineapple—that juicy, aromatic and delicious, large, yellow-orange 'fruit' crowned by a tassel of leaves. In knowing the pineapple, they therefore also know one of the most important representatives of the pineapple family, the *Bromeliaceae*, which was so named by the French botanist Charles Plumier in honour of his friend, the Swedish doctor Olaf Bromel. The pineapple, with its scientific name *Ananas comosus*, is important in many ways: firstly, it is one of the few utilitarian plants of the *Bromeliaceae*. It is grown on plantations in all tropical areas, and it has become a plant of great economic importance. Secondly, the pineapple was the first representative of the family to reach Europe—as early as 1690. The plant had already been depicted in 1535 in the work *Historia General de Las Indias*. The interest, however, was directed, as it is now, less to the plant than to the edible fruit spike. Not until much later did people begin to appreciate other bromeliads, especially the bright parrot-like colours of their flower spikes, which made such an exotic impression with the dazzling red of the prominent bracts and the contrasting, glowing colours of the flowers. Exotic tropical plants came into fashion around the middle of the nineteenth century in Europe. In 1864, the famous Kew Gardens in England was cultivating 100 different bromeliads: today the number of known species is around 2,000.

Although the era of great discoveries is past, every research and collection trip into inaccessible and remote mountain areas of Central and South America constantly reveals new, hitherto unknown bromeliads. This fact is attested to by the numerous publications and new descriptions by the man who is currently the world's top ranking bromeliad authority, Lyman B. Smith in the U.S.A.

Towards the middle of the last century, Belgian commercial growers were especially active in the cultivation of bromeliads and were constantly striving to introduce new species into Europe. By the turn of the century, a number of bromeliads were in cultivation which even today are still among the most cherished trade plants; these include *Vriesea splendens*, *V. tesselata*, *Aechmea fasciata* and *Tillandsia lindenii*. While the Belgians were busy with the care and culture of bromeliads, in Germany the botanist Carl Mez was examining their scientific make-up, and he published his findings in Latin in the 1935 work *Bromeliaceae**, which is still considered one of the standard works of bromeliad literature.

In the U.S.A. it was primarily Mulford B. Foster who, beginning in the

*Engler *Das Pflanzenreich*, Vol. 32, 1935.

1930's, was responsible for the popularizing of bromeliads. He not only founded the American Bromeliad Society, but also undertook several collecting trips to Central and South America and thus discovered a number of new species, several of which bear his name, e.g. *Aechmea fosteriana, Dyckia fosteriana, Cryptanthus fosterianus*. He was also active in cultivation, and we are indebted to him for numerous beautiful hybrids. Since the death of Carl Mez, the scientific study of this plant group has rested in the trustworthy hands of Lyman B. Smith.

Not only in the U.S.A., but also throughout Europe, the number of bromeliad fanciers is on the increase. This is not surprising as these plants—even when they are not in bloom—make ideal decorations for modern homes. Our collaborator, the late Dr. R. Oeser, who dedicated his life to the culture of rare and beautiful bromeliads, was able to demonstrate that the grey-leafed, aerophytic tillandsias can be grown even by the amateur plant lover (see page 58).

Most good florists now feel obliged to carry at least some bromeliads among their stock (*Vriesea splendens, Aechmea fasciata,* among others) as many people want to try their hand with bromeliads and to enjoy their bizarre forms and colours. Of course, due to space limitations, one cannot collect all species and genera and from the very beginning, one should specialize in certain categories which are most suitable for one's home surroundings. There are still difficulties in collecting rare and beautiful species apart from the usual plants offered by nurseries, but we live in an age in which distances are shrunk to a minimum. For most of us, it is only a few hours by jet to the homeland of bromeliads.

Acknowledgements

The descriptions of plants were prepared from specimens grown in the Botanical Gardens of Heidelberg University by the Chief Inspector of Gardens H. Lehmann and are so written that even the absolute beginner can identify the plants. Details of the care and culture of the atmospheric tillandsias stem from the pen of Dr. R. Oeser, who was involved with the culture of this plant group for several decades. The short reports of collecting expeditions and the chapters on the structure and life of bromeliads are the contribution of Professor W. Rauh, who has made several trips into the native areas of bromeliads. He has studied their morphology and ecology, and has collected an abundance of living material that is being cultivated in the Botanical Garden at Heidelberg.

Plants have been identified from imported stock, the authors making full use of the available literature. The original and often incomplete original descriptions have been considerably expanded and dubious instances were checked by L. B. Smith, to whom special thanks are due.

Great value is placed on the large number of illustrations and thanks are due to the expert photographer Antje Buhtz for the production of most of the colour photographs; all were taken with a Leica and the appropriate accessories. Almost without exception, the plants illustrated are imports.

Thanks also to the scientific illustrator I. Gegusch for the drawings and to our secretary I. Henninger for the preparation of the manuscript. Our very special thanks, however, go to the bromeliad specialist L. B. Smith of the Smithsonian Institution, Washington, D.C., U.S.A., who reviewed our deliberations where we were doubtful and who took responsibility for the descriptions of the new species discovered by Professor Rauh. However, we must also thank our friends A. Blass of Munich, L. Califano of Naples, and especially J. Marnier-Lapostolle of St. Jean Cap Ferrat, France, the owner of the world renowned botanical garden 'Les Cedres' with its great bromeliad collection. They all placed the extensive materials of their collections and photographs at our disposal. We are grateful also to H. von Appen of Lima, Peru, who unselfishly supported Professor Rauh on his travels in Peru. Many friends in South and Central America, who helped us in the acquisition of rare and nearly inaccessible species, who must remain unnamed, also deserve our thanks.

Sadly, Dr. Oeser, L. Califano and J. Marnier-Lapostolle are now deceased.

Abbreviations and Dimensions

In order to simplify the descriptive notes for each of the plants, an abbreviated system of notation is used in Part II. This is set out below:

Pl Plant shape, growth form

Fo Foliage

LS Leaf Sheath—the broad base of the leaves, usually more or less sharply defined

LB Leaf Blade—lamina (without the wide base)

Sc Scape (the shaft of the inflorescence)—the flowerless portion of the inflorescence

I Inflorescence—the fertile, flower-bearing segment; its axis is called a rhachis or spindle.

SB Scape Bracts—the leaf organs of the scape which lack flowers in the inflorescences of plants with compound (segmented) inflorescences (Fig. 12)

PB Primary Bracts—occurring in compound inflorescences and having segmented inflorescences in their axes (Fig. 13)

Sp Spikes

FB Flower Bracts—scape bracts with single flowers in their axes (Fig. 12)

Se Sepal—when we speak of 'posterior' sepals, we mean those turned toward the axis of the inflorescence

Fl Flowers

Pe Petals

St Stamen

Pi Pistil

Ov Ovary

Fr Fruit

Ha Habitat (and range)

CU Culture and Use

Syn Synonym (see page 81)

Throughout the book, the metric system of measurement is used for altitude, plant dimensions, bloom diameter, etc. In the descriptions of plants in Part II, measurements (cm and mm) stated are inevitably average values as these vary with location.

All temperatures are given in degrees Centigrade (°C).

Notes on Illustrations

The book contains three types of illustrations. The colour photographs are referred to as 'Plates' (Plate 1, Plate 32, etc.), the black and white photographs as 'Illustrations' (Ill. 1, Ill. 32, etc.) and the line drawings as 'Figures' (Fig. 1, Fig. 32, etc.). The colour plates and black and white photographs will be found in special sections of the book and the line illustrations are fully integrated with the text.

For details of the sources of illustrations, see page xii.

Part I
Growth and Culture

1 Bromeliads in their Native Habitat

The family is exclusively of the New World, with the exception of *Pitcairnia feliciana*, which is of African origin*. In the Americas, we find bromeliads all the way from Virginia in the north to southern Argentina. At the southernmost limits are the genera *Greigia* and *Fascicularia* in southern Chile, approximately on the 44th parallel. The family is not distributed at all evenly over the whole area, but certain points of concentration are apparent. The places particularly rich in bromeliads are Mexico, the Antilles, Costa Rica, eastern and southern Brazil, the Andes of Colombia, Peru and Chile.

If we look at this distribution on a map of the Americas, we see that the area where bromeliads grow stretches over approximately 80 degrees of latitude and this means that the plants are subjected to the most varying climatic conditions. Their distribution north and south of the equator means that plants receive rainfall at differing times of the year; as a result they have widely differing life patterns. In home collections, however, they must be cultivated under more or less one set of conditions. In addition, the plants, in purely tropical areas, range from the plains of the tropical rain forests to the icy regions of the mountains. In Peru, tillandsias are found at altitudes as high as 4,000 m!

From the huge bromeliad area from where we collected, just two regions are singled out to demonstrate the variation in climate and their living conditions: Mexico and Peru.

Mexico, stretching from the Rio Grande del Norte (30° latitude North) to south of the Isthmus of Tehuantepec (15° latitude North) with the provinces of Chiapas, Tabasco, Campeche and Yucatan, is essentially highlands averaging 2,000 m (6,500 ft). It is bordered in the east by the chain of the Sierra Madre Oriental and in the west by the Sierra Madre Occidental with an average height of about 3,000 m (10,000 ft); the latter continues to the south as the Sierra del Sur and the Sierra de Chiapas. The highest elevations in the east are the Cofre de Perote at 4,820 m (15,814 ft)

---

*It is worth noting that among the American cacti, with which some bromeliads are commonly allied, there is also only one species found in Africa, *Rhipsalis cassytha*.

2

and the Pico de Orizaba at 5,747 m.

The temperature distribution comprises three large climatic zones: the *tierra caliente* (hot zone, from sea level to 800 m), the *tierra templada* (moderate zone, from 800 to 2,000 m and the *tierra fria* (cold zone, from 2,000 to 5,700 m).

Even though Mexico has generous summer rains, especially from May to September or October, the upper temperatures and the distribution of rainfall on the Gulf of Mexico, especially on the east coast, are strongly influenced by the trade winds, which, laden with moisture, break on the east slope of the Sierra Madre Oriental and rise up along the mountain range. The air masses then cool, and between 1,500 and 2,000 m the moisture begins to condense. Heavy cloud banks, which release generous amounts of rain, collect in the form of a broad belt around the mountains. Then the trade wind blows into the central highlands as a relatively dry airstream, bringing precipitation in the form of strong storms in the rainy season. Temperature and rainfall are primary factors in the wide distribution of vegetation, not only in Mexico but throughout the whole world.

On a trip form Veracruz, located on the Atlantic coast, up across the Sierra Madre Oriental to Mexico City (2,400 m) we crossed a series of plant formations overlapping each other at increasingly greater heights. The hinterland of Veracruz is composed of a savannah, dotted with swamps and clusters of wooded areas. It is hot and humid, especially towards the end of the rainy season. This is the *tierra caliente*. We collected a number of noteworthy tillandsias in this hot zone. On the calabash tree *Crescentia,* one of the most common trees in the coastal savannah, we found the pretty, little, mounding *Tillandsia ionantha* (Plate 1), whose scape bracts become a vivid red during the blooming period. We noticed further the prominent *T. streptophylla*, whose broad leaf sheaths form a big, swollen bulb covered by the curled leaf ends; in the branches of small trees and large bushes, we observed a small form of the Medusa's head tillandsia *T. caput-medusae, T. balbisiana* and many others.

An evergreen, tropical rain forest connects to the narrow strips of coastal savannah ascending to heights of 600 – 800 m. This rain forest once covered the whole Yucatan peninsula and its luxuriant growth and its abundance of plant species can be compared to the rain forest of the Amazon basin—the Hylaea. As a result of constant high humidity and even temperatures, the trees display a rich growth of epiphytes, mosses, lichens, orchids—and also bromeliads, of which *Tillandsia leiboldiana* and *T. flabellata* especially captured our interest.

Above 800 m we entered the *tierra templada*. The floral composition of the forest changes; the tropical, evergreen rain forest gives way to a mountain forest, in which are mingled nut trees, alders, oaks, magnolias and also tree ferns. High humidity and a yearly average temperature of 15°C (59°F) produces a richly developed epiphytic flora. The branches of the trees are enveloped in thick coats of *Tillandsia punctulata*, which because of its short runners, tends to form big, loose, clusters. Of special beauty, however, is *Tillandsia multicaulis*, which with its numerous, flaming red 'swords' is one of the most beautiful sights at heights of 1,000 to 1,400 m. It is also one of the few bromeliads that bear multiple inflorences (Plate 8). Among the most graceful phenomena in these heights are without doubt *Tillandsia argentea* (Ill. 35) and *T. filifolia* (Ill. 99), two avidly collected species with hair-like leaves sticking out in all directions and with their thick bases (Ill. 36) forming bulbs. *Tillandsia juncea* (Ill. 109) is also widespread at these heights. It is a very variable species whose native habitat extends from Mexico to Peru and whose compact mounds can completely envelop the branches.

Higher up, it not only becomes cooler but it also becomes much more damp; this is the area of the fog banks caused by the trade wind. The forest here seems dismal and eerie, standing on the steep slopes. The ragged trees protrude like ghosts from the fog; trunks and branches are overgrown with mosses and lichens, which hang down in long beards and are witnesses to the high humidity; epiphytic peperomias, orchids, ferns and aroids keep company with the rosettes of great funnel-shaped bromeliads. Often the trees are so thickly covered with epiphytes that one would think they were about to collapse under the weight of the *parasitos*, as the natives call the epiphytes. For the first time we encountered some great bromeliad species, whose leaves form funnel-like rosettes with a diameter up to 1 m and sometimes more. These funnels are capable of collecting substantial amounts of water for dry periods; these pools teem with an animal world of frogs, lizards, and mosquito larvae, among other things. It is advisable to keep clear when tearing down the plants from the trees to avoid an involuntary shower.

In the cloud forests of the Sierra Madre Oriental, at heights between 1,800 and 2,000 m, there occurs a very special, but frequent combination of bromeliads, especially tillandsias, which one recognises as typical plants of the cloud forest and are strictly limited to these heights: here grow two of the most beautiful of Mexican tillandsias, *T. imperialis* (Plate 14) and *T. prodigiosa* (Ill.1). The lower range limit of *T. imperialis* is about 1,800 m; it is very common at 2,000 m and disappears again at 2,400 m. *T. imperialis* forms large funnel-shaped rosettes which bloom to produce a thick, club-shaped inflorescence covered tightly with red primary bracts, which support flower spikes with luminous, blue-violet blooms (Plate 14)—really an 'imperial' sight! In Mexico, the plant is brought to the market places in great numbers at Christmas time—it blooms precisely at this season—and serves as a substitute for Christmas candles. Because of the specific conditions of its native habitat, it is difficult to grow in cultivation. The same is true for *T. prodigiosa*, the 'marvellous one', very beautiful but completely different from *T. imperialis*. In contrast to most other tillandsias, the metre-long inflorescences, which are equipped with white or pink scape and flower bracts, do not grow upright but hang limply and for nearly half a year (Ill. 1). *T. prodigiosa* can hold its own with the most beautiful of the Mexican orchids. Wherever this tillandsia grows, it never appears alone but always in masses, primarily on evergreen oaks, but also on conifers. It is found not only in the dank mountain forests of Chiapas but also in central Mexico in the cloud forests above Teotitlan, between Mexico City and Cuernavaca, as well as near Oaxaca. The long inflorescences are very remarkable and make in one's memory, a deep impression. They are collected by the natives who use them to decorate religious shrines, churches and chapels. According to T. MacDougall, a noted collector of Mexican plants, this custom seems to be especially common in Chiapas.

Usually, *T. prodigiosa* is found with another tillandsia which also forms great rosettes of grey leaves, which are vegetatively indistinguishable from those *T. prodigiosa*. The differences are not apparent until the plant blossoms, for the inflorescences are not pendulous but upright, and the leaves are bent backwards (Ill. 2). We have determined that this plant is *T. bourgeaei*. All the tillandsias named here occur only in the cool, damp cloud forests, at heights of between 1,800 and 2,200 m. Among other tillandsias at these heights we saw *Tillandsia butzii* (Ill. 81), a pretty but variable species whose leaf sheaths, like those of *T. caput-medusae*, form a hollow pseudo-bulb (Ill. 33b). We also found the interesting *T. chaetophylla* (Ill. 3), which appears in thick bundles and whose narrow leaves look more like grasses or sedges than bromeliads. Without the inflorescences, one might believe that grasses were growing epiphytically on trees. Of the same growth habit are *T. festucoides* (Ill. 98) and *T. linearis* (Fig. 38). Both occur at the same height. Climbing higher and leaving the cloud zone behind, the vegetation scene changes once again; it becomes noticeably drier. Oaks were already a part of the cloud forest, but now they predominate and grow in extensive groups. In a great abundance of species (Mexico has around 110 species of oak), they cover the mountains between 2,200 and 2,500 m. Most of them are evergreen, hard-leafed and show only an insignificant deciduousness. These oak forests are abundant with succulents, especially agaves, species of *Dasylirion*, and even cacti. There is still a relatively high humidity, which is indicated again by the thick growth of tillandsias. But they have a completely different appearance from those of the cloud zone. Their leaves are grey to almost white and thickly covered with large, protruding trichome hairs, which are able to absorb atmospheric moisture. One of the most prevalent species of these dry oak forests is *T. usneoides*, the 'Beard-Moss Tillandsia' or 'Spanish Moss' (Ill. 4), which we already observed along the upper limits of the dank cloud forests. It occupies an enormous distribution area horizontally: it is found from Florida to southern South America, and in growth form differs in many respects from all other bromeliads. Its phytomers (plant segments between joints) are very thin, extremely elongated, often twisted in a corkscrew shape (Fig. 4). They form strands of 1 m, which are completely rootless and hang in thick folds, not only from the trees but also from the faces of cliffs somewhat like the beards of moss in the upper reaches of the middle ranges and in the cloud zone of the European Alps. Wherever *T. usneoides* appears, it occurs in profusion because its method of proliferation is very simple: pieces broken off and spread by the wind grow in any medium— even on telephone wires. In cultivation, one can achieve great success with *T. usneoides* when designing floral displays for window greenhouses. Its large, snow white variety *major* is especially suitable for this sort of effect; we were able to collect it most readily in the vicinity of Oaxaca.

Other collectable species worthy of cultivation which we came across were the snow white stars of *Tillandsia plumosa* (Plate 23) with its silvery trichome hairs; the filament-like, white *T. magnusiana* (Ill. 5); *T. matudai* (Ill. 6); *T. seleriana* (Ill. 30); and a large form of *T. caput-medusae*.

Carefully we removed a large specimen of the Medusa's head tillandsia, *caput-medusae*, from the crumbly oak bark in order not to harm it. In doing so, however, we make an unpleasant discovery; *T. caput-medusae* proved to be a rather unpleasant 'ant plant', one which sheltered whole nests of biting ants in the bulb formed by the broad, radiating leaf bases. With remarkable speed these ants attacked and bit us so severely that we felt the after effects for days!

T. seleriana is also an 'ant plant'. As we examined it closely, we discovered at the base of the exterior leaves a conspicuous hole (Ill. 30), through which the ants enter and leave. Upon bisecting such a bulb, we observed that a whole ant nation had established itself in the hollow of the base. Having been warned by this discovery, we hereafter left large plants of the pseudo-bulb tillandsias on the tree, and we collected only small examples—which, to be sure, were also inhabited by ants. It should not be assumed that the tillandsias benefit from the association with the ants. As in other plants, the ants just inhabit the cavities formed by the leaf bases. Among the 'ant tillandsias' in Mexico are *T. circinnata, T. butzii, T. utriculata* and *T. xerographica*. The latter two, especially *T. xerographica*, with its great, silvery rosettes and the splendidly colourful inflorescences (Ill. 7), are among the most impressive sights in the dry (xerophilous) oak forests. *T. utriculata* (Ill. 8) was also encountered occasionally growing on cacti.

The ecological counterpart of *T. plumosa* is *T. atroviridipetala*, a pretty, little, feathery-scaled tillandsia of the dry but very light oak forests on lava flows called Pedrigalles. Driving from Mexico City down to Cuernavaca, we encountered it in masses. It grows here not only on oaks, but also on *Commiphora* species (Fam. *Burseraceae*), usually with the crown pointing downwards (Plate 22). *T. atroviridipetala*, which is spread over a very small area, is closely related to *T. plumosa*. Both have intense green flowers, but in *T. atroviridipetala* the flower is hidden in the rosette (Plate 22) while *T. plumosa* has a clearly developed inflorescence shaft (Plate 23). Both are especially suited to the grower who has limited space, since the compact, silver or grey rosettes reach a diameter of only 10 to 15 cm.

At 2,400 m occasional conifers appeared among the oaks; above 2,600 m the oaks disappear and conifer forests cover the upper regions to nearly 4,000 m. Many Europeans could feel quite at home here in these surroundings if the tree trunks were not occasionally covered from top to bottom with tillandsias and other bromeliads (*Catopsis*). While the tillandsias of the cloud forests are distinguished by the formation of large funnel-shaped rosettes with broad, rigid leaves, the leaves of the species in the conifer forests are supple, grey and covered with felt-like trichomes. At heights of 2,400 to 3,000 m, a miniature version of the impressive *T. prodigiosa* is found. It is *Tillandsia macdougallii,* whose inflorescence grows to 30 cm and which hangs limply with its glowing red bracts (Plate 18). Closely related to these but considerably smaller are *T. benthamiana*, with white flowers (Plate 17), and *T. andrieuxii* (Plate 16), with beautiful blue-violet flowers. Also, *T. oaxacana* from the vicinity of Oaxaca is related to these species (Plate 19). A splendid but essentially larger species is *T. carlsoniae* (Plate 25), which resides especially in the conifer forests in Chiapas near San Cristobal de las Casas, where it occurs along with *Tillandsia fasciculata* (Plate 34), a species which is just as decorative as it is variable. Its habitat ranges from Florida to South America, and it thrives at almost all heights.

At heights greater than 3,500 m we have only seldom observed bromeliads in Mexico—usually in the form of a *Catopsis* on conifer trunks.

After crossing the Sierra Madre Oriental, the landscape changes decisively. The forests disappear; extensive corn fields cover the highlands of Mexico. Scattered among them are plantations of agaves, whose 'hearts' are cut out to make the Mexican national drinks of tequila and pulque. In the dry season, it is very monotonous landscape! But with the beginning of the rainy season the highlands are transformed into a thriving garden. Huge patches of the native, glowing red *Cosmea bipinnata* cover the land, mixed with the yellow of wild sun flowers, coreopsis, etc. (favourite garden plants elsewhere).

Descending into the basin of Tehuacan, located at about 1,600 m, there is another change of vegetation. It is noticeably drier, and we entered a landscape which most Europeans have in mind when they hear the name Mexico— dry semi-deserts with scattered tufts of mesquite (*Acacia*). This is the realm of succulents and of the numerous Mexican cacti. Very impressive is *Echinocactus grandis*, whose 2-m high and 1-m thick shoots remind us of gigantic barrels; *Myrtillocactus geometrizans* with its regularly branched candelabras along with the upward striving columns of the white-spined *Cephalocereus hoppenstedtii* and the 6-m high neobux trees form regular forests and undergrowth with tree-shaped opuntias, spiny agaves and mighty, tree-shaped lilies of the species of *Yucca, Beaucarnea* and *Nolina*. The latter seems strangely primeval. Its ridged, outsized base resembles an elephant's foot and, like the cacti, is able to store great amounts of water for dry periods.

Even in these dry regions, we encounter bromeliads, especially members of the species *Hechtia* and *Tillandsia*. The terrestrial hechtias cover large parts of the meagre ground in great variety and usually in masses. Yet there are also tillandsias, which grow as epiphytes on the great columnar cereus plants. They are of a different form from those of the wet mountain slopes. Their leaves are rigid, hard and stiff and occasionally close into tube-shaped vases. There are *Tillandsia dasyliriifolia* (Ill. 90), *T. makoyana* (Ill. 9) and *T. utriculata* (Ill. 8), all relatively large species. Of the 'dwarfs', only *T. recurvata* is at all common; it grows in masses on mesquite bushes as well as on cacti.

The wealth of species is by no means exhausted by the above mentioned bromeliads; Mexico alone has around one fifth of all the known tillandsias!

This short travelogue serves only to show that Mexico is a country full of geographical and climatic contrasts; but in all areas, from the coast to the high mountains and even in the dry semi-deserts, interesting and collectable tillandsias grow. That makes their cultivation elsewhere rather difficult, of course. Even if special care is taken to re-create the natural circumstances of the various habitats, we can never do justice to all species: those from the *tierra caliente* require warmth and humidity; those from the *tierra fria* want to be cool and have periodic dry spells. Therefore, one should cultivate bromeliads, especially tillandsias, like orchids—under various conditions and even in greenhouses of various temperatures. But what amateur can afford that? Even most botanical gardens keep all their bromeliads in one greenhouse, and even among the professionals there is the common (mistaken) belief that bromeliads are exclusively inhabitants of hot and humid climates. In order to dispel this opinion completely, let us make a jump to South America, to Peru, which is perhaps even richer in bromeliads than Mexico.

What makes Peru so interesting in comparison with all other South American countries is the great variety and contrasts in its topography and the closely related climate and vegetation zoning. In Peru, we find nearly all climates and geographical features: the sand desert; the hot, dry and sun drenched rocky desert; the high snow covered mountains, whose peaks in the 'White Cordilleras' tower to 6,700 m; the periodically dry high steppes, the *puna*; the cool, wet cloud forest of the eastern Andes, the *Ceja de la montaña*; and the hot, humid rain forest of the Amazon plains. And in all these regions, even far above 4,000 m, we encounter bromeliads in abundance.

Peru is dominated by the powerful Andean mountain range, the longest mountain range on earth, also called the Sierra or Cordilleras. The Andean wall, which in a very short distance (100 km) rises from sea level to nearly 7,000 m, is bordered by a narrow strip of sandy desert, stretching as far north as the fourth parallel north of the equator. This desert strip is wider in the *Desierto de Sechura* (northern Peru between Chiclayo and Piura) and in the south stretches far on toward Chile in the Atacama Desert. Like the African Namib Desert, the Peruvian coastal desert is a so-called cold air or fog desert. It owes its origin to a cold sea current, the Humboldt or Peruvian Current, which rises in the Antarctic, runs parallel to the coast, and, at about the fourth parallel south of the equator, deflects and joins the warm equatorial current coming from the north and influencing the coastal areas to the north. The cold Peruvian Current then causes the formation of fog, the so-called *garuas*, which in the winter months (April to October) enshrouds the whole coastal area up to the fourth parallel in a thick cover reaching approximately 500 to 600 m. Above the cloud cover, between 600 and 2,000 m, there is sunshine the year round. Although it very seldom rains in the coastal desert (Lima records a yearly rainfall of 21 mm), the moisture precipitating from the fog cover is nevertheless so great that the streets are made wet and slippery and a remarkable amount of vegetation can thrive. The desert, especially at heights of 300 to 600 m, is covered in the *garua* season with the fresh greenery of annual plants, which in 'good' *garua* years appear in such great number that the sand appears at a distance to be green meadows. This impression is further enhanced by the fact that the highland Indians bring their cattle down from the dry Sierra to the coastal land to graze on the *lomas* (low hills, undulating ground). Among these annuals are tuberous and bulbous plants including the splendid *Hymenocallis amancaës*, a kind of narcissus, which makes a fine impression with its large, yellow, bell-shaped blossoms. In the vicinity of Lima, it appears in such great masses that the people celebrate a narcissus festival in its honour.

As soon as the hot season begins at the beginning of November, when the clouds begin to lift and the sun begins its six month reign in a cloudless sky, the *loma* vegetation disappears and the desert during the summer months presents itself as a real sand desert. But even then it is not completely without vegetation, for between Trufillo in the north and the Chilean border, on flat sands as well as on the slopes of the lower Andes foothills one finds great stretches of tillandsias (Ill. 13—16), although they may be separated by broad areas without vegetation. Their extent coincides precisely with that of the *garua* fog banks, for these desert tillandsias live exclusively on the humidity in the air. Most of them produce no roots, or very few, and lie on the bare sand so that one can easily pick them up. Some of them give rise to miniature dunes (Ill. 16).

There is sufficient humidity for these desert tillandsias not only during the *garua* season, but also during the fog free period which is accompanied by sea breezes moist enough to sustain the plants and to protect them from the intensive sun rays. The growing tips of the tillandsia growths are all pointed toward the moist breeze and the sea (Ill.13 and 14). Such tillandsias, living exclusively on humidity, were called atmospheric (aerophytic) tillandsias by Carl Mez. The mechanism of water absorption will be dealt with later.

As soon as we left the capital Lima, driving on the Carretera Central into the valley of the Rio Rimac, we first met the desert tillandsias at the Inca ruins of Cajamarquilla. A most interesting species is *T. paleacea*. Its elongated, usually rootless sprouts tangle into long strands, in which all the growing tips form a more or less compact 'front' and point in one direction—to the sea, into the direction of the life-giving moisture. While the strands develop new growth on the seaward front, the sprouts in the rear die and rot into a black raw humus (Ill. 13), which is blown away by the wind. The year-round high humidity is also evidenced in the one-cell green algae, which lives on humidity alone and finds a home between the water-absorbing scales on the leaves. As a result, the older leaves of a plant take on a grass-green colouration.

T. paleacea appears in huge masses in the vicinity of Cajamarquilla, and at a distance, its strands pointing to the sea, are reminiscent of a rough sea crowned with whitecapped waves (Ill. 13). In its habitat, this species prefers to reproduce vegetatively by detached sprouts blown by the wind to a suitable place, from which new strands grow. Blooms and seed pods are seldom observed.

T. paleacea was found by us in great amounts only near Cajamarquilla; two closely related species, *T. straminea* and *T. purpurea*, have a considerably broader distribution than *T. paleacea*. Both of them appear either in great masses, or the individual plants intertwine like *T. paleacea* into long strands (Ill. 38), which in areas with shifting sand—as near Trujillo in northern Peru—give rise to small dunes (Ill. 16). In contrast to *T. paleacea*, these two tillandsias bloom very freely, and in August and September their white blossoms, edged in violet, emit a strong fragrance like that of stock.

Great expanses of bare desert are also decorated by *Tillandsia latifolia*, a rosette-shaped species found in many areas and in many forms, of which several varieties are known. The two varieties *major* (Ill. 17) and var. *latifolia*, distinguished by their difference in size, are strictly desert tillandsias. They settle in thick masses on the great stretches of the bare desert, while var. *divaricata* (Ill. 56) grows epiphytically on cacti in the dry valleys of the inner Andes and on well nigh inaccessible cliffs in the higher regions of the Andes.

Of special interest is the reproductive manner of the desert varieties of *T. latifolia* and its colony-like appearance. Similar to the pineapple, the end of the inflorescence remains vegetative and develops into a large rosette (Ill. 57). Under this weight, the relatively thin scape bends to the ground so that the offspring can take root and thus carry out vegetative reproduction. Since the inflorescence scapes are woody, hard and tough, and deteriorate slowly, several generations are often connected; of these, however, only the youngest is alive, since the old rosettes die after blooming. (For details see page 45).

The *garua* fog bands extend about 30 km into the oblique valleys of the Andes accompanied by the atmospheric tillandsias stretching into the rocky desert-like slopes of the rainless Andes foothills, where they usually are found in cohabitation with cacti. One of the most beautiful and widely spread rock-wilderness tillandsias which also grows in the dry valleys of the Andes is *T. tectorum**, a plant with an elongated scape, whose broad, recurved leaves, similar to the Mexican *T. plumosa*, are covered so thickly with large trichome hairs that in their dry state they appear snow white. Like *T. latifolia, T. tectorum* forms extended colonies between the rocks (Ill. 18).

There are a great number of terrestrial bromeliads, especially members

*The plant in W. Richter *Bromeliaceen (Bromeliads), Houseplants of Today and Tomorrow*, page 345. Fig. 4 and captioned *T. tectorum* Morren, is in fact the central American *T. argentea* Griesb.

of the genera *Bromelia, Puya, Hechtia, Pitcairnia, Deuterocohnia*, among others, which live exclusively on the ground. Terrestrial tillandsias, on the other hand, occur only in dry areas with high humidity, as in Peru, northern Chile, Brazil, Argentina, Uruguay and Paraguay. Normally, however, tillandsias are epiphytes living on trees or on cliffs.

Northern and central Peru are especially rich in tillandsias. Here, as a result of the increasing dryness as one proceeds southwards, bromeliads disappear on the eastern slopes of the Andes, and they are found only in the dry inner valleys, where they appear in an astonishingly great number of species.

Moving into the northern Peruvian Andes, into the valley of the Rio Saña, up to the Hacienda Taulis, the area has not only an enticing landscape, but is also botanically very interesting. There is a nearly untouched mountain forest, which covers the western slopes of the Andes between 1,800 and 2,800 m.

Leaving the small provincial town of Chiclayo and crossing the extensive sugar cane fields of the Hacienda Pumalca, we approached the Andes foothills, which lay before us in the brilliant sunlight, brown, burnt and completely without vegetation. Relentlessly the sun bears down day after day all year long. It seldom rains in this zone. Green is found only along the banks of the broad river bed of the Rio Saña in the form of rice and cotton and clumps of Humboldt willows (*Salix humboldtiana*), reminiscent of pyramid poplars.

About 70 km inland, at a height of 200 to 300 m, the landscape changes startlingly. The weathered granite of the Cerros carries orderly forests of imposing columnar cereus (*Neoraimondia gigantea*) up to 8 m high and little ball cacti of the genus *Melocactus*. In clefts in the rocks, the first bromeliads appeared, these being the terrestrial *Deuterocohnia longipetala*, one of the few bromeliads whose shafts last several years and produce new blooms again and again.

At approximately 350 m the vegetation changes again. *Neoraimondia gigantea* is no longer seen, and in its place appear other cacti, white, felt-like columnar cacti of the genus *Espostoa*. They are accompanied by shrubs and little trees, which also bear the first epiphytic tillandsias, small species such as *Tillandsia capillaris* and *T. recurvata*, which are spread far throughout Peru and appear in a wide range of communities and climates. In addition, we noted the narrow-leaved *T. juncea*, which we had encountered in Mexico, and *T. disticha*, a little species with narrow, grass-like leaves, which are broad and spoon-shaped at the base and combine into an onion shape. We also observed *T. ebracteata*, which is very common in northern Peru and southern Ecuador and is easily recognized by its yellow-green leaves and funnel shape. It occurs at altitudes of 500 to

1,000 m in the crowns of deciduous trees (especially of the genera *Ceiba* and *Erythrina*) so thickly that the trees even in their leafless state give the impression of a dense roof of yellow-green leaves (Ill. 19). *T. ebracteata* is not limited exclusively to the crowns of trees, but also grows epiphytically on the white, felt columns of *Espostoa*, a common columnar cereus at this height (Ill. 20).

Of special interest is *Vriesea* (*Tillandsia*) *espinosae**, a characteristic plant of the rainy savannah forests of northern Peru and of southern Ecuador, whose growth habit forms large, loose tufts which cover the branches of the host plant in thick mantles (Ill. 21). The mature rosette measures 5 to 8 cm in diameter and at its base it occasionally develops several rosette offshoots, which begin with a runner-like section 5 to 10 cm long covered with hard scale leaves (Ill. 155). The mature rosettes, with their hard, stiff leaves remain for a long time, giving rise to great, loose clumps, in which only the youngest rosettes are alive. In spite of its abundance, *V. espinosae* is relatively seldom cultivated, since in contrast to many species from the same area, this plant is an unreliable grower and is difficult to grow.

On the Puente Papayo, which crosses the Rio Saña gorge, we encountered one of Peru's largest tillandsias, *T. rauhii*, discovered by Rauh as late as 1959 and described by Lyman B. Smith. It grows at 600 m on steep cliffs and its large funnel-shaped rosettes, up to 1.5 m wide, are comparable in size to those of *T. grandis* and *T. maxima*. The ribbon-shaped rosette leaves, up to 10 cm wide, are of a reddish-violet colour and frosted with grey; when they bloom a huge arched, double-spiked inflorescence about 1.5 m long, rises from the middle of the rosette (Ill. 22).

There were tillandsias of similar size and with pendulous inflorescences of several metres in the dry inner Andean valleys that cut deep into northern Peru, e.g. in the valleys of the Rio Utcabamba, Rio Ucayali, Rio Marañon. On the way from Bagua to Chachapoyas the cliffs from afar appeared green from the giant rosettes of *Tillandsia ferreyrae* (Ill. 23), and in the dry valley of Huanuco there were, on the steep cliffs, mass stands of a giant tillandsia later identified by L. B. Smith as *T. fendleri* or a similar species (Ill. 24), but to return to the valley of the Rio Saña.

Above the bridge, in the direction of the little village of Florida at about 1,500 m, the natural vegetation had been severely disrupted; *chacras*, the natives' fields of sugar cane, bananas, pineapple, corn and other tropical

---

*Because of the scale formation at the base of the sepals, *Tillandsia espinosae* is now placed in the genus *Vriesea*. In other respects, the Ecuadorian plants are much larger than the Peruvian. The typical formation of runners in *V. espinosae* is not mentioned in L. B. Smith's original diagnosis.

crops, stretch to the mountain slopes; at their far edge were the remains of the former forest, whose remaining trees bore a rich growth of bromeliads. Here we collected *Guzmania monostachya* with its candle-shaped inflorescence covered with black-veined, red bracts (Plate 9), and a strikingly wide-leafed form of *Tillandsia complanata*, which decorated the crowns of the trees with its glowing red or green funnels. As in the Central American *T. multicaulis*, the inflorescences are lateral and are rather numerous in each rosette. However, they are much less colourful than *T. multicaulis*. The heads are considerably smaller and are on elongated, thin, pliable and arched stems hanging downwards (Ill. 47). Also, *Tillandsia usneoides* was very common at this height and hung from the trees in long beards.

Above Hacienda Taulis, which lies at 1,600 m, we found a unique phenomenon for the western side of the Peruvian Andes, a mountain forest, or cloud forest, almost undisturbed by human activity. It was rich in epiphytes, mosses, lichens, ferns as well as orchids, remarkably mounding *Ericaceae* and bromeliads, which, as expected at these heights, are funnel-shaped. An exception was the abundant *Tillandsia adpressa* at the edge of the forest at 2,800 m. Its broad leaf sheaths form a large, loose, water-holding bulb (Ill. 73).

Many tillandsias, as well as vrieseas and aechmeas, were collected in the forest of Taulis; some of them have as yet not been identified.

We made a further journey over the western cordillera, into one of the largest inner-Andes dry valleys of northern Peru—called the Huancabamba Valley, after the village of the same name. About 50 km south of Piura, we left the coast, again crossing the deciduous dry forests with the enormous green-barked kapok trees *(Ceiba)*, which in September are laden with the white puffs of the ripe fruit. The monotonous grey-brown of the leafless forests was punctuated with thorn-covered *Erythrina* trees (*Papilionaceae*) with their crowns of thousands of bright red blossoms, and in the underbrush the delicate red-violet of the blooms of *Bougainvillea peruviana* brought vivid colour into the picture. We noted essentially the same forms of tillandsias which we encountered on the way to Taulis. The vegetation did not change until we crossed the West Cordillera and descended into the valley of the Huancabamba. Above the little village of Canchaque in the highest reaches of the Andes' west side, between 2,500 and 3,000 m, there was an impenetrable mountain forest in which grew the splendid *Guzmania variegata* (Ill. 153), with its long spike decorated with glowing red, bowl-shaped primary bracts, and the prominent *Tillandsia tetrantha* (Ill. 132 and 133). But the landscape became drier and drier as we descended into the valley of the Rio Huancabamba. Before we reached the village of the same name, we encountered the first stands of the white, felt-like *Espostoa lanata* at an altitude of 2,000 m on the east side of the Andes, and the decorative grey-green *Armatocereus laetus*, whose columns of 3 to 4 m display a notable segmentation of seasonal growth patterns. Both of these cacti bore a rich growth of epiphytic tillandsias.

Tillandsia recurvata appeared at 1,700 m in such masses, that the young growths of the cacti were completely covered with them (Ill. 12); *T. usneoides* hung in long strands from acacias (*Acacia macracantha*), cliffs, and also from the columnar cacti. One of the most beautiful examples of tillandsias in this dry valley is *T. cacticola*, a species eagerly sought by collectors. It is of medium size and has white, felt-like leaves and a panicle inflorescence 30 to 50 cm high. The individual branches of the inflorescence bear lilac-coloured flower bracts in a tight shingle-like arrangement. In their axes are sweet-scented cream-coloured flowers edged in violet (Plate 15). Even the natives are enthusiastic about the beauty of *T. cacticola*. They pick the spikes and sell them not only in the market at Huancabamba, but also take them down to the coastal cities.

The species name *cacticola*, the cactus dweller, is actually misleading. Occasionally, this beautiful free bloomer is found on cacti (Ill. 11), but it thrives just as well on trees or on cliff sides. Its habitat stretches from Huancabamba to the Rio Chamaya, Rio Marañon and Rio Utcabamba. Also in the dry valleys around Chachapoyas we found substantial stands of *T. cacticola*. There seems to be intermediate forms between it and the forms of *T. purpurea*, which are to be found in the inner Andes dry valleys.

It is not only the dry valleys of northern Peru that are rich in epiphytic and terrestrial bromeliads—especially the grey, aerophytic tillandsias—the dry valleys of central and southern Peru, the valleys of the Rio Santa and the Mantaro, the dry valleys of the Vilcanota, the Urubamba and the Apurimac and its numerous, sometimes impenetrable tributaries also abound in bromeliads. How many unknown species grow here in the steep cliffs of the deep valleys and await discovery?

Again we went into the cordillera, departing from the little village of Casma, about 330 km north of Lima. Thick fog lay over the coastal desert; it was damp and cold, but just 20 km inland we penetrated the fog cover, and a beaming blue sky stretched out over the barren rock desert. Only the wide valley floor had a light covering of dry forest growth, *Prosopis* and *Acacia*. Approximately 30 km inland, at 600 m, we saw the first cacti; the great columns of *Neoraimondia*, *Armatocereus*, *Haageocereus* and the white, felt-like *Espostoa melanostele*, which forms such thick growth on the steep cliffs between 1,200 and 1,500 m, that from a distance the cliffs appear almost white. But tillandsias also inhabit the steep, almost perpendicular valley walls in great number. Some are species already known to us from the desert, grey tillandsias with large water-holding trichomes such as *T. latifolia* and *T. purpurea*, growing here, however, on dry and steep cliffs. Joining them at altitudes up to 2,800 m are mass stands of *Tillandsia tectorum*, in an especially small, nearly rosette form, which looks remarkably like the Mexican *T. plumosa*. Between 3,000 and 3,300 m we collected *T. sphaerocephala* and *T. oroyensis*, two other grey tillandsias.

We then arrived at the *Cordillera Negra*, the Black Cordillera, at about 4,000 m, with its monotone brown steppe, which being winter time looked dead. The only exception to the monotonous brown bleak waste of the *puna*, as the grassy landscape is called, were the large, white puffs of the wool cacti, *Tephrocactus*, which from a distance made one think of flocks of sheep. The young growths of this cactus are protected from the cold by wraps of thick wool.

They reach a diameter of 2 m and more and, at the beginning of October, are covered heavily with numerous, bright red or lemon-yellow blooms. A splendid sight! Although during this season the night-time temperatures sink far below freezing, and the delicate flower petals are completely unprotected from the frost, not a single blossom shows any sign of frost damage.

In the late afternoon we reached the Carhuas Pass and looked down into the heavily populated Santa Valley; this was an indescribable picture. Deep below us lies the little city of Huaraz, the capital of Calleón de Huaylas. Above it rises an immense wall of ice-capped mountains, the White Cordillera, the *Cordillera Blanca*. One six-thousander after the other; the highest of them is the double-peaked Huascarán, at 6,778 m, lying to the north above the little city of Yungay. Icy peaks as far as the eye can see!

We descend to Huaraz. An unforgettable trip, with the entire range of the mighty mountains of the White Cordillera stretched out before us. From the centrally located Monterrey, a pretty little hotel with thermal springs, we were able to research the entire Santa Valley. Our first trip took us northwards toward Yungay, Peru's Zermatt, from where we used a new, narrow, but otherwise good road far into the *Cordillera Blanca* to the Yanganuco glacier at 3,800 m.

On a previous visit, we had a strenuous ride on mules for a whole day; but this time we reached the glacier comfortably by car in an hour. It is rimmed with glacial mountains and splendid mountain forests, in which *Polylepis* is predominant—a tree-form *Rosaceae* whose knobby stem is covered with a red, paper-thin, peeling bark. Even at 4,000 m, the trees are covered with bromeliads, especially large funnel-shaped tillandsias. We collected approximately ten different species from the trees and steep cliffs of the former glacial valley, now filled with three lakes. Very decorative but too large to collect was the *Puya rauhii*, which establishes itself with its hearty stem between cliff rocks at the edge of the lake. Its hard, thorny leaves form great rosettes, from whose midst the 1.5 m inflorescence ascends, covered with large, ice-blue blooms.

Another trip took us up into the grandiose Santa Valley to the little villages of Recuay and Catac. Here we left the main road and tried to make headway on a very bad road, barely navigable with a Volkswagen, into the *Quebrada Pachacolo*. At the bottom of the valley, surrounded by glacial mountains, at 4,000 m, there was supposed to be a large stand of South America's most famous bromeliad, the gigantic *Puya raimondii* (*Pourretia gigantea*). We were skeptical about the ability of our little vehicle to get through in the bad weather; it was beginning to rain and snow. But after about 10 km we were able to discern the first stands across the rocky and partially marshy *puna*. From a distance, the blooming specimens resemble giant candles rising into the snowy sky. But not until we are standing directly before them does one feel their primitive power. We felt like dwarfs next to them (Ill. 25). A thick, leafless stem of 1.5 to 2 m height carries a leaf tuft of approximately 2 to 3 m diameter, with edges carrying sharp thorns and resembling daggers pointing in all directions.

Puya raimondii is most imposing at its bloom time, when a thick club-shaped spike 5 m high rises from the middle and is covered with thousands of yellowish-white flowers (Ill. 25). Although most of the specimens had already bloomed and faded (there is no information on the number of months required for the plant to form the gigantic inflorescence), we were fortunate to be able to see some specimens in full bloom and thereby to see at 4,000 m the rare spectacle of the pollinating birds fluttering around the blossoms to suck the abundant honey. *P. raimondii* is not only the largest, but also the most interesting bromeliad since it blooms only once in its life and then dies; in contrast to all other bromeliads, *P. raimondii* is propagated exclusively by seed. Warming, the Danish botanist, called such plants hapaxanthic. In the course of many (presumably 50 to 70) years, the plant produces a thick stem topped by a crown of rigid, upright leaves. The stem stops growing as the great terminal spike forms. Thereupon the growth material deposited in the rosette leaves is exhausted, and the leaves on blooming plants turn yellow and hang limp (Ill. 25). As soon as the seeds have been dispersed, the plant dies completely. Although one specimen produces a million, perhaps even a billion, tiny, winged seeds (Fig. 17, right), only a few of them will find conditions favourable to germination, so that in a stand of old specimens of *Puya raimondii* only a few seedlings will be found. The existence of *P.*

raimondii is especially threatened by the Indian herdsmen, who graze their herds at these heights and burn the faded plants before their seeds are spread, in order to warm themselves by the fire.

The plants are rich in gum-resin and burn like torches. The fire is spread by the wind to nearby, non-blooming plants, which are thus destroyed or severely damaged. *P. raimondii* is therefore threatened with extinction in Peru and can be found in only a few stands. In addition to the *Cordillera Blanca*, it may be seen also near Aija in the *Cordillera Negra* and near Lampa in southern Peru along Lake Titicaca. These may be the last stands of a formerly much larger, connected area, which once stretched from central Peru to Bolivia. Surely it is time for the Peruvian government to take measures to protect this really rare monument of the Peruvian Andes and to punish their burning with severe penalties. Only then can this unique natural phenomenon be preserved for posterity.

A few hundred metres lower and a few kilometres from the *Puya raimondii*, in the valley of the Rio Santa, grows the dwarf *Tillandsia bryoides*, reminiscent of moss, in clumps in the branches of shrubs (Ill. 26). What a contrast! It is hard to believe that *Puya raimondii* and *Tillandsia bryoides* are representatives of the same plant family.

A review of the results of our trips in Mexico and Peru—other countries would display a similar zoning—show that the *Bromeliaceae* is an exclusively New World family, but is very wide spread; we find examples from the southern U.S.A. almost down to the southern tip of South America. They grow epiphytically on trees, on cacti, on steep cliffs or terrestrially. They inhabit the sunny, sandy deserts of the coast as well as the steamy, hot, damp, rainy forests, the cool cloud forests, the deep, dry valleys and the high steppes up to 4,000 m. But they are especially plentiful in the cool, constantly damp cloud forests of the tropical high mountain ranges. Only orchids compete in their horizontal and vertical spread.

The next chapter deals with the growth forms of the bromeliads and how they differ according to their habitat and their growing conditions.

2   Morphology of Bromeliads

Growth Form, Branching Habit (Ramification) and Life Cycle

In regard to size, and as briefly noted in the preceding chapter we find within the family of bromeliads and even within a single genus, such as *Tillandsia*, an entire range from tiny, moss-like species to gigantic forms. The individual offshoots of *T. bryoides* join to form a turf that is reminiscent of billowing moss mounds; in *T. tricholepis* (*T. polytrichoides*) (Ill. 135) the species name is reflective of the fact that the plant resembles the Great Star Moss, *Polytrichum*. In contrast, there are the great funnel rosettes of *Tillandsia grandis, T. rauhii, T. maxima* measuring 1.5 m in diameter. The giant among the bromelaids is doubtlessly the high Andean *Puya raimondii* (*Pourretia gigantea*), which even at 4,000 m reaches a size of 10 m at bloom time.

The life span of bromeliads ranges from several years to perennial. Except for *Pitcairnia pusilla* there seem to be no one-year, i.e. annual-hapaxanthic species—plants which have a life cycle from germination to bloom and seed formation and death within one vegetation period and propagate solely from seed. *Puya raimondii* is pluriennial-hapaxanthic—it blooms and bears fruit only once; of course it needs many years (estimated at 50 to 70) to reach maturity. After dispersing the numerous, small, winged seeds, the plant perishes.

All other bromeliads, though, are long-lived or perennial plants with sympodial dichotomy*, explained in Figs. 1–3.

In most bromeliads, the dying process does not occur suddenly, but gradually, because the nutriment stored in the primary shoot is transferred to the developing offshoots (Figs. 1–3, Ek). These are buds that arise in the leaf axes of the mature plant and take over the progress of the offshoot system. Colloquially these buds are sometimes referred to as pups. They are the major means of vegetative propagation (for details see page 66), because from these 'pups' one gets bloom-size plants much more

*SYMPODIAL DICHOTOMY—Where at each forking of the plant, a shoot continues to develop and the primary shoot decays.

19

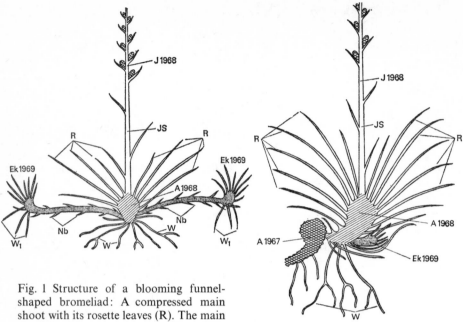

Fig. 1 Structure of a blooming funnel-shaped bromeliad: A compressed main shoot with its rosette leaves (R). The main shoot has elongated into the scape (JS), which terminates in the inflorescence (bloom spike) (J). Offshoots (Ek) have developed in the axes of the basal rosette leaves. The offshoots begin with a runner complete with cataphyllary leaves (Nb). As these mature, they also develop a leaf rosette and adventitious roots (W₁).

Fig. 2 A blooming funnel-shaped bromeliad with sympodial-monochasial branching. A (1967) dead axis. A (1968) successive axis. J bloom spike. Ek offshoot for coming year. Remainder as in Fig. 1.

quickly than through the cultivation of seeds. Usually, several offshoots appear at the base of the dying mature plant (Fig. 1, Ek 1969; Ill. 27a, Ek).

Frequently, however, only two offshoots develop (Fig. 3, Ek 1969; Ill. 27b) forming a forked ramification (branching) (Fig. 3; Ill. 27b, Ek). This is referred to as a sympodial-dichasial ramification. If only one offshoot develops at the end of the bloom period (Fig. 2, Ek; Ill. 28a, c), then a sympodial-monochasial dichotomy results. Many *Vriesea, Guzmania, Aechmea* and *Neoregelia* species (Fig. 10) show this phenomenon as do some tillandsias (Fig. 11).

In species with elongated axes, especially in tillandsias, the offshoot

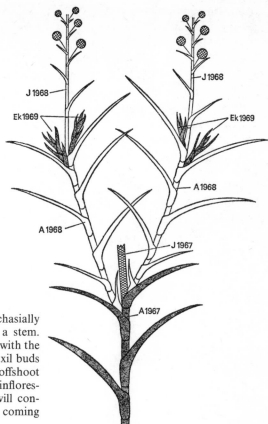

Fig. 3 Offshoot structure of a dichasially branched tillandsia which forms a stem. The axis (A) has ended its growth with the inflorescence spike (J 1967); two axil buds at its base have continued the offshoot system and have matured into inflorescences (J 1968). The buds (Ek) will continue the offshoot system in the coming year.

arises at the base of the inflorescence spike, i.e. in the axils of the uppermost live leaves (Fig. 3; Ill. 27b–c). The lateral branching is thus directed toward the apex, i.e. acrotonal. But in most of the funnel-shaped bromeliads such as *Tillandsia, Vriesea,* and *Guzmania* species, etc., the offshoots appear at the base of the mother rosette in the axils of the basal rosette leaves. The dichotomy is then basitonal (Fig. 1; Ill. 27a, Ill. 28).

Although the leaves of the generation that has bloomed dry up and decay relatively quickly, the dead axes of past years frequently remain preserved and joined for a long time (Ill. 28c). In many species, however, the pups detach themselves from the dying mature plant as soon as they have reached a certain size and then become independent. They quickly grow into new plants, which in turn repeat the behaviour of the mother plant, i.e. they too form a bloom spike and die, and their offshoot system is continued by axil buds. In this manner a species can theoretically live on forever.

Of fundamental importance for the growth forms of bromeliads is the length of the phytomers or internodes (segments between two joints), as well as the type of dichotomy. *Tillandsia usneoides* has extremely elongated internodes, thereby differing from all other species in the genus (Fig. 4). It is well-known that *T. usneoides* hangs in beard-like bundles and strands from tree limbs, cliff sides or telephone wires (Ill. 4) and is therefore called Old Man's Beard, or in Florida is known as Louisiana Moss or Spanish Moss; it takes its species name, *usneoides*, from its similarity to a local beard moss *Usnea*. By the thickness of the axes, the length of the grey-white leaves and the internodes, several varieties are distinguished: the most delicate is var. *minor* (Fig. 4, left); the most sturdy is var. *major* (Fig. 4, right).

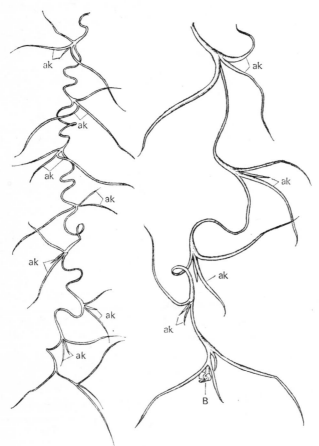

Fig. 4 *Tillandsia usneoides* var. *minor* (left) and var. *major* (right), showing long shoots with dormant axil buds (ak) and bloom (B).

Aside from its growth form, *T. usneoides* is notable in another respect; normally it is considered to be completely rootless. This fact is true, however, primarily only for old plants, since seedlings, as pictured in Fig. 5, are definitely provided with a root (W). Although this root is not well developed and has a short life, it nevertheless serves the purpose of attaching the seedling to a foundation. The seedling plants also differ from the older ones by the fact that no internodes have yet developed. The two foliage leaves (Fig. 5, B) that follow the cotyledons (Kb) cluster in a rosette. Elongation into internodes takes place with the formation of approximately the fifth leaf. The internodes themselves are sometimes bent into an S-shape, twisted like a corkscrew (Fig. 4) or rolled like a clock spring. Since they are extremely thin and essentially get no support through the long leaf sheaths and, in addition, show no geotropic (orientation to gravity) sensitivity, the offshoots hang limply from the branches of trees (Ill. 4) or from cliff sides. In the axils of the remote leaves, buds appear, whose axes are compressed and normally produce only two or three leaves; they are therefore designated as short shoots or brachyblasts (Fig. 4, ak); only upon damage to the offshoot system do

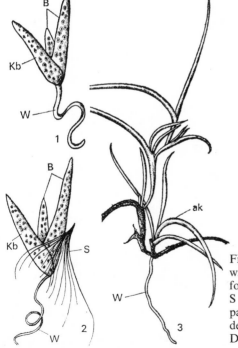

Fig. 5 *Tillandsia usneoides*: 1, 2 seedlings with the cotyledons (Kb) and the first foliage leaves (B); W single root; S remainder of the seed capsule and the pappus; 3 Piece of an older plant with the developing axil bud (ak) and root (W). Dead portions are shown in black.

they grow into long shoots. Once the plants have reached maturity, the primary offshoot concludes its elongating with the formation of a terminal, single-blossomed inflorescence (Fig. 6), and axil buds take over the continuation of the offshoot system. Thereafter sympodial dichotomy occurs.

The older parts of the plant remain for a long time, but then gradually die and are held together by the horsehair-like vascular bundles which are surrounded by dead sclerenchyma sheaths (tissue of thick-walled cells). *T. usneoides* is accordingly also called 'vegetable horsehair' and like real horsehair has been used by the natives to fill mattresses; it is also quite suitable as packing material in which to transport living plants. Each year, in the Heidelberg Botanical Garden, we noted how *T. usneoides* is 'stolen' in the spring by birds, especially by blackbirds, who use it to build their nests. The long, flexible strands are especially suitable for this purpose.

The propagation of *T. usneoides* by seed occurs only to a limited degree. Vegetative propagation is much more common; it occurs by the wind carrying off plant scraps which then use their twisted internodes to hook on to branches of trees and bushes and onto cliff sides. Since *T. usneoides* lives completely from the atmospheric moisture, it also colonizes telephone wires, fences and walls.

T. cauligera (Ill. 80), *T. purpurea* (Ill. 38), *T. straminea* (Ill. 134), *T. paleacea* (Ill. 115), *T. araujei* (Ill. 74), *T. decomposita* (Ill. 29), *T. duratii* (Ill. 95). *T. tectorum* (Ill. 129) and many others have elongated, often

Fig. 6 *Tillandsia usneoides* shoot with the single-flowered inflorescence.

low and rootless offshoot axes. They are primarily inhabitants of areas with little rain. Even though the plant parts of these species are not so extremely elongated as those of *T. usneoides*, they are nevertheless so long that the leaves stand remote from each other, and their more or less long sheaths only partially cover each other (Fig. 3).

Of special interest in this regard are the epiphytic *T. decomposita* and *T. duratii*. Both have elongated (1 to 2 m) axes that clamber over the host shrubbery and are primarily rootless. Their newest leaves are directed upward; the older ones, on the other hand, point downward, and their ends, as soon as they come into contact with a firm object, such as the branch of the host, begin to roll inward and twine around the branch, thus giving the epiphyte a firm hold. The tips of the older leaves function thus as tendrils. In cultivation, too, the leaf tips roll in a spiral without having been touched (Ill. 29; Ill. 95). Since the 'tendrils' are hard, but at the same time elastic, they can continue to make holds in the branches of their host. Also, in this way the plant achieves the necessary support and can therefore dispense with the formation of adhesion roots.

As the above mentioned species show dichotomy, which happens usually in connection with the formation of terminal inflorescences, the new offshoots develop at the base of the inflorescence, i.e. in the axils of the uppermost foliage (Fig. 3, Ek). This lateral formation has already been described on page 21 as acrotonal (directed toward the apex). Thus a sympodially constructed offshoot system occurs, in which the previous generations remain for a few years (Fig. 3), but then gradually decay.

This acrotonal type of dichotomy has resulted in two very interesting growth forms of bromeliads—the strand forms of the Peruvian desert tillandsias and the mounding form of the genera in the high Andes.

First, a word about the mounding growth, which is the typical growth form of many plants of the high mountains as well as of tropical and non-tropical regions. Within the family of *Bromeliaceae* this growth form occurs in its most beautiful manifestation in the representatives of the genus *Abromeitiella*. These are terrestrial bromeliads from the high mountain ranges of Argentina and Bolivia. The best known and most cultivated species is *A. brevifolia*. In its habitat it forms very hard mounds up to 0.7 m high and 1.5 m in diameter. The leaves are toothed and sharp pointed (Ill. 37); Fig. 7 shows that the mounding growth is the result of an acrotonal dichotomy repeated during every bloom period. Since every offshoot has approximately the same length of growth, and because of its position in the mound, grows more or less upward, a compact mound results in the course of many years, whose upper surface is formed from the tightly squeezed growth tips (Ill. 37). The older generations in the interior of the mound wither gradually into humus (in Fig. 7, stippled) in

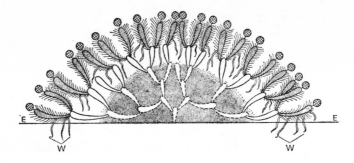

Fig. 7 Growth of a mound forming bromeliad (*Abromeitiella*). The inflorescences are shown in doubly hatched circles. The living shoots (stippled) are shown with leaves. The dying shoots are shown in white with solid edges. Portions already dead are shown with broken lines. The larger stippled area represents old growth that has withered into humus. W adventitious roots. E Ground level.

which the living offshoots are rooted. Physiologically such an *Abromeitiella* mound is comparable to a cactus: the surface of the mound corresponds to the green rind of the cactus because both serve the purpose of assimilation. The central, water-storing tissue of the cactus is functionally the same as the humus filling the interior of the mound, for the humus is capable of absorbing water during the rainy season and thus forms a water reservoir for rainless periods.

Abromeitiella is of special interest not only to the bromeliad collector but also to the cactus fan because the mounds serve quite well to add interest to the cactus collection, especially since this genus lives alongside of cacti in its native habitat. But it reaches its typical growth form only when it is planted in the open.*

Other species, such as *Tillandsia ionantha*, are also inclined to form mounds when they are cultivated as free hanging plants. In the course of many years, because of continuous dichotomy, a compact, nearly round ball of *T. ionantha* is formed (Ill. 31).

Basically, the same type of dichotomy is demonstrated also by the interesting strand-forming tillandsias of the Peruvian deserts (*T. paleacea, T. straminea*). As we follow the development of a strand, which we can pick up in its entirety from the bare desert sand, we note that it has begun from a single seedling or a little piece of a plant which has continued to branch acrotonally and laterally; thus a flat cushion comes into being. In contrast to *Abromeitiella*, the growth pattern is not upward, but the elongated axes more or less recline; but the growth tips are

*Or in a very bright light. *Ed.*

pointed upward. Very early a growth pattern occurs in the lateral edges of the cushion, which leads to a considerable spread perpendicular to the growing front (Fig. 8). Thus the formation of strands takes place, which in the course of years can reach a breadth of several metres (Ill. 13 – 14). The fact that their diameter remains approximately the same (50 to 80 cm) throughout their lifetime is based on the fact that the old segments wither more rapidly than is the case in *Abromeitiella* (Fig. 8; Ill. 14). In the 'fronts', which always point towards the sea, the intertwined growth tips are much more loosely arranged than in the mound surface of *Abromeitiella*, since the leaves are somewhat longer than those of the *Abromeitiella* (Ill. 37). Once such a strand has reached bloom size, all of the sprouts form inflorescences and then new sprouts at nearly the same time. This phenomenon is especially evident in the strands of *T. purpurea* and *T. straminea* (Ill. 38).

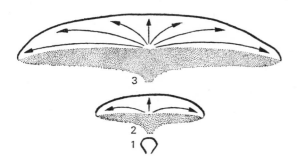

Fig. 8 Development of a strand-forming tillandsia: 1 Seedling; 2, 3 strand formation. Arrows indicate the direction of growth; their length represents the vigour of the growth. Withered sections of the strands are indicated at the base of each strand.

The propagation of the strand types, especially of *T. paleacea*, takes place mostly vegetatively. Pieces torn loose by the wind then grow in suitable places and become new strands.

A further growth style is demonstrated by the miniature tillandsias *T. andicola, T. capillaris, T. recurvata, T. bryoides, T. pedicellata* and *T. tricholepis*. Their elongated offshoots, although often only a few centimetres long, form a loose moss turf, which evolves through the individual plants forming new sprouts at their bases but also branching out acrotonally around the time of formation of bloom spikes (Ill. 27b–c).

The resemblance of some of these species to moss comes from the type of foliation and the shape of the leaf. In *T. tricholepis,* the leaves are very much abbreviated and jut out sparsely along the thin axis (Ill. 27b–c; Ill. 135). The plant therefore resembles the Great Star Moss, *Polytrichum* and formerly bore the species name *polytrichoides*. In *T. bryoides* and *T. pedicellata*, the scoop-shaped leaves point upward; they cling closely to the axis and completely cover it (Ill. 26; Fig. 28; Fig. 43). These species, too, have a great similarity to moss but look more like the young shoots of certain lycopods, especially the club moss, *Lycopodium clavatum*.

In most bromeliads, however, the axis of the young shoot is short and compact; often it is somewhat knotty or club-shaped and is visible only in cross section since it is almost completely enwrapped by the leaves, which are closely arranged in a rosette or in a bundle. Within the group of rosette-shaped bromeliads we can distinguish two growth types.

The most widely spread growth type is represented by the vase or funnel-shaped bromeliads. We find this form not only in the genus *Tillandsia* but also in the closely related genus *Vriesea* and in *Guzmania, Aechmea, Nidularium, Neoregelia, Billbergia, Wittrockia, Catopsis*, etc.

The vase-shaped bromeliads live epiphytically on trees, on cliff sides or terrestrially. In the large species their leaves can become longer than one metre; they are strap-shaped and either pointed upward throughout their entire length (Ill. 1) or are recurved from the middle on (Ill. 23). The (usually) broad sheaths of the leaves overlap each other to a large extent and lie so close together that they form a watertight funnel, which is capable of collecting generous amounts of rain water for the dry periods. The amount of trapped water depends on the size of the funnel. In very large rosettes the amount may be several litres. The prerequisite for holding water is the condition that the funnel-shaped rosette grow upright, even if it means that it clings sideways to its foundation (steep cliff sides, for example).

It is notable that the tanks hold not only particular flora, but also particular fauna. In the vases we find not only flowerless growth such as algae (*Chlorophyceae* and diatoms), but even blooming plants such as small *Utricularia* (Bladderwort). C. Picado was able to identify no less than 250 animal species in the vases of bromeliads in Costa Rica alone. Some are completely peculiar to bromeliads. We have observed tiny animal life along with tree snakes, salamanders and numerous, sometimes very colourful tree frogs. Although we had always shaken out our collected plants thoroughly, at night we were often deprived of sleep by loud, disharmonic frog concerts at our quarters where we deposited the day's collection! In contrast, when we collected in the dry areas, we often experienced scorpions marching around our room.

One further type of growth is represented by the bulbous or onion-shaped tillandsias. These live exclusively epiphytically in dry areas as well as in rainy regions. In contrast to the funnel-shaped bromeliads, which have negative geotropism (their axes always grow away from the earth), almost all bulbous forms are ageotropic; their offshoots are not affected by gravity and grow in the direction of the spot where the bud first

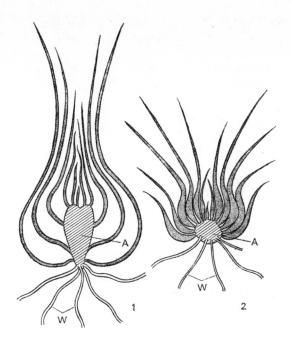

Fig. 9 Bulb formation in tillandsias: 1 pseudo-bulb;
2 true bulb (in longitudinal
cross section); A axis of
shoot; W roots. Leaf tissue is
stippled.

appeared. The plants, whether they are positioned on the upper or lower
side of their host branch, are directed sometimes horizontal, sometimes
downward. This phenomenon is especially beautiful in *Tillandsia
atroviridipetala* (Plate 22), *T. bulbosa* (Ill. 32), *T. caput-medusae, T.
seleriana,* among others. Examples of bulbous bromeliads, especially
tillandsias, are *T. argentea* (Ill. 35–36), *T. atroviridipetala* (Plate 22), *T.
balbisiana* (Ill. 79), *T. baileyi* (Ill. 76), *T. bulbosa* (Ill. 32), *T. butzii* (Ill. 33),
T. caput-medusae (Ill. 82), *T. disticha* (Ill. 92), *T. seleriana* (Ill. 30), *T.
streptophylla,* among others.

At a very early date, Troll suggested that the unbranched plant of a
bulbous tillandsia gives the impression of an onion. Indeed, the broad
base of the leaves of the tillandsias fit tightly together like the leaf organs
of an onion, but succulence typical of the onion layers is not present in the
leaves of bromeliads. Instead, there are large hollow spaces between
them, which serve to hold water (Fig. 9, 1 and Ill. 33b). Schimper (1888)
described the manner in which water reaches the hollow spaces between
the leaf sheaths and is prevented from escaping from the 'onions' that are
pointed upward. He uses the example of *Tillandsia bulbosa*:

'whose leaves are spoon-shaped at their sheath-like bases, while the leaf blade is
cylindrical and either gutter-like with a narrow slit or reed-like. The leaf edges are

sometimes located closely together or sometimes overlap. The leaf blade is always more or less recurved and twisted around its axis. The sheaths form a nearly completely tight onion-like structure, which, because of their spoon shape and their close proximity, contains large hollow spaces continuing upward into the hollow of the reed-like leaf blade and having only a very narrow opening to the outside at the transition point between sheath and blade. If one drops water onto the edges of the blade—no matter whether they overlap or are only close—it is quickly sucked up by capillary action. The same phenomenon occurs at the edges of the sheaths and at the narrow opening at the base of the blade. In this manner one can soon fill the hollow spaces, and the same thing takes place in nature during rain and dew. The fact that water does not leak out, even in an upside down position, needs no explanation since each chamber, with the exception of the little opening at the top, is tightly closed. In their innermost chambers the bulbs always contain water as well as dirt and dead insects, while the outer chambers have no water and are inhabited by ants.'

However, on our trips through Mexico, we discovered that *Tillandsia caput-medusae*, *T. seleriana* and *T. butzii*—all members of the bulbous group of tillandsias—did not contain a single drop of water between their leaf sheaths even during the rainy season (July), but housed colonies of large, biting ants.

Schimper recognised the symbiotic* nature of these tillandsias and ants when he says:

'The plants get their nitrogen supply from the bodies of the ants, that are not satisfied living just in the dry, peripheral hollow spaces, but also make fatal excursions into the water-holding chambers. The ants, of course, use the narrow opening at the base of the blade as their entrance'.

It is not clear from Schimper's remarks what he means by the 'narrow opening at the base of the blade'. We, however, discovered a large hole at the base of the outer 'onion' leaves, even in young plants, as can be seen clearly in Ill. 30. The ants enter and leave through the hole. It has not yet been explained whether the hole is made by the ants themselves or whether the leaf structure decays at the appropriate spot, as is the case with the orchid *Schomburgkia tibicines*, which also houses ants and lives in close proximity with *Tillandsia caput-medusae*. In these orchids the elongated stems (bulbs) are hollow and have an entrance at the base for the ants. In young bulbs an early spot anticipates a later hole.

Oeser also made observations regarding the problem of myrmecology†

---

*SYMBIOSIS—the living together of disimilar organisms with benefits to one or both organisms.
†MYRMECOLOGY—the inhabitation by ants of plants, offering specialised shelters or food.

of the tillandsias in cultivation:

'It has always been a joy for me to observe the ants that have made the trip with freshly imported plants. They have an extraordinary orientation ability and immediately find the wire from which the plant hangs. They apparently eat insects and allow no aphids to thrive. After a few weeks they had disappeared again. The tillandsias do not seem to mind the disappearance of their tenants; with sufficient fertilization they grow without ants'.

In time, though, we developed an enormous respect for the ants, as we collected *T. caput-medusae, T. seleriana*, and others.

In *T. streptophylla*, which, by the way, also belongs to the tillandsias housing ants in their bulbous bases, the leaf blades are spiral at their tips, rolled up and flattened (Ill. 128), a phenomenon which contradicts Schimper's theory of capillary water absorption.

In contrast to the 'onions' of the above named species, there are also tillandsias with genuine bulbs. These likewise have an onion-like swelling, which occurs because the broadened leaf sheaths are not only tightly packed together, but also are succulent and form a thick, water-holding tissue, as is the case in the culinary onion. This onion quality is especially evident in cross section (Fig. 9, 2 and Ill. 36).

Examples of tillandsias with true bulbs are, among others, *T. argentea* (Ill. 35 – 36), *T. filifolia, T. atroviridipetala* and *T. plumosa*—all relatively small, rosette types with narrow, sometimes grass-like (*T. filifolia*) leaf blades, which are frequently heavily covered by large trichomes, so that the plants appear snow white or grey (*T. plumosa, T. atroviridipetala*).

Between the true bulbs of the type of *T. argentea* and the hollow pseudo-bulbs, there are, of course, transitional forms. Thus, in *T. disticha* (Ill. 92), the bases of the scoop-like leaf sheaths are succulent, have a water-holding tissue and lie tightly packed together; the non-succulent upper sheath sections, on the other hand, form hollows inhabited by ants (Ill. 34).

The propagation tips of the rosette bromeliads begin at the base of the mother plant either with a short, more or less upright axis (most common), or they begin with an elongated, thin, runner-like section.

This runner first grows horizontally (plagiotropicly), is equipped with scaly (cataphyllic) leaves and later becomes a rosette. In doing so, the axis changes its direction of growth and becomes more compact and decisively stiffer (Fig. 1). We find the formation of runners or stolons in epiphytic as well as terrestrial bromeliads. In this regard *Vriesea espinosae* is of special interest. It is a characteristic plant of the rainy, green savannah forests of southern Ecuador and northern Peru. At altitudes of 300 to 600 m it forms large, loose clumps, which cover the branches of trees in thick cloaks (Ill.

Fig. 10 *Neoregelia ampullacea*. Multiple-shoot segment with s y m p o d i a l - monochasial branch-ing. Each shoot has a horizontal runner adorned with cata-phyllary leaves (A_1 to A_3) and an upright funnel rosette (R_1 to R_3). Arrows indicate reorientation in the direction of growth. Ek_4 offshoot bud for the fourth year; W adventitious roots.

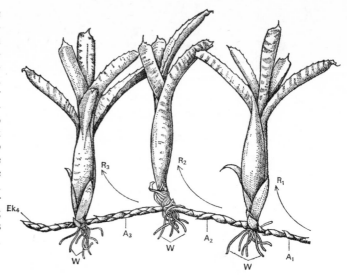

21). If one detaches one of these clumps from the host plant, one discovers that the 'cloak' structure is based primarily on the formation of stolons: every blooming mature rosette produces at its base several offshoots, each of which begin with a stolon about 10 cm long and are surrounded by robust, cataphyllic leaves before they take on the form of a rosette (Fig. 1; Ill. 155). The dead rosettes with their woody stolons remain alongside the newest, living rosettes for many years.

The genus *Tillandsia* has very few stoloniferous representatives; *T. tricolor* and *T. punctulata* have short stolons. The formation of stolons is more common in the representatives of the genera *Aechmea, Nidularium, Pitcairnia, Hechtia* and *Neoregelia*, of which *N. ampullacea* is an example (Fig. 10).

With the help of these stolons, some species are even capable of climbing trees: for example, *Neoregelia ampullacea, Nidularium bracteatum, Pitcairnia sceptriformis* and *P. scandens*.

Species of the genera *Bromelia, Pseudananas, Pitcairnia*, etc. reproduce by means of subterranian stolons and frequently appear in massive stands in their native habitat.

Leaf Arrangement and Leaf Form

Leaf arrangement is called phyllotaxis. In contrast to most monocotyledons, very few bromeliads show a purely two-part (distichous) phyllotaxis, i.e. the successive leaves are arranged on the axis in two rows and stand at an angle of 180° to each other. This is the case with a few tillandsias; *T. usneoides, T. capillaris* with its numerous forms (Fig. 29), *T. recurvata, T. crocata* (Fig. 31), *T. gilliesii* (Fig. 11), among others. In the vegetative area of most bromeliads we find spiral phyllotaxis. It can happen, however, that in the blooming phase a transition to distichy occurs so that in the inflorescence area the leaf organs are arranged in two lines (details on page 64 *et seq.*).

The form and size of the leaves are extremely variable. In length, they vary from a few millimetres (*Tillandsia bryoides, T. tricholepis*) to 1 m and longer; but the width, too, is quite variable: there are all kinds of intermediate stages from filiform, grass-like examples (Ill. 3) to wide, strap-like blades (Ill. 25).

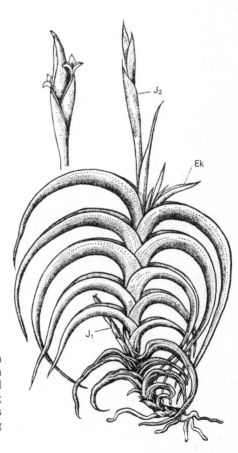

Fig. 11 *Tillandsia gilliesii* Baker as an example of a stem forming tillandsia with distichous leaf arrangement and sympodial-monochasial branching; J_1 last year's inflorescence; J_2 this year's inflorescence; Ek offshoot for the coming year.

In many species, especially in the vase and bulb forms, the base of the leaf is distinctly broadened and more or less sharply distinguished from the actual blade; it not only fulfills the water storage needs of the plant, as already mentioned, but also is active in the role of nutrition absorption (see page 56).

Seldom do the leaves show a petiole-like narrowing; however, some *Pitcairnia* species and *Cryptanthus beuckeri* have this characteristic.

In all of the *Tillandsioideae*, the leaf edge is smooth and entire; the *Pitcairnioideae* contain both smooth and spined edged leaves, and in *Puya* species, the leaf edges bear stiff, hard, curved thorns; frequently the spines near the base are directed downwards while those on the blade are pointed upwards. In the Bromelioideae the leaves of *Bromelia, Fascicularia, Aechmea, Billbergia* and *Neoregelia* species are also equipped with spines.

Not only the form, but also the colour of the leaves varies greatly in the bromeliads. Often one can deduce the climatic conditions under which a species lives by observing the leaf colour. The white or grey colour of many tillandsias comes from the fact that the leaves are heavily covered with absorbent scales (page 54), which hold air between them and cause a total reflection of light. If one dips these plants briefly in water or sprays them, the leaves become green since the air between the scales is compressed and the assimilation parenchyma* shines through. These so-called grey-leafed tillandsias inhabit exclusively areas with little rainfall but with high atmoshperic humidity. We will look again at the fact that it is precisely these scales that provide water absorption and are capable of drawing moisture from the air. Green-leafed tillandsias, on the other hand, belong almost exclusively to the vase type bromeliads, which inhabit areas with much rainfall. In these, the formation of scales is limited to the leaf sheaths, while the blades are nearly scaleless.

The banding and diagonal striping of many species, e.g. *Billbergia zebrina, Aechmea fasciata, Cryptanthus zonatus*, etc. comes from the fact that zones or strips of heavy scale alternate with zones of less scale (Ill. 39). Stripes, spots and irregular patterns can also be the result of colour. The green-white longitudinal striping of many species such as *Neoregelia carolinae* var. *tricolor* (Ill. 40), *Nidularium innocenti* var. *lineatum* and the so-called 'variegata' forms of pineapple comes from the fact that stripes of tissue with chloroplast alternate with chloroplast-free stripes. Red colouration is based on the presence of red pigment, anthocyanin; in brown leaves, the green chlorophyll is blanketed with red anthocyanin.

An especially beautiful leaf marking is displayed by *Vriesea splendens,* a splendid and very popular commercial plant, in whose leaves sharply

*ASSIMILATION PARENCHYMA—Food-converting tissue.

delineated, jagged, dark purple bands with anthocyanin alternate with green bands with no anthocyanin (Plate 63). The leaves of the Peruvian *Guzmania lindenii* (Ill. 152) show a similar, perhaps even more beautiful configuration. In *Vriesea hieroglyphica* (Ill. 165) and *Guzmania musaica* the bands dissolve into irregular squiggles (Ill. 41), which look like script, or more precisely, like hieroglyphics. The leaves of *Vriesea gigantea* (*V. tesselata*) (Ill. 42) and *V. fenestralis* display a grid-like configuration. It comes from the fact that chloroplast-bearing tissue covers the leaf veins, while the tissue between the veins bears less chloroplast; the grid configuration reflects a faithful image of the arrangement of leaf veins, which is especially clear when the plant is held to the light (Ill. 42).

Otherwise, the colouration of leaves is heavily dependent upon the intensity of the sunlight. The leaves of many bromeliads take on a brilliant colouration in the bright summer months, not only in their native habitat, but also in cultivation. During the duller winter months, they revert again to grass green. Varying light intensity also explains the observation that in a single tree the same species can appear in a red and in a green modification, depending on the individual plant's position in the shadows or in full sun on the crown of the tree.

As a rule, the leaves of a mature plant (except for the scape bracts and the cataphyllary leaves of the pups) are all structured alike. A true heterophyllia (leaves of different forms) in mature plants is found only in some *Pitcairnia* species, e.g. in *P. palmeri* and *P. pungens*; in *P. heterophylla* the heterophyllous quality is reflected in the species name: during the dry period, in which the plant blooms, it produces brown leaves with long spines, followed by normal green leaves in the rainy period (Ill. 43). The formation of the different leaf forms is determined by the seasons in this case.

Roots

In general, the roots of bromeliads do not play the same role as roots do in other flowering plants; in most plants roots serve the primary purpose of absorbing water and dissolved mineral nutrition; secondarily they also support the plant. Bromeliads are different. Since many species take in atmospheric humidity and are capable of taking mineral nutrition with their trichome hairs (details on page 54 *et seq.*), the roots, at least in the epiphytic species, are almost exclusively anchoring devices. They are, therefore, very hard, surround their foundation and excrete a rubber-like substance from their usually flat underside (according to the investigation by Chodat and Vischer, 1916), which connects them so firmly to their foundation that they can be removed only with much force.

Anyone who has collected epiphytic bromeliads from trees or rocks knows the difficulties of removing the plants from their host undamaged. In the vase-shaped bromeliads, it is easier to break off the main axis than it is to loosen the roots. It is better to collect tree-growing types with a bit of the branch since their growth pattern will then be less disturbed. Bromeliads that need no foothold dispense with roots altogether as they mature, or have only a weak root system. On page 23 we indicated that *T. usneoides* has roots only in the seedling stage (Fig. 5, 1–2, W); older plants are usually rootless. Only under special cultural conditions, e.g. very high humidity, can weak roots be developed, and they are of no significance either as nutrition absorbers or anchoring devices. But nevertheless, *T. usneoides* has not completely lost the ability to form roots. In Fig. 5, a piece of an older plant (3) with an already dead axis has been depicted. A side shoot has developed, which has grown a wretched little root (W). Also the plants of *T. duratii* and *T. decomposita* are usually rootless in maturity since their leaf tips function as tendrils (Ill. 29) or hook on to branches and thus provide an anchor for the plants.

On the other hand, the large terrestrial bromeliads such as *Puya, Hechtia* and *Bromelia* species have especially strong root systems. In these plants, the roots serve not only to anchor the plant but also absorb water and mineral nutrition.

Also, in the so-called atmospheric tillandsias, i.e. tillandsias that live primarily from humidity and are equipped with a thick trichome mantle, at least the young roots might function as water absorbers. In imported plants it sometimes takes months before new roots are produced. But then they appear suddenly and in great numbers; with sufficiently high humidity they stick out in all directions from the main axis without at first clinging to the mount. It is therefore to be assumed that their tender, soft tips are thoroughly capable of absorbing moisture that condenses on them and conveying it to the plant. In any case tillandsias that are rooted in this manner and kept sufficiently moist grow considerably better and more rapidly than others, as can be proven by the large collection of A. Blass in Munich. Blass cultivates his plants on pieces of cork oak bark and hangs them in close proximity over large, open, heated water basins in his spacious greenhouse. The atmosphere is thus nearly saturated by the constant evaporation. This is favourable to root development and considerably speeds the growth of the plants.

Since the roots of most monocotyledons are short-lived, even the initially tender and succulent air roots of the bromeliads soon become brittle, hard, woody and wire-like. If these have not already established themselves on their mount, they are useless to the plant.

The origin of the roots is of special interest. Since bromeliads are monocotyledons, they soon lose their prominent main root; this is visible only in the young stages, it soon dies and is replaced by axis roots, i.e. roots which emerge from the plant axis. In the large, stemmed species they are located in the vicinity of the plant apex, the vegetation point, specifically in the tissue of the cortex; they do not immediately penetrate the cortex, but often grow some way lower down and then emerge below the place of their origin (Ill. 46, top). A cross-section of the stem of a *Puya*, showing the circle of roots in the outer cortex is shown in Ill. 46, (bottom).

Especially notable in this regard is a 'nameless' *Hechtia** frequently cultivated in the gardens of the Mediterranean. It forms large stoloniferous rosettes whose leaves in intensive sun show a decorative red-green banding. The roots arising from the growing tip of the plant first run downwards for a stretch in the stem cortex, then enter a leaf, travel through the entire leaf blade, emerge at its tip and enter the soil.

Inflorescence

Every bromeliad grower, of course, likes to see his plants bloom. But in comparison to the generally plain leaves of orchids, there are among the bromeliads many species which even in their non-blooming state are the pride of any collection and evoke the delight of their owner. Among them are *Guzmania musaica* (Plate 56), *G. lindenii* (Ill. 152), *Vriesea hieroglyphica* (Ill. 165), *V. gigantea* (Ill. 162), *V. fenestralis* (Ill. 161), and others. In contrast to other plants, a bloom is often quite undesirable in bromeliads because, as explained on page 19, this often means the death and the loss of the plant if it does not 'pup'. It is precisely the beautifully marked 'foliage' bromeliads often show little inclination to form pups, so that the preservation of the species is guaranteed only through propagation through seed. Yet one can obtain seed only if one has two simultaneously blooming specimens of the same species but of different origin.

As a rule, the bromeliad rosette ends its growth with the formation of a terminal bloom spike (inflorescence). The plant axis stops producing foliage, and its vegetative tip engages in the formation of an inflorescence. This means that every plant, every rosette, produces only one branched or unbranched inflorescence (Fig. 12). The only exceptions are *Tillandsia complanata* (Ill. 47) and *T. multicaulis* (Plate 8), in which the inflorescences—simple spikes—appear multi-laterally to the axis of normal rosette leaves. The fact that the inflorescences here are actually equivalent to true side shoots derives from the fact that they, like

---

*The plant is known under the name *Hechtia glomerata*, but is not the true *H. glomerata* in many respects. It cannot be definitely identified, since the plant has never bloomed.

monocotyledons in general, begin with a bracteole equipped with leaves with two keels.* The axis end of the mother rosette remains vegetative, and its vegetative tip produces normal, green leaves subsequent to the inflorescence. In further behaviour, however, *T. multicaulis* and *T. complanata* differ from each other: in the former the blooming mature plant dies after the seeds ripen and reproduces by pups forming at the base of the old rosette (Ill. 44); in *T. complanata*, on the other hand, pup formation does not take place. The vertex of the mature plant continues to produce leaves and seems to produce inflorescences for several years in a row. But there is no documentation on *T. complanata* on the frequency of blooming since this plant is difficult in cultivation and does not last long in the collection.

There are articles published citing lateral inflorescences for *Quesnelia lateralis*, but it may be a matter here of variant behaviour. *Q. lateralis* produces lateral branches which normally elongate and are used for vegetative propagation. Occasionally, however, they remain very short and end their longitudinal growth with the formation of inflorescences, which then seem to spring from the axis of the rosette foliage and thus produce the deceptive effect of true lateral inflorescences.

Also *Tillandsia lindeni* var. *luxurians* is said to produce lateral inflorescences in addition to the terminal inflorescence. It has not been possible to determine whether this is a case of true lateral offshoots from the mature rosette or basal branching of the terminal inflorescence.

Except for those examples described above, the inflorescences of all bromeliads are terminal. From the outset to complete unfolding of the individual blooms, weeks may pass—in the giants like *Tillandsia maxima*, *Vriesia gigantea*, *V. imperialis* and the mighty *Puya raimondii* it may take months or even years. There is little literature and documentation on this subject. According to reports by Ule, the complete formation of the inflorescence of *Aechmea sphaerocephala* takes more than a year. F. Müller (1896) reported that a *Canistrum* needs 2 to 3 months for the complete development of its inflorescence.

Once the plants have reached maturity, their subsequent behaviour varies in the different species: for example *Tillandsia brachycaulos, T. ionantha, Guzmania sanguinea* show their readiness to bloom by the fact that the upper leaves of a rosette—otherwise indistinguishable in size and form from the green foliage—take on a vivid, usually red colouration and become an attraction apparatus for the flowers, which hardly rise above the rosette (Plates 1–3).

*We will not discuss here the question of how this deviant positioning of the inflorescence fits into the basic structure of bromeliads.

In most bromeliads, however, the initiation of the bloom phase results in an elongation of the axis, which in the vegetative stage is compact. In the rosette types, the rosette loses its form and changes into a half-rosette pattern (Fig. 12).

As to the direction of growth in the bloom spike, generally it grows upwards, i.e. it is orthotropically oriented. But there are exceptions to this behaviour. We have already mentioned that the axes of *T. seleriana, T. bulbosa, T. butzii, T. caput-medusae*, etc. do not react to gravity (geotropism); they are ageotropic. Consequently, the inflorescences are also ageotropic and grow in the direction of the vegetative axis without bending upwards (Ill. 81). *Tillandsia atroviridipetala* grows preferably

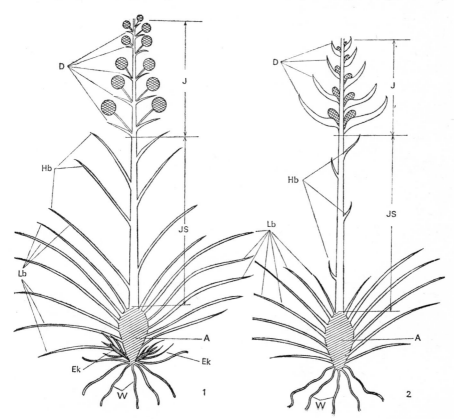

Fig. 12 Blooming rosette bromeliad with racemose inflorescence (1) and with spicate inflorescence (2): A axis of shoot; W adventitious roots; Ek offshoot; Lb foliage leaves; JS scape (flowerless segment of the inflorescence) adorned with scape bracts (Hb); J fertile, flower-bearing segment of inflorescence; D flower bracts, where individual flowers are found in axes.

with its rosette in a hanging position; and its very short inflorescence is therefore likewise pointed downwards (Plate 22). The inflorescences of *Tillandsia tenuifolia* (*pulchella*), *T. araujei, T. meridionalis* (cover picture), *T. aeranthos* (Plate 24) and many others also seem to be ageotropic since they are sometimes directed upward, downward or horizontal. The inflorescences of many large Vrieseas and Tillandsias, such as *T. maxima, T. rauhii* (Ill. 22), *T. ferreyrae* (Ill. 23), and several *Aechmea* and *Billbergia* species are pendulous.

The behaviour of *Tillandsia prodigiosa*, the 'marvellous one', which grows in the cloud forests of Mexico at altitudes of 1,500 to 2,000 m primarily on oak trees, is interesting. It forms large funnel-shaped rosettes from whose midst the inflorescence emerges. Its axis is initially directed upwards, but then it undergoes a geotropic orientation in the course of its development and elongation. Then it grows downwards and later hangs limple approximately 1 to 1.5 m (Ill. 1). If for no other reason, the name *T. prodigiosa* is justified.

Other Mexican species (usually on conifers) with pendulous inflorescences are *T. andrieuxii* (Plate 16), *T. macdougallii* (Plate 18) and *T. oaxacana* (Plate 19). In *Aechmea filicaulis* (Ill. 184) the thin, thread-like inflorescence axis reaches several meters in length.

Every developing bromeliad bloom spike shows distinctly recognisable parts: a basal, bloomless part and an upper, flower-bearing, fertile segment. The former is called the 'scape' (in Fig. 12, JS). It usually bears leaf organs in a loose arrangement, called 'scape bracts' (Fig. 12, Hb). Also the axis (rhachis) of the fertile section, which is the actual inflorescence (Fig. 12, J), is equipped with leaf organs, which, in the case of unbranched, simple spikes as depicted in Fig. 12, bear flowers in their axes. These leaf organs are called flower bracts (Fig. 12, D). In simple inflorescences the flower bracts follow immediately after the scape bracts (Fig. 12, D; Ill. 53–55); but in compound inflorescences, which are treated on page 42 *et seq.*, leaf organs follow the scape bracts; they have entire segmented inflorescences in their axes instead of individual flowers. In the species descriptions they are designated as primary bracts (Fig. 13, T).

In most bromeliads the scape is more or less elongated and can vary in length from a few centimetres to a metre or more. In sessile inflorescences, the scape is completely missing. Occasionally it develops only after bloom (*Tillandsia pedicellata*, Fig. 43).

Inflorescence diameter varies from hair-thin to several centimetres, from species to species.

The foliation, the form and the colour of the scape bracts, present an astonishing variety;

1 The transition from rosette leaves to scape bracts takes place gradually and smoothly (Fig. 12, 1, Hb). The basal scape leaf organs have typical foliage characteristics, and the following leaves become successively smaller—at first as a result of a gradual reduction of the blade; a shortening of the leaf base also occurs in the upper section of the inflorescence shaft. In their development the scape bracts are retarded leaves, which they frequently resemble in their green colouration. Examples of this behaviour occur with *Tillandsia latifolia* (Ill. 48), *T. polystachya* (Ill. 116), *T. plumosa* (Plate 23). In these, the scape bracts gradually blend into primary bracts of the fertile region. Frequently, the scape bracts which follow the green rosette leaves are distinguished by vivid colouration, such as with *Tillandsia intumescens* (Plate 5).

2 The transition from rosette leaves to scape bracts also takes place gradually and smoothly, but the scape bracts increase in size towards the tip. Also they are brightly coloured and form the actual attraction apparatus for the relatively unattractive flower spike. In many *Aechmea* and *Billbergia* species the usually bright red scape bracts are followed directly by tiny flower bracts, in whose axes insignificant flowers appear. The splendid *Aechmea mariae-reginae* is a representative species (Plate 4).
   In *Guzmania* species, for example in *G. lingulata* and *G. minor*, the bright red scape bracts increase in size toward the tip (Plates 6 and 7) and with their flat blades form a kind of involucre. But then they gradually change into the similarly coloured primary bracts or flower bracts of the inflorescence, as a result of increasing foreshortening (Plate 7).

3 The transition from rosette leaves to scape bracts takes place more or less abruptly. The rosette leaves become gradually shorter toward the centre of the rosette, but still belong to the vegetative axis, while the actual scape bracts of the inflorescence shaft itself appear as small, frequently scoop-shaped, pale green organs closely clinging to the axis. In this respect the bloom spike differs sharply from the vegetative region (Fig. 12, 2). Examples of this behaviour are many tillandsias with sword-shaped inflorescences such as *Tillandsia lampropoda* (Plate 39), *Vriesea carinata* (Plate 11), *V. splendens* (Plate 63) and *V. ×morreniana* (Ill. 49), among others.

4 In some *Nidularium, Neoregelia* and some *Canistrum* species, as well as *Guzmania sanguinea*, an elongation of the scape does not occur. On the contrary, the scape here remains so short that the compact inflorescence is sunk into the centre of the rosette (Plate 3), even under water, and only its flowers are raised above the surface of the water. The fact that the young flower buds do not rot is explained by the fact that rot-retarding

substances are emitted from the sheaths of the inner rosette leaves. This type of inflorescence is known as a 'nest' and is exemplified notably in the genus name *Nidularium*. In nearly all species with nest-like inflorescences, the inner rosette leaves, so-called heart leaves, are distinguished by vivid red, rust-red, bluish or sometimes white colouration. They are constructed similarly though varying in form (homologous) to the coloured scape bracts of an elongated inflorescence and for months make up the charm of these plants. *Cryptanthus*, some *Neoregelia* and *Wittrockia* have scape-less inflorescences without colourful heart leaves.

Inflorescences are either simple or compound, maturing into racemes or spikes.

A raceme has stalked flowers sitting in the axes of the flower bracts (Figs. 12, 1; 13, 1), a spike has flowers without stalks (Figs. 12, 2; 13, 2). Most common of the simple inflorescences is the spiked form (Fig. 12, 2 and Ills. 53 – 55).

A simple inflorescence is unbranched or single.

Compound inflorescences can reach enormous size as in *Puya raimondii* (Ill. 25) and appear in a few cases as multiple racemes (Fig. 13, 1a) or as panicles but more often as multiple spikes (Fig. 13, 2a) or as spike panicles. Multiple racemes are made up of a number of simple

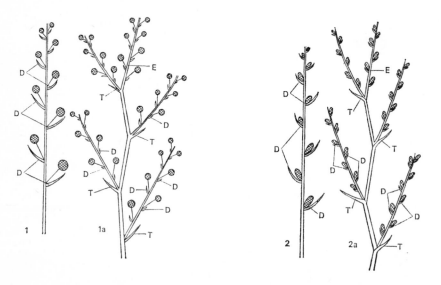

Fig. 13 Bromeliad inflorescences: 1 simple raceme; 1a compound raceme; 2 simple spicate inflorescence; 2a compound spicate inflorescence. Single flowers are cross-hatched; D flower bracts; T primary bracts, segmented inflorescences (not individual flowers) are found in axes; E terminal inflorescence.

racemes arising from the axes of the primary bracts (Fig. 13, 1a, T) below the terminal raceme (Fig. 13, 1a, E) (First order side branches). *Fosterella penduliflora* (Ill. 50) and *Aechmea ferruginea* (Ill. 51) are examples. Further branching of the simple racemes (second or higher order) brings about a panicle inflorescence, very rare in bromeliads. Multiple spikes (Fig. 13, 2a; Fig. 14, 2; Ills. 52 and 56) follow the branching pattern of multiple racemes; further branching of the simple racemes forms a spike panicle as in *Tillandsia guatemalensis* (Ill. 107).

A single spiked inflorescence is made up as follows: along the axis (also called a rhachis or a spindle), of the inflorescence bracts are set in either spiral (Ill. 53) or in distichous (in two vertical ranks) (Ill. 54) arrangement which bear the flowers in their axes (Fig. 12, D). In the distichous arrangement of the bracts, which are common in the vegetative regions of bromeliads with spiral leaf arrangement, a change of leaf arrangement occurs as the fertile section matures.

If the internodes of the inflorescence axis are elongated, then the flower bracts are widely spaced and the rhachis is visible (Fig. 12, 2; Ill. 54; Ill. 85 to 88). If the internodes of the rhachis are very short, then the bracts succeed each other closely and partially overlap in a distichous arrangement; this is an imbricate arrangement of the bracts (Fig. 14 and Ill. 55). When these are greatly enlarged and strikingly coloured, the inflorescence is known as sword-shaped. The 'Flaming Sword', *Vriesea splendens*, is one of the most beautiful examples (Plate 63), and other sword examples are *Tillandsia multicaulis* (Plate 8), *T. anceps, T. cyanea, T. lindeni* (Plate 30), *T. lampropoda* (Plate 39) and many *Vriesea* species.

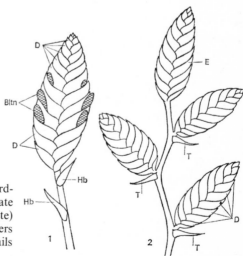

Fig. 14 1 sketch of simple spicate, sword-shaped inflorescence; 2 compound spicate inflorescence with overlapping (imbricate) arrangement of the flower bracts. Flowers (Bltn) in 1 are cross-hatched. Other details as Figs 12 and 13.

Form, size and colour of the floral bracts can vary greatly:

1   They may be small and unnoticeable and appear as tiny scoops, e.g. in many *Aechmea* species, and in *Catopsis* (Ill. 52).

2   The sometimes coloured scape bracts may continue on into the inflorescence as floral bracts that become smaller and smaller toward the tip (Fig. 12, 1). This behaviour is found in a number of *Aechmea, Vriesea, Billbergia* and *Guzmania* species.

3   The scape bracts which follow the foliage rosettes may be small and insignificant, but continue into the floral bracts, which are vividly coloured and become larger toward the tip. Examples are *Vriesea incurvata* (Ill. 55), *V. splendens* (Plate 63), *Tillandsia multicaulis* (Plate 8), and *Guzmania monostachya* (Plate 9).

4   In comparison to the scoop-shaped, insignificant scape bracts, the floral bracts are greatly enlarged and usually also vividly coloured. Frequently, the transition from scape bracts to floral bracts is abrupt so that the flower-bearing segment is sharply distinguished from the scape. An example of this behaviour is shown in *Vriesea carinata* (Plate 11).

In the compound inflorescences, not only the floral bracts of the segmented inflorescences are distinguished by vivid colouration, but frequently the scape bracts are also colourful, e.g. *T. cacticola* (Plate 15). Often it is the scape bracts alone, which make the inflorescence attractive. This is the case in *Tillandsia leiboldiana* (Plate 12), *Guzmania variegata* (Ill. 153), *Tillandsia imperialis* (Plate 14), and many others. In many bromeliads this interplay of colour between the scape bracts and the floral bracts, lasting weeks or even months, is the source of the beauty of their inflorescences and makes them valuable in commerce.

The number of flowers in a racemed or panicled inflorescence can vary greatly generally numerous blooms are carried, but also there can be but a single flower of which examples are *Tillandsia usneoides* (Fig. 6), *T. bryoides, T. pedicellata, T. capillaris*. This is however not a single flower, but a single-flowered inflorescence, for above the flower there is a short, sterile tip of the inflorescence axis and the bloom sits in the axis of a floral bract.

In segmented inflorescences also, the number of flowers can be as few as one. Thus *T. atroviridipetala* (Plate 22) carries a compound inflorescence composed of two to six single-flowered spikes giving the illusion of a simple inflorescence with two to six flowers.

Another peculiarity of all bromeliad inflorescences is the fact that they are always 'bare' at their tips, i.e. they have no terminal flowers. In some species the inflorescence end develops into a tuft of sterile bracts as with *Aechmea hystrix* and *A. comata* (Pl. 67) or leaves. With *Ananas comosus* the terminal leaf rosette is capable of vegetative reproduction; the simple-panicled fruit spikes are sterile and form no seeds. Another example of leaf formation at the ends of the inflorescence is the Peruvian desert tillandsia, *T. latifolia*. Its inflorescence is composed of spikes, and its terminal spike as well as the side spikes are capable of leaf tuft formation. Usually, however, only the terminal spike forms a tuft, which, even while still attached to the mother plant, forms a sizeable rosette (Ill. 57). As a result of its increasing weight, the elongated scape arches to the ground so that the young rosette can take root and grow into a new plant. Since the scape becomes woody, it remains connected to the old rosette for a long time. This abundant vegetative propagation also explains the mass stands of *Tillandsia latifolia* in extensive parts of the Peruvian coastal desert (Ill. 17). Similar behaviour is described by Harms (1928) for the south Peruvian desert tillandsia, *T. werdermannii*, by Herzberger (1910) for *T. tenuifolia* and by L. B. Smith for *T. baileyi*.

In another species, designated by L. B. Smith as *T.* aff. *denudata*, which we collected in central Peru near Tarma (at about 2,000 m), we observed brood shoots not at the tips but at the bases of the spikes on an inflorescence composed completely of compound spikes (Ill. 58, ak).

Some tillandsias have yet another kind of vegetative reproduction; the reproductive offshoot does not develop near the actual inflorescence, but in the axes of the scape bracts. Such a tillandsia, *Tillandsia somnians* appears in mass stands in northern Peru between Piura and Ayabaca at altitudes between 1,800 and 2,000 m and completely envelops the crown of small trees. From the middle of the rosette, which is about 50 cm in diameter, yellow-green or dark wine-coloured, a thin, tough, arched scape 1.5 to 2 m long arises, bearing a few small spikes at its tip (Ill. 60, J). As soon as the plant has finished blooming and the mature rosette (Ill. 60, R) dies, offshoots (ak) appear from the axis of the scape bracts, and because of the effect of gravity, all seem to spring from the upper side of the scape (Ill. 60). We have observed that the mass stands of this species result exclusively from the generous vegetative reproduction, whereby lignification of the scape frequently allows up to four generations to remain attached.

The same kind of reproduction is described by André (1890) in *Tillandsia flexuosa* var. *vivipara, T. incarnata* and *T. secunda*. In the unidentified Peruvian tillandsia depicted in Ill. 59, new rosettes develop in the axes of the basal scape bracts of the severed inflorescence while the mature plant dies without sending out offshoots.

Aside from the above exceptions, the inflorescences of bromeliads die after the distribution of seed (frequently accompanied by the withering of the old rosette). An exception to this behaviour is *Deuterocohnia longipetala* (Ill. 246) found in masses at 600 m in the cactus desert of northern Peru. Its paniculate, slightly woody inflorescence of 1 to 2 m lasts for several years, and the panicle branches produce new bloom growth at their bases at the beginning of the next growth period. There is no documentation on the possible number of repetitions of this process. In blooming greenhouse plants, it was determined that the inflorescences can produce flowers for approximately three years.

Flowers

The flowers of bromeliads are typical monocotyledon flowers, i.e. all the flower whorls are of three parts. Exceptions appear only in abnormal flowers. The three sepals are followed by three delicate petals, then stamens arranged in two circles of three each, and finally the hypogynous (superior or inferior) ovary formed of three fused carpels.

Before describing the flower organs, let us look briefly at the botany of the flowers. They are short-lived in comparison to their primary bracts. Their bloom period ranges from a few hours to several days. Some *Vriesea* species, e.g. *V. fenestralis, V. gigantea, V. racinae*, are exclusively night bloomers; their flowers begin to open in the late afternoon or later at night and are already closed again by morning. The remarkable scent during the blooming process indicates pollination by nocturnal moths or butterflies. In general, however, since they are pollinated primarily by humming-birds, the blooms of bromeliads are usually odourless. The flowers of the Peruvian desert tillandsias *T. straminea* and *T. purpurea* are distinguished by a pleasant scent (like stock flowers), as are those of *T. cacticola, T. duratii, T. xiphioides, T. dyeriana, T. hamaleana*; the flowers of *Aechmea cylindrata* are said to smell of hyacinths and those of *Vriesea regina* are said to smell of jasmine. The nocturnal flowers of *Vriesea jonghii* give off a replusive smell of opossums.

The succession of individual flowers in the simple as well as in the segmented inflorescences is acropetal, i.e. in succession from bottom to top—in the head-shaped inflorescences from outside to inside. However, there are exceptions to this behaviour.

The flowers are generally radial. In some *Billbergia* and *Pitcairnia,* species they are slightly zygomorphic because of twisting of the perianth tube. The same is true in *T. multicaulis*, in which at the unfolded stage the flowers are unilaterally split.

The sepals are usually clearly distinguished from the petals by size, colour and consistency; they are usually firm, greenish and shorter than the petals (Fig. 15, κ), although many Vrieseas, Pitcairnias, Billbergias and Aechmeas have coloured calyxes. In *Guzmania musaica* (Plate 56), however, the vividly coloured calyx supersedes the short flower.

The sepals may be separated from each other or fused to some degree from bottom upward; they are either symmetric or asymmetric. In *Puya* and *Canistrum* species the sepals are covered with a loose woolfelt.

The petals are usually many times taller than the calyx and are distinguished by vivid colouration (Fig. 15, B). The colours blue, red, violet, white, yellow, and more rarely brown (*T. aureo-brunnea*) or green are predominant. The petals of certain *Puya* species are strikingly metallic blue, blue-green or aqua. Frequently, the petals fade after bloom and display a completely different post-floral colour.

In many *Puya* and *Pitcairnia* species, the petals twist like a cork-screw after blooming; in billbergias of the subgenus *Helicodea*, they form a spiral from outside to inside.

Generally, the petals are separated; only in some species of *Navia, Guzmania, Greigia, Cryptanthus, Neoregelia, Nidularium* are they fused. The form and the size of the petals are extraordinarily variable. There are flowers that are so small that one nearly overlooks them, e.g. *Tillandsia bryoides* and *T. pedicellata*; on the other hand the flowers of *Pitcairnia ferruginea, P. mirabilis* and *Tillandsia grandis* as well as certain billbergias of the subgenus *Helicodea* are up to 12 cm long.

The petals themselves are longitudinally either lanceolate, lineal, or oval, entire margined, seldom finely dentate on the margin (*Tillandsia xiphioides*) and almost always with a 'claw', i.e. they have a narrow, more or less elongated basal segment (claw) surrounded by the sepals and gradually or suddenly forming into a broad and often angular blade (lamina).

At the base of the petals there are sometimes outcrops in the form of small scales, which are also called ligulae (Fig. 16, 2, L). *Petala ligulata* and *petala eligulata* are the features which distinguish the genera *Vriesea* and *Tillandsia*: the petals of *Vriesea* are ligulate (Fig. 16, 2), those of *Tillandsia*, are eligulate (Fig. 16, 1). Since the scales of dried specimens are fragile and not always visible, many species of *Vriesea* were once assigned to the genus *Tillandsia*. Vegetatively, the members of these two genera can hardly be distinguished from one another; names such as *Vriesea tillandsioides* are indicative of this fact. These scales, which can be simple, deeply separated (double), entire margined, dentate, fringed or split, have been interpreted to be 'devices to prevent the escape of the honey. This makes sense only in the pendulous flowers (e.g. in *Billbergia bakeri**). The interpretation is not plausible in many cases' (Harms, 1930).

Generally, bromeliad flowers open to differing degrees; the plate, (the

---

*Billbergia bakeri is now *B. distachia*.

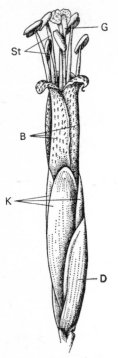

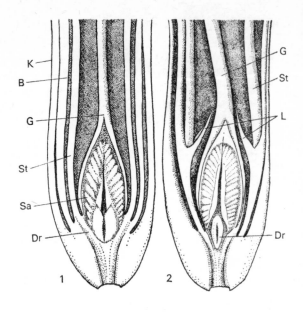

Fig. 15 Individual flower of a tillandsia: D flower bract; K sepals; B petals; St stamens; G pistil.

Fig. 16 Longitudinal cross section of the base of the flower of a *Tillandsia* (1) and of a *Vriesea* (2): L scales (ligulae); K sepals; B petals; St stamens; G pistil. The ovary has been bisected to show seeds (Sa) and glandular tissue (Dr).

tips of the petals) bend back, flatten out or even roll into a spiral. Observations by Ule indicate that in some nidulariums as well as one *Canistrum*, one *Quesnelia* and *Aechmea fasciata*, the dome-shaped corolla remains completely closed. It has not yet been proven whether these species display a true cleistogamy (self pollination). Ule (1898) cites only one case of cleistogamy—in *Aechmea sphaerocephala*, a large terrestrial bromeliad of eastern Brazil. In this plant, the flowers are said to remain completely closed so that self pollination results. The lack of nectar in the flower also indicates this fact.

The relationship of the length of the stamens (Fig. 15, St) to the corolla varies and is significant in the identification of the subgenera of *Tillandsia*. The stamens may be shorter, exactly as long as or longer than the corolla. If the petals are fused at their bases, then the filaments are all united with the perianth tube at varying heights. In separated corollas, only the filaments of the inner stamen whorl are attached to the petals at varying heights. In *Tillandsia monadelpha*, the filaments of all stamens are fused (Fig. 40).

The three-part ovary, consisting of three fused carpels, in the *Tillandsioideae* and *Pitcairnioideae* is superior or half-inferior; in the *Bromelioideae*, it is inferior. In the latter, above the actual ovary, there is a long or short tube for the acceptance of honey, which is secreted by special glands located in the partitions of the ovary. Septal glands (Fig. 16, Dr) are present in all bromeliads and secrete a thick or thin slimy liquid of clear or dark brown colour, which seeps through special openings in the space between the perianth and the ovary.

The majority of bromeliads have hermaphrodite flowers. Incomplete hermaphroditic flowers—that is those in which one sex is greatly suppressed—are found only in *Hechtia* and *Catopsis*. In the latter (specifically in the subgenus *Tridynandra*) the male inflorescences are much more richly branched than the females. The male flowers are smaller than the females.

Pollination and Fruit Formation

Few precise obversations of the process of pollination in bromeliads have been made since studies can be made only in the natural habitat. The vivid colouration of the scape bracts and the flower bracts, the production of sticky pollen and the often heavy secretion of honey indicate cross-pollination—primarily by birds, but also by moths and butterflies. Relatively few species emit any scent—unlike the Peruvian desert tillandsias *T. purpurea* and *T. straminea* (with the scent of Stock).

The heavy secretion of nectar in bromeliad flowers, the prevalence of 'parrot' colours i.e. the effect of contrasting colours such as red, yellow and blue-violet—and especially the glowing red of the scape bracts, the primary bracts and the flower bracts—would indicate that many bromeliads are ornithophilous (pollinated by birds). The brilliant red not only stands out as a complementary colour to the surrounding green, but is also the colour to which a bird's eye is especially sensitive. Blue, a colour least sensitive to birds, is very rare in bromeliads. Whenever blue tones are present, they are tinged with red or violet.

Among the pollinating birds, the nectar-sucking hummingbirds play a major role. On our trips into bromeliad territory it was again and again an inspiring, beautiful and impressive experience to see the sometimes very colourful hummingbirds as they hovered before the bromeliad flowers, dipping their long, arched bills into the blooms to quench their thirst with the secreted nectar and thereby causing pollination. Even at elevations of 4,000 m, we observed birds swarming around the gigantic inflorescences of *Puya raimondii*. According to Johow (1898) *Puya chilensis* is pollinated by the Chilean Starling *Curacus aterrimus*, 'which uses the sterile ends of the segmented inflorescence as a perch and drinks the heavily secreted, sugar-poor nectar from the odourless, greenish-yellow flowers and powders his forehead with bright yellow, very sticky pollen.'

A peculiarity of the ornithophilous bromeliads is the protandrous* quality of their flowers: the anthers appear at a time when the stigmata of the same flower are not yet mature. If, for example, a nectar-seeking hummingbird touches the ripe pollen sacks, it picks up pollen. If it then flies to another flower with ripe stigmata, it may wipe off some of the pollen and so effects pollination. The flowers that are visited by birds—particularly by hummingbirds—are recognizable by the fact that their perianth has a relatively long tube, from which the stamens and the pistil and stigmata protrude. Short-tubed flowers seem to be sought out by butterflies. It is not yet clear whether the bumblebees, honeybees and ants seen on bromeliad flowers play a role as pollinators.

In addition to flowers that require cross-pollination, there are bromeliads that are self-pollinating. This is the case with *Guzmania* and certain *Aechmea* species. *Vriesea splendens* displays a special form of self-pollination. The pollen sacks initially stand above the stigmata; if cross-pollination does not take place, then the pistil and stigmata grow into the anthers and self-pollination occurs.

Fruit and Seed

The type of fruit formation (fructification) is closely related to the position of the ovary. On page 49 it was noted that the members of the subfamily *Bromelioideae* have inferior ovaries while those of the subfamilies *Pitcairnioideae* and *Tillandsioideae* have superior or half-superior ovaries. From the latter, a hard capsular fruit develops and grows within the partitions (septa) (Ill. 61), to form a septate capsule. All genera with inferior ovaries have berries (Ill. 63 and 64), which are sometimes vividly coloured (white, red, blue) and perhaps entice birds which then distribute the seeds. In *Ananas* and *Pseudananas*, the entire inflorescence becomes fleshy and juicy, and the bracts (except for their papery terminals) unite with the ovary to form a club-shaped fruit.

In the *Pitcairnioideae*, the capsules open along the septa, and also split in the loculus—generally in the upper third.

The *Pitcairnioideae* seeds are usually small, triangular to lanceolate,

---

*PROTANDROUS—the anthers mature before the pistils in the same flower, the pollen being dispersed before the pistils are receptive.

Fig. 17 Winged seeds of *Pitcairnia ferruginea* Ruiz et Pav. (left) and *Puya raimondii* Harms (right).

winged with a thin tissue (Fig. 17). They are scattered by the wind and blown into clefts in the rocks, where they find suitable germination conditions. Nearly all *Pitcairnioideae* are terrestrials. We found young plants of *Puya raimondii* only on boulder slopes among the rocks.

The seeds of the berry-forming *Bromelioideae* are small, spherical, and generally pear-shaped. They have a grey, brown or black husk (Ill. 64), which is covered by a jelly, which helps the seeds to stick to trees— primarily to those with a rough bark. Since the berries are eaten by birds, the seeds may also be spread through their droppings.

The seeds of the capsulate *Tillandsioideae* are of quite a different construction. The fruits need about 6 to 12 months to mature. They are mostly cylindrical or have three blunt edges and are up to 6 cm long. Sometimes in *Catopsis* the fruits are small, short and oval.

The seeds of *Tillandsioideae* are 1 to 7 mm long, slender and spindle-shaped. In their mature state they are attached to a parachute-like construction of hairs at their base (Fig. 18, 1 and 2) originating from the cells of the outer layer of the seed covering (integument) and separating from top to bottom in the form of hair-like strands attached to the basal end of the spermatic cord and the inner layer of seed covering, whose cells likewise develop into hairs and surround the spermatic cord (Fig. 18, 1 and 2). The hairs themselves are segmented and composed of elongated cells, which are joined either in a simple fashion or, more commonly, dovetailed with their fork-like ends (Fig. 18, 2a).

At the tip the seeds of *Tillandsioideae* have tufts of varying size attached to the husk (Fig. 18, 1 and 2, Sa).

Müller* cites the formation of a double parachute (Fig. 19) in *Tillandsia gardneri*:

*Müller: 'Die Keimung einiger Bromelien' ('Germination of some bromeliads'). Bulletin of the *Deutschen Botanischen Gesellschaft*, vol. XIII, 1895, pp. 175–182.

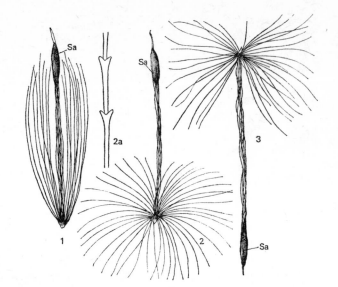

Fig. 18 Seeds of *Tillandsia utriculata*: 1 and 2 unfolding of the parachute (pappus); 3 seed (Sa) in flight (the heavy seed hangs downward); 2a segment of a pappus strand (enlarged).

'The outer seed covering breaks down into hair-like cell strands. Some of these detach from top to bottom and remain attached at their ends to the seed strand. Others separate from bottom to top, where they (not always at the same elevation) remain attached to the seed. Thus results a lower and an upper umbel, which are attached by the seed strands'

In the fruit, the seeds lie tightly pressed to each other at the top of the capsule; the hairs of the basal appendage are not yet separated and form a stem at whose tip sits the brown, spindle-shaped seed (Ill. 62; Fig. 18, 1). Once the capsule breaks open, the basal flight hairs open out like a parachute (Fig. 18, 2; Ill. 65). Hanging freely in the air, the unit makes a 180° turn; the heavy seed is then on the bottom and is borne by the parachute (Fig. 18, 3). The tuft serves not only as a flight apparatus, but it

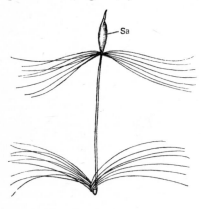

Fig. 19 *Tillandsia gardneri* Lindl. Ripe seed with a double pappus.

is also capable of sticking permanently to tree bark, rock cliffs, etc. However, our own observation has been that trees with smooth bark have essentially no bromeliads while trees with rough bark as well as rough cliff sides are ideal locations for attaching the parachutes.

Wind is the sole means of the distribution of the seeds; in rainy weather the flight hairs of all the seeds of one or several capsules can become so matted that hundreds of seedlings unite to form a clump. We have observed such behaviour in nature especially in *Tillandsia prodigiosa, T. imperialis, T. rauhii* (Ill. 68). In such a cluster, for unknown reasons, one seedling usually grows more rapidly than the others (Ill. 68) and develops into a mature rosette, while the others are retarded in growth.

With normal development of the seed, its flight hairs are distinguished by a beautiful silky, white sheen and open up as the capsule splits during dry weather, a sign to the grower that the seed is viable. Frequently in imported plants a premature ripening occurs, resulting in straw-like seed flight hairs, and seeds not capable of germination. During long rainy periods, naturally ripened seed cannot be spread by the wind, and germination frequently occurs on the old fruits (Ill. 66).

A number of studies have been made concerning the germination of bromeliad seeds. They all agree that the main root either does not develop at all, or, as is the case generally in monocotyledons, withers after a short time and is replaced by adventitious roots. Rootless tillandsias, such as *T. usneoides*, only in their youngest stages have a weakly developed root (Fig. 5), which soon withers and the plant grows on without roots.

3 The Living Bromeliad

Of greatest importance to the life of bromeliads are the scales or trichomes which cover the entire surface of the leaf, or appear in bands and stripes (Ill. 39). In tank bromeliads the scales are generally located on the upper side of the broad basal parts of the leaves. In the grey tillandsias the whole leaf blade is covered with the scales. There is a direct correlation between the whiteness of the leaves and the density of the trichomes. The leaves appear to be so white only because the scales, during dry weather, hold much air, which reflects the light well.

The scales absorb atmospheric water through capillary action like blotting paper and delivers it to the leaf tissue, where it is stored in a special water tissue (parenchyma) that varies in density in the various plants. As the plant becomes wet, the air between the trichomes is squeezed out and the white colour of the leaves gives way to green.

The thickness of the scale covering provides a clue to the ecological conditions of the habitat and the amount of precipitation available to the plants, and from this important information for the cultivation of bromeliads becomes available: the grey and white bromeliads, especially the tillandsias, live in areas of little rainfall but with high humidity. Frequently they appear along with cacti (Ill. 9–12). The green tillandsias, on the other hand, and all tank bromeliads are inhabitants of areas with relatively high precipitation—cloud forests and evergreen rain forests (Ill. 1 and 2).

Every trichome consists of two parts, the shield or trichome covering (Fig. 20, D), and the water absorption cells (Fig. 20, A).

In *Tillandsia* or *Vriesea*, with their highly developed trichomes, four shield cells are in the covering (Fig. 20, 1–3, Z). These are usually surrounded by a ring of 8 and then an additional ring of 16 cells (Fig. 21). The latter are surrounded by a wing (Fig. 20, F1), which normally is composed of 64 cells (there are exceptions), so that we have the following cell formula for the trichome covering: $4 + 8 + 16 + 64$ (Fig. 21).

In the dry state, the scale shrinks as shown in Fig. 20, 3; as soon as water is present, the 'accordion' walls of the shield cells expand (Fig. 20, 2). Then the water absorption process sets up a suction, and the cells

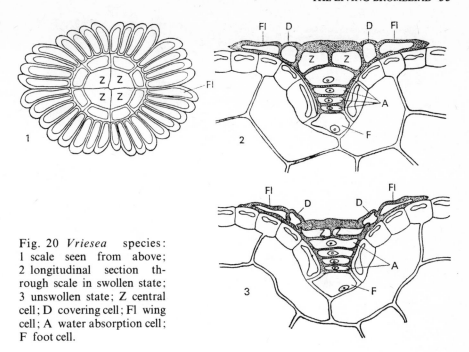

Fig. 20 *Vriesea* species: 1 scale seen from above; 2 longitudinal section through scale in swollen state; 3 unswollen state; Z central cell; D covering cell; Fl wing cell; A water absorption cell; F foot cell.

are filled with water (Fig. 20, 2A). Now the absorption cells can reassume their turgor and relay the water to the leaf tissue osmotically. Here, the water is then stored in special water tissues ready for the dry seasons.

The scales also act as a protection against transpiration and they reduce water evaporation considerably during dry periods; they may also act as light filters. The simplest scale development is found in the *Pitcairnioideae*, in the genera *Navia*, *Dyckia*, *Pitcairnia*, *Puya*, *Hechtia* and *Fosterella* (Fig. 22, Z). The trichomes are located primarily on the undersides of the leaves and have in the main no water-absorbing function, but rather they serve exclusively as protection against transpiration.

All species of the above named genera are terrestrials, with a highly developed root system that serves to absorb water. The rosette leaf blades are narrow and have only a slight sheath, in which no water collects.

A number of members of the Bromelioideae such as *Greigia*, *Fascicularia*, *Cryptanthus* and *Ananas* usually have only one central cell. But the scales are of such great number that they also contribute to the absorption of atmospheric moisture, even though these plants are equipped with a well-developed root system that absorbs water and leaf

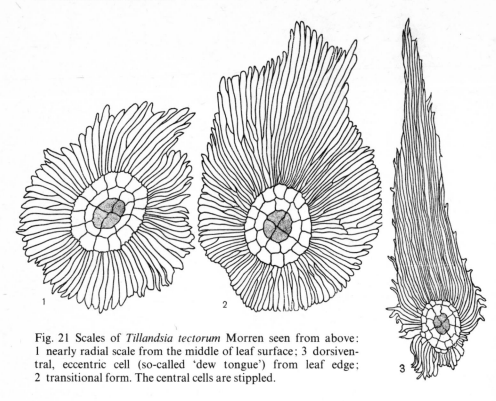

Fig. 21 Scales of *Tillandsia tectorum* Morren seen from above: 1 nearly radial scale from the middle of leaf surface; 3 dorsiventral, eccentric cell (so-called 'dew tongue') from leaf edge; 2 transitional form. The central cells are stippled.

blades that are deeply channelled, with their broad sheaths. Every leaf base has many trichomes and forms its own little tank, a so-called water niche.

Another group of bromeliads are the 'epiphytes of low development'. Among them are the genera from the subfamily of *Bromelioideae* that are not terrestrial but grow epiphytically, such as *Nidularium*, *Canistrum*, *Aechmea* and *Billbergia*, as well as a part of the *Tillandsioideae*, i.e. those members of the genera *Tillandsia*, *Vriesea*, *Guzmania* and *Catopsis*. The common feature of all of these forms is their watertight tank (page 45). In these tank bromeliads the water-absorbing trichomes are gathered at the base of the upper side of the leaves from which the plant draws the necessary water. The tank forms an aquarium for a fauna and flora that is

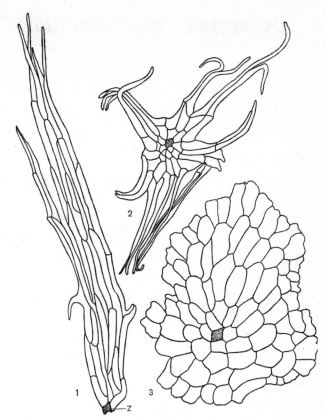

Fig. 22 Absorption scales of various 'primitive' bromeliads: 1 *Pitcairnia heterophylla*; 2 *Lindmania* (*Fosterella*) *penduliflora*; 3 *Hechtia tillandsioides*. Central cells are stippled.

specially adapted to bromeliads. Decomposition of material falling in the water helps to provide nourishment.

The highest level of development is found in those bromeliads which are all extremely xerophytic; white and grey tillandsias, and some vrieseas, such as *V. tillandsioides*, *V. rauhii*, *V. olmosana*, etc. All are inhabitants of areas with little precipitation but with high humidity.

In all atmospheric tillandsias, the whole leaf blade—the underside as well as the upper side—is equipped with large, water-absorbing trichomes. Special water absorbing tissue is highly developed in the bulbous tillandsias, *T. argentea*, *T. filifolia*, *T. plumosa* and *T. atroviridipetala*, in which the thick leaf bases function as reservoirs (Ill. 36).

4   Bromeliads in Cultivation

Propagation from Seed

Propagation of bromeliads is possible in two ways: vegetatively through the offshoots (pups), and by seed. Since bromeliads bloom only once in their life and then die, it is safest to propagate rare species vegetatively because the plant cannot be relied upon to set seed; many bromeliads are self-sterile and produce no seed from pollination with their own pollen. Mutations and good hybrids are also best propagated vegetatively. Propagation by offshoots is possible only in limited numbers, therefore seed growing is the only commercially practical propagation method.

It is not difficult to get seed on those tillandsias which bloom in our cultural conditions. Seed set is easiest to achieve in the self-fertile species such as *T. schiedeana, T. butzii, T. tricholepis*, etc. which pollinate themselves and produce seed.

For successful seed set of self-sterile species, one needs two plants of different origin blooming at the same time, i.e. they must not be from the same clone. Two pups from the same mother plant, even if they bloom simultaneously and are pollinated, will not produce viable seed. There are now dealers who import bromeliad seeds, but anyone who has collected bromeliads in their native habitat knows that the harvesting of seed of wild plants is often by pure chance.

Growing Atmospheric Tillandsias from Seed

The late Dr. R. Oeser, who for years worked on the cultivation of grey and white tillandsias and whose cultural methods found recognition, employed the methods described below.

Store the ripe seeds in an open glass container—*not* a closed one, since the germination process begins immediately. If you interrupt this process by shutting off the air, the seeds die. This also accounts for the rapid loss of their growth potential. In the fresh air, where oxygen is present, the seeds remain viable for a long time even in the direct sun. Seeds of *T. bergeri*, without contact with water in droplet form, absorb so much moisture from the air that they swell visibly.

Dr. Oeser used bundles of conifer twigs on which to sow seed of the atmospheric tillandsias. The branches of *Thuja* and *Juniperus* species (thuja and juniper), from which the young twigs have been only partially removed, proved especially suitable. Two pencil-thick twigs about 30 cm long, with all their side branches squeezed together, make up the basis of the seed bed. They are then wrapped with more thin branches until the bundle is 2 to 4 cm thick. In the middle and at the ends, it is tied tight with a wire, and the ends are trimmed.

Distribute the seed evenly but just loosely attached to the rough surface of the bundle, and then wrap the bundle with nylon thread. The actual securing of the seed comes about, by spraying the whole bundle—at first very carefully, later normally. Then the seeds and their parachutes lie tightly against the mount and will not fall off even when the bundle is dipped in water. Finally, attach to the upper end a strong wire hook that will allow the bundle to be hung up in a semi-shady place in the greenhouse.

No fungicides (chemicals that prevent damping off) are necessary, and sterile conditions are not required.

At least once a day (more frequently in warm, sunny weather), the bundles are dipped into clean rain water to which a weak solution of a general fertilizer has been added. The greenhouse windows are opened whenever the outside temperature permits; the humidity ranges from 90 to 95 % with the windows closed, to 20 to 30 % with the windows open. On the whole, the seedlings are cultivated under the same conditions as apply to older plants; they are, however, better protected from the sun's rays and are dipped more frequently. Seedlings treated in this manner reach maturity in approximately three to five years.

The success of germination on the *Thuja* bundles can perhaps be explained as follows: tillandsia seeds in nature stick to tree trunks and branches. The rain moistens them on the underside of the branches, where they are protected from the hot sun and where they remain moist for a longer period. The sap flow in the living branch prevents its surface from becoming so hot even in full sun. This is not the case on a dead branch. This cooling sap flow in the living branch is now replaced by our bundle of twigs: it has the effect of a wick, whose centre retains a certain amount of moisture long after being dipped. Cooling is accomplished by evaporation and has a favourable effect on the growth of the seedlings. There is an additional important factor: tillandsias are capable of withstanding long periods of repeated drought, but algae and fungi cannot stand long dry periods. Therefore, the surface of the bundle hanging freely in the breeze dries quickly and remains free from algae and fungus growth, while seeds sown in a pot on soil, peat, moss or fern root are quickly covered with a thin film of fungus and algae. This film prevents the penetration of moisture into the deeper layers and also prevents respiration in the young seedlings. They begin to wither and finally perish. No transplanting is necessary at this stage, and very little damping off occurs. In nature, no plants are transplanted. Either the plant survives where the seed landed, or it perishes.

Thuja and the other *Cupressus* (cypress) contain resins and essential oils that have a lasting effect; perhaps they also have a fungicidal and anti-bacterial effect without harming the tillandsia seedlings. It is also certain that the degree of acidity in the mount is important to the success of germination.

In the first year, the growth of the seedling hardly exceeds three to ten leaflets. During the autumn and winter months, growth stops. In the following spring, however, an astonishing increase in size is seen. In the summer, the bundles must be hung so that they receive no direct sun. During the darker winter months, however, they should be kept in a light position; they will not be burned by the sun since it shines at an oblique angle and has a weak effect. However, in spring the sun's rays can cause much burn damage!

Artificial lighting in winter will cause no harm. One should cover the bundle completely and for this reason it is a good plan to mix the seeds of fast-growing and slow-growing species and sow them together; the fast-growing *Tillandsia brachycaulos*, for example, not only protects the slow-growing *T. streptophylla* and *T. bulbosa*, but also creates a favourable micro-climate for them.

In the third and fourth year, the bundle is so quickly overgrown (Ill. 67) to necessitate thinning and transplanting. This often presents difficulties because the root development in young plants holds them so tightly to their mount that it may be necessary to dismantle the bundle. Along with their strong and usually lignified roots the seedlings are then attached with nylon thread to a fresh, thicker bundle of branches.

For small tillandsia seedlings that are hard to tie with nylon, grasp the roots of the seedling with the two arms of a double pine needle and prevent the seedling from falling out by wrapping the double needle with a thin florist's wire; the seedling and the double needle can then be tucked into a twig bundle.

The experience gained from the cultivation of innumerable tillandsia seedlings has also contributed to the improvement of the care of imported plants. As long as they are not too large, they are mounted in the fork of a rather large *Thuja* branch, and the point where the plant is attached is covered with thin branches that serve to retain moisture; fern fibre or moss is not used. If the point of attachment needs to be broadened for better distribution of the roots, wire on an old pine cone from which some scales have been removed. The tip should point upwards so that it can catch and retain water for a while. Not only are the cones durable, but they also serve the needs of the tillandsia roots.

If tillandsia seedlings are left on the material for several years, it is subject to natural aging; the thin branches become brittle. If the bundle becomes unsightly it can be renewed by inserting new twigs into the gaps and tying with wire. For such renovating branches of *Juniperus* and *Thuja* species with tightly pressed, spoon-shaped foliage, which can be pushed into the old bundle are used.

Already, in the introductory chapters, we alluded to the fact that most members of the subfamily of tillandsias are epiphytes and are falsely referred to by natives as *parasitos*. They colonize trees, cliffs, wires, even telephone wires and need only air and humidity in the form of dew, fog or rain. Their real habitat, however, is trees—living trees with rough or wrinkled bark. What is more appropriate than tree bark for growing epiphytes, especially the grey-leafed tillandsias, the decorative bark of the cork oak is particularly suitable.

It is useless to hope that the bromeliads will develop strong roots in the usual orchid mix because fungi and algae soon cover the surface of this plant material and prevent the absorption of water. Also, since peat moss is very hard to soak again once it has dried out, it is best not used for bromeliads at all. It is much better to attach the plants to bare cork bark. Astonishingly good root growth of the plants (the roots creep into all the crevices of the cork bark) is obtained when they are hung over heated water tanks and thereby are exposed to a constant high humidity.

Massive oak branches are used as bromeliad 'trees' in conservatories and florists' show windows and are very beautiful indeed, but they do not permit individual care of the individual species, such as frequent dipping. When such epiphyte 'trees' are freshly planted (Ills. 69 and 70), they look very fine but then after a few months of insufficient light and humidity the epiphytes wilt and fail. In botanical show houses, on the other hand, such epiphyte trees with the necessary care can convey a good impression of the living habits of tropical epiphytes.

It is not necessary that cork bark alone be used for mounting; the bark of some native trees is also suitable and decorative. The bark of old oak trees or the sprung bark from the base of old birches is very durable.

Highly recommended as the most suitable and especially the most beautiful mounting material for mature bromeliads and especially for the atmospheric tillandsias are grape vine stocks. They last about six to eight years before rotting. Oak branches, for example, in the warm, humid greenhouse would rot in only two to three years. Fast-growing wood such as spruce, fir or pine would deteriorate even more rapidly. The wood of the California *Sequoiadendron giganteum*, on the other hand, is very durable, but in other parts of the world *Cupressus* (cypress) branches for the growing of seedlings and grape stocks and cork bark for the further culture of plants and have provided the most success.

With age the quality and acidity of the mounting material declines as does fallen trees and branches in the jungles and cloud forests—bromeliads, orchids ferns and other epiphytes soon die under these conditions. In cultivation it is advisable to replant or remount after about four to six years.

On trips through the native habitats of bromeliads, one sees again and again that most of the grey-leafed, atmospheric tillandsias come from areas of little rainfall: they colonize trees of the savannah forests, which are foliated only during the short rainy season but are otherwise bare; they grow on dry cliffs and also appear as desert-like companion plants with xerophytes. As a result of their extreme locations, the atmospheric tillandsias can survive great drought, considerable temperature extremes between day and night and high light intensity. They live under conditions similar to those of cacti, with which they not only associate, but on which they often perch (Ill. 9 to 12). That means that one can easily grow cacti and atmospheric tillandsias together.

Like cacti, the atmospheric tillandsias can be taken outside in the spring after the danger of frost has past and left until autumn. People who have gardens can hang their plants from the branches of trees. At the Botanical Gardens of the University of Heidelberg a structure was built just for the outdoor culture of the grey-leafed tillandsias (Ill. 71). In the first few days and weeks, the plants are shaded to avoid burning, but are gradually accustomed to the full sunlight. As a result, the leaves of many species take on the glowing red colouration that they possess in their native habitats.

In the early morning and in the late evening, the plants are sprayed with rain water, to which a complete nutrient fertilizer is added every four weeks. The morning and evening sprayings correspond to the nightly dew in the native habitats. After only a few weeks the vigorous development of offshoots and roots tells us that the plants feel much better outdoors than in the clammy air of the greenhouse; not only do they have better colour, but their new growth is much more robust. Protection against long, hard rain is necessary only if the plants have not yet been sufficiently hardened.

In the famous family-run botanical garden 'Les Cedres', in St. Jean Cap Ferrat (southern France)—whose owner, the late J. Marnier-Lapostolle, was an enthusiastic collector of bromeliads and had the largest collection in Europe—the epiphytic tillandsias are so accustomed to outdoor culture that they self-sow. Of course, tillandsias and other xerophytic bromeliads from the genera *Hechtia, Bromelia, Aechmea* and *Puya* can remain outdoors all year only in favourable climates. In the European climate, the plants must be returned to the house or greenhouse as soon as frost threatens. To preserve their beautiful colour, they should be given the lightest spot possible. As house plants, we put them in windows which receive the most light; in the greenhouse we hang them on wires directly beneath the glass (Ill. 72). No artificial lighting is needed. There is no danger of burning because the plants have been 'hardened off'. Not until spring, when the sun's rays become more intensive and the plants have become 'soft' because of the poor winter light, is caution prescribed.

During the winter months, spraying can be reduced but not completely withdrawn because many species retain their native rhythm even in cultivation, i.e. during the winter months they prepare their inflorescences and finally reveal them in the spring. The winter temperature should be about 15°C.

Whenever climatic conditions permit, the greenhouses should be briefly aired during the winter months. In the cultures of the Botanical Garden at Heidelberg, automated air humidifiers and ventilators have stood up well when applied to the culture of atmospheric tillandsias. The humidity is distributed equally, and the plants dry relatively quickly again, thus reducing algae growth and also reducing the danger of the plants rotting at the bases.

The cultural measures described serve as a guide and should be altered to fit local conditions. Everyone who studies his plants intensely, will soon develop his own cultural recipes.

Growing Green-leafed Tillandsioideae and Other Bromeliads from Seed

W. Richter has thoroughly studied the culture of the soft, green-leafed Tillandsioideae—*Tillandsia, Vriesea, Guzmania* and *Catopsis*—as well as other bromeliads, and his experience still forms the basis for commercial bromeliad growing.

Normal rectangular or square clay seed dishes are carefully sterilized; their bottoms are covered with a layer of shards, which in turn are covered with 2 cm of coarse, steamed peat and a thin layer of pulverized fern root (*Osmunda*). There should be 2 to 3 cm of space below the·rim of the dish. The seeds are distributed evenly on the *Osmunda* fibre, and, since bromeliads germinate in the light, should not be covered. Finally, they are sprayed with a 1:1,000 solution of Chinosol to prevent the growth of fungi and algae. In place of fern fibre, use a 1:1:1 mixture of pine needles, peat and sand and sow the seeds on this. A one to one mixture of peat and sand is also very good.*

Until germination the dishes must be kept covered with glass and held at a temperature of 22–25°C. Germination begins after about 10 to 25 days †. Now light shading with thin paper is necessary. Richter does not recommend fertilizer for the young seedlings; additional lighting is to be used only during the winter months: 'You should leave the seedlings in the seed trays as long as possible with even temperature and humidity, as long as algae or moss do not constrict the development. Do not transplant too soon.' Usually, transplanting takes place four to five months after germination. The winged seeds of the genera *Puya* and *Pitcairnia* are treated in the same way.

The seeds of the berry type bromeliads—*Aechmea, Billbergia, Neoregelia, Nidularium* and *Cryptanthus*—are sown after removing the sticky remains of the sweet jelly and after thoroughly disinfecting the seeds with Chinosol. Sow the seeds on leaf mould mixed with milled peat moss and sand. The planting medium must be loose, well-drained and slightly acid. The pH value (degree of acidity) should be 4. The seeds of this kind of bromeliad, too, should not be covered with soil. Germination of the berry type bromeliads occurs at a temperature of about 20°C and constant high humidity after about 8 to 10 days† —sooner than those of the capsule type. In the beginning, you should cover the seed dish with glass, which can be raised as soon as the first leaflets appear.

Once the seedlings have grown a few leaves, they are separated and replanted. As a medium for the young seedlings, we use a mixture of milled leaf mould, milled peat and sandy humus in the ratio 2:2:1, i.e. slightly acid. Polyester granules which absorbs no water but loosen the medium for better root formation may also be added. For the first transplanting it is advisable to place the seedlings close enough together so that their leaves touch, in order to create a favourable climate for further growth.

As the seedlings grow, keep the dishes at a relatively high temperature and even humidity and spray the seedlings several times a day with rain water to provide the necessary moisture. The fast-growing seedlings of the berry type bromeliads can be transplanted again after 8 to 10 weeks into a coarser mixture of peat, soil and sand; the seedlings of the capsule-forming bromeliads do not require transplanting until after 3 to 4 months.

Potting-on should correspond to the small root system of bromeliads. A mixture of coarse pine needle chaff, sandy soil, peat, decayed compost,

*Of course, there are many other mixtures. *Ed.*
† With really fresh seed, germination begins much earlier. *Ed.*

some moss and sand makes a good medium. In any case it should be loose and well drained. If using pure peat, you must be careful not to let the surface become dry, since it is difficult to saturate again.

In most terrestrial bromeliads, the roots play an important role, not only as anchors but also as water absorption organs. The growth of the young plants proceeds more rapidly than in the atmospheric tillandsias, but on the whole they grow more slowly than other house plants. Repotting need occur no more than once a year. If the plant becomes too large for the pot, loosen it carefully. If the roots have grown rapidly to the walls of the pot—as is often the case—it is better to smash the pot than to disturb the root ball. The new pot should also not be too large. Use the same medium and pack it around the plant.

The best time for replanting in Europe is spring. Newly repotted plants should be watered only slightly at first but should be sprinkled frequently with room-temperature water if conditions permit. In the greenhouse, you can grow the plants in saturated air, which furthers new root growth.

Usually, it takes several years from seed to inflorescence, but if you notice an inflorescence appearing in the centre of the rosette, do not stop watering the centre of the plant. A lack of moisture endangers the further development of the inflorescence—you need not fear rot (see page 41). From the time the developing inflorescence is first noted until the first flowers appear can take several weeks or months. By warmth, light and chemicals (see page 71) you can speed up the development of the inflorescence. By cooling, on the other hand, the actual flowering time can be postponed and lengthened.

It is not absolutely necessary to grow the green-leafed Tillandsioideae in pots, for in their native habitats most of them do not live in the ground, but epiphytically on trees. Like the grey-leafed tillandsias, you may grow them on bark, branches or larger epiphyte 'trees', but it must be understood that the green species require much more careful attention than the grey-leafed tillandsias if they are to live long and not look shabby after a short time. Since the roots of the green-leafed bromeliads also serve to absorb water, you must take care to keep the root balls slightly damp.

To prepare such an epiphyte tree, remove the plants from their pots, shake the root balls well, wrap them in the roots of the fern *Polypodium vulgare*, which is also used as a medium for epiphytic orchids, and tie the whole mass to the branch with a thin wire or nylon thread (Ill. 70, b).

Peat moss (*Sphagnum*), used for wrapping the root ball, dries too quickly and is then hard to saturate.

Another way to attach the plants is to form little pockets out of the leathery, fibrous blades of older palm leaves, nail them to the branch, set the plant in with a little potting mix, and wrap the whole thing with a thin wire (Ill. 70, a). The fibres of the palm leaves not only look good, but they are also quite durable.

So that the epiphyte tree might give the impression of a natural combination of plants, the bromeliads may be accompanied by other plants that normally grow alongside bromeliads in their natural settings. These plants could be those which can stand the atmosphere of the home, e.g. *Cattleya* or *Rhipsalis* species, epiphytic *Araceae*, ferns, etc. (Ill. 69).

When creating an epiphyte tree, you must know where it will be placed and whether you can give the mounted plants the proper care; it is not easy to water plants mounted on a tree. The tree can be set into a large pot or tub and filled with concrete (for stability) which also allows the entire construction with its plants to be moved at will.

Epiphyte trees are best kept in a special, closed flower window or in a conservatory. For direct southern exposures (north of the Equator), northern exposures (south of the Equator), xerophytic tillandsias, hard-leafed aechmeas, billbergias and neoregelias are appropriate. If you want green-leaved Tillandsioideae, you must either arrange shading or plant them near the floor in the background in the shadows of the more robust, hanging species. For windows with east or north exposure (west or south in the Southern Hemisphere), the green, soft-leafed tillandsias, vrieseas, guzmanias, etc. are recommended. They cannot tolerate the direct rays of the sun. There are today small, transportable room greenhouses and portable glass houses in many models with sprinkling systems, electric heat and artificial lighting. They seem to be very useful for the culture of bromeliads since an ideal micro-climate can be created in them. (See also pages 76–78.)

Vegetative Propagation

Although the mature plants bloom only once and then die, they arrange for their continuance not only by the production of seeds, but they also reproduce vegetatively by the formation of offsets or 'pups'. This process, known as propagation, renewal or innovation (Ill. 28), repeated with every bloom period, makes the plants well nigh immortal. There are few exceptions to this behaviour: *Puya raimondii*, which is exhausted upon the formation of its gigantic inflorescence and dies without having offshoots; also *Tillandsia complanata*, which blooms for several years in a row before dying, but also produces no offshoots.

Aside from these exceptions, the nutrition deposited in the mother plant is used up by the offsets. They grow quickly while the old plant is dying. Offshoots develop in several ways:

1   In the stem-forming species, especially in tillandsias, one or more pups appear directly at the base of the inflorescence and continue the organism sympodially. This is the case in the extreme atmospheric tillandsias such as *T. tricholepis* (Ill. 27, b), *T. andicola* (Fig. 24), *T. capillaris*, *T. gilliesii*, *T. recurvata*, *T. mallemontii* (Fig. 39b), *T. decomposita*, *T. duratii*, etc. as well as in the clumping bromeliad *Abromeitiella* (Ill. 37).

In *T. decomposita*, *T. mallemontii*, *T. capillaris* and *T. gilliesii* the single pup develops so quickly and so strongly right from the beginning that the inflorescence is pushed to the side and thus gives the impression of being a lateral inflorescence (Fig. 11, J1).

2   In most bromeliads, especially in the funnel-shaped bromeliads, the pups appear in the axes of the basal rosette leaves of the strong axis segment. They either sit directly on the axis (Ill. 27 a), or they begin with a thin, elongated, stoloniferous section equipped with rudimentary leaves. This is the case with *Vriesea espinosae* (Ill. 155), many *Aechmea* species, *Neoregelia ampullacea* (Fig. 10), some pitcairnias, etc. The stoloniferous segment grows either horizontally (plagiotropically), as in *Vriesea espinosae* and *Neoregelia ampullacea* (Fig. 10), or it is directed diagonally upwards and towers over the mother plant (Ill. 28 b).

An exception to pup formation at the base of the mother plant is the beautiful *Guzmania sanguinea*, whose foreshortened inflorescence is situated under water (Plate 3). Its single pup appears at the base of the inflorescence, i.e. in the centre of the rosette. As the offshoot develops, the flower head is pushed to the side so that the seed capsules develop out of the water and can be released dry.

3   One further form of pup formation is observed in *Tillandsia viridiflora*, *T. rauhii*, *T. prodigiosa*, *T. imperialis*, *T. multicaulis*, etc. Here in the early stages of development, long before the mother plant is mature, a tuft of pups appears at the base of the rosette (Ill. 45). These are called adventitious offshoots and are small plants that are only loosely attached to the mother plant and can easily be removed and treated as seedlings. The cause of the formation of these adventitious pups is unclear, especially since by no means all species are capable of similar functions (Ill. 44).

4   Finally there are the offshoots which appear on the inflorescence (*Ananas comosus*, *Tillandsia latifolia*, *T. somnians*, etc.) as described before.

In all cases, pups are a means of vegetative propagation with the object of forming mature plants quickly. While a plant grown from seed blooms at the earliest in 3 to 5 years, a well cared for pup can reach the size of the mother plant in 12 months and thus be capable of bloom.

With a sharp knife, separate the offshoot directly at its point of attachment; with the stoloniferous type, remove the lignified runner with a garden shears since this runner is worthless for further purposes. It will not produce new roots.

Removing the pup soon after its appearance causes it to take much more time to reach bloom size, but by so doing the mother plant will often be encouraged to produce more offshoots. As they develop, the pups assume most of the nutriments deposited in the mother plant. Then the leaves of the mature rosette begin to yellow and finally to wither after the seeds mature. *Catopsis*, closely related to *Tillandsia*, demonstrates these processes in a striking way: after the mother plant has fulfilled its duty, its leaves become noticeably yellow, papery, and crisp and soon fall off entirely as if the plant had 'amputated' them itself.

In other species, especially in the clumping or turf-like species such as *Tillandsia ionantha*, *T. caput-medusae*, *T. pruinosa*, *T. festucoides* and *T. chaetophylla*, the mother plants do not die for years and because they continue the processes of assimilation, they continue to transfer nutriments to the young plants. It is, therefore, advisable to leave the pup attached to the mother plant at least as long as the parent is capable of supplying nutriments to the offshoots. They will then far exceed the mother in growth, size and beauty.

Many species of *Cryptanthus* show a very peculiar form of pup formation. Here, numerous offshoots appear in the axes of the rosette leaves, and as they grow they force themselves out and fall off. On a suitable medium they again take root.

With the grey-leaved tillandsias, it is better to leave the pups attached to the mother plant in order not to disturb their growth, and carefully to remove the old, dead leaves of the mother plant as they form; by this time the offshoots themselves have formed roots.

Treat the pups of the geen-leafed tillandsias like older, 2-year-old seedlings, i.e. pot them into larger dishes or suitable seed beds. Set them rather close so that their leaves touch and create a favourable micro-climate.

As a planting medium, use a peat/sand mixture with an addition of an aggregate to keep it loose. In the first few days keep the new sets dry and only mist the room or the greenhouse, then water regularly; the tanks of the funnel-shaped bromeliads should also be filled with water. The temperature should be between 20–25°C; furnish sufficient shade.

Plants grown from offshoots differ from seedlings only in the fact that their leaves are wider and taller right from the beginning while the primary leaves of seedlings even of wide-leafed plants, e.g. *Vriesea splendens*, are generally narrow and grass-like. With good care and sufficient fertilizer you can expect to have mature plants within one year.

In Europe, only the darkest winter months (November to January) are unfavourable for vegetative propagation.

Care of Imported Plants

Once we have made ourselves familiar with the care of atmospheric tillandsias and are able to cultivate and propagate the green, funnel-shaped bromeliads, it will not be difficult to grow bromeliads that have recently been imported from their native lands. The author's expeditions to Peru, Ecuador and Mexico not only furthered the study of native growing conditions, but also served to introduce living plants and their culture to Europe. His collecting activities are responsible for the fact that numerous rare species are now found in cultivation in Europe. In many areas of South and Central America there are collectors who continually export bromeliads to Europe. Unlike cacti, bromeliads should be sent only by air. Although the 'tough' tillandsias from the rocky deserts may survive for several weeks on board ship, the losses during the growing on period are relatively large. However, specimens of *Tillandsia usneoides*, that had laid for nine months in the herbarium pressed between newsprint, were in full bloom when removed; but this is certainly an exception, because when tillandsias (*T. latifolia*, *T. tectorum*, *T. recurvata*, *T. capillaris*) have been used as packing material for the transport of cacti by ship, more than half of them die later as a result of the long trip.

In their habitats, bromeliads, especially tillandsias, seldom grow alone, and usually form large masses. Remove the plants carefully from their supports. Although the old roots will no longer grow and therefore are *not* capable of taking up a new mount, they should not be removed for fear of damaging the rosette axis to which they are attached. At the very least you will reduce the risk of damage to the plant's growth if you leave it attached to the branch. In the case of little species such as *T. atroviridipetala*, *T. argentea*, *T. filifolia*, *T. ionantha*, which colonize thin twigs, this is easily possible. Dead mother plants and the dead basal leaves of the living rosettes should be carefully removed. Funnel-shaped bromeliads should be well shaken, so that no animals and other pests receive a free ride. Pay special attention to the 'heart' of the collected plants. Pull gently on the innermost leaves. If these are tight, then the plant is healthy. If they pull out easily, then the plant is worthless, even if it otherwise appears to be healthy. The cleaned plants are then given a collection number and data on the location, the exact elevation and the surrounding plant life. This information is of great importance for later scientific study. If there should be a new species among the collection, it could not be described without such information.

The collected and cleaned plants are then wrapped in dry newspaper and packed. But it is possible to ship the plants loose and not packed in cartons or baskets. This method is especially recommended for species with hard leaves; fewer leaves will be broken and the plants require less space when arranged in loose layers.

Equipped with a 'health certificate' supplied by the Ministry of Agriculture in the homeland, the plants can now begin their trip to their new home. After their arrival, they are again sorted and thoroughly cleaned. Old, damaged and unattractive leaves are removed and damaged roots are cut off. As a precaution, the imports should be dipped in a suitable insecticide in order to kill unseen pests. Although the plants were heartily shaken earlier in their homeland, one is always amazed at the abundance of beasties that still show up! Finally, every plant is registered, i.e. it gets an index card on which is written the exact collection information, the name of the collector and later the bloom time, bloom colour, etc. In addition, you can give each species a number for your own collection. Only collections organized in this manner have scientific value.

The further treatment of the imports depends on the kind of cultural requirements. The grey-leafed, atmospheric types are immediately tied onto wood or bark with nylon thread in an arrangement similar to the way they grew in their native habitats. Many species, such as *Tillandsia atroviridipetala*, hang from their natural mount. The neck of the root system is also surrounded by the planting medium, recommended for offshoots. In the first few weeks the plants are watered lightly. As soon as they show signs of life—as can be noted when the leaves become stiffer—they can be dipped and fertilized regularly. If it is suspected that grey-leafed import tillandsias are not entirely healthy and strong, it is recommended that they be placed on wet, loose peat in a bright location. Then they need be sprayed very little, but they profit from the air rising from the wet peat. The danger of rotting is then considerably lessened. This treatment can be extended to many weeks.

The soft, green-leafed *Tillandsioideae* are cleaned and put into a mixture of peat, sand and aggregate. Special attention should be given to watering during the first few weeks so that the plants will not rot at the base. Only after they begin to form new leaves, can they receive copious amounts of water.

The best times to import plants are spring and summer. If plants arrive in the winter, they should be kept as dry as possible until spring so that they will not rot.

Bloom Promotion

With the spread of bromeliad cultivation, their commercial development and their acceptance into collections of house plants, there has also been an increasing interest in experiments directed toward influencing the bloom of bromeliads in order to gain a certain amount of control over the bloom period. For the commercial grower, cultivating large numbers of bromeliads is profitable only if he can market several thousand specimens of, say, blooming *Vriesea splendens*, all at one time.

As early as the 1930's, it was noted that pineapple plants reached bloom earlier when the surrounding brush was burned. It was determined that the ethylene gas (C_2H_4) in the smoke was the factor that instigated bloom. A short time later it was discovered that acetylene gas (C_2H_2) would also influence the formation of fruit. Just as in the old fashioned carbide lamps, it is produced by the 'dissolution' of calcium carbide (CaC_2) in water:

$$CaC_2 + 2H_2O = Ca(OH)_2 + C_2H_2.$$

The practice was to place calcium carbide powder into the rosette after a rain, resulting in the quick formation of acetylene and subsequent formation of the inflorescence.

In 1962, such experiments were carried out with ornamental bromeliads, first using guzmanias. Today this method is used predominantly with *Aechmea fasciata*, *Vriesea splendens* and other funnel-shaped bromeliads. The empty tanks are filled with a solution of 50 g of carbide in 10 litres of water. During the next two weeks, the concentration of the carbide solution should not be changed. Watering with care, development of the inflorescence then takes approximately 6 to 7 weeks.

The use of calcium carbide in water has the disadvantage that the tank easily becomes blackened. Therefore growers have switched to using acetylene gas directly from a cylinder. By means of a depressurizing valve, the gas is directed into the tank water at a pressure of 0.2 atmospheres and thus is dissolved in the water. According to the experiences of the late Mr. Gülz, commercial bromeliad grower in Bad Vilbel, Germany, the application of the gas should last for some 8 to 12 seconds. For the treatment of great amounts of plants, Gülz recommended 'covering them with sheets of plastic film stretched on metal frames has proven to be useful. We introduce the properly calculated amount of gas into the air space under the film, preferably in the evening, and let it sit overnight.'

Before the acetylene treatment can be effective, the plants must already have reached bloom size. With the proper nutrition, sufficient fertilizer and extended lighting during the winter months, you can speed the growth and considerably shorten the wait for the acetylene treatment of marketable bromeliads. Such experiments have been carried out only on the funnel-shaped bromeliads. There are no studies on the behaviour of the tank-less atmospheric tillandsias.*

Diseases and Pests

In contrast to many other house plants and tropical greenhouse plants, bromeliads are relatively free from diseases and pests.

Botanical Diseases

The major botanical pests are fungi, particularly the fungus which covers seed pots with a fungal film, and which not only prevents the penetration of water into the potting medium, but also prevents an exchange of gases with the outside world. It is best to sterilize the pots and the medium before sowing and to wash the seeds in a weak solution of suitable fungicide to kill any clinging fungus spores. Do not sow the seeds too closely; water carefully and illuminate and aerate the seedlings well. If the formation of a fungus film cannot be prevented, then the seedlings can sometimes be saved only by transplanting them immediately. When *Tillandsia* seeds are sown on *Thuja* or *Cupressus* branches as described on pages 58–62, they are usually not attacked by fungus, since the cultures are aired and lighted from all sides.

Commercial growers of *Aechmea fasciata* have become concerned in recent years about a fungus disease known as Aechmea wilt, which has completely devastated *Aechmea* cultures in many nurseries. The cause of the disease is a fungus from the genus *Fusarium*, whose members can cause great damage by stopping up the water routes in aechmeas and in other cultivated plants (carnations, gladioluses, tomatoes, etc.).

The fungus is first noticed in the form of a brown, quickly spreading spot at the base of an exterior leaf, which soon becomes limp, withers, dies and falls from the rosette. Gradually the inner rosette leaves are attacked and killed by the fungus. High humidity and temperatures over 25°C favour the spread of the disease. A cure with chemicals is not yet possible. Therefore, as soon as the disease is detected, the affected plants should immediately be taken out and destroyed; the temperature should be lowered and the cultures well aired. In any case the greenhouses in which *Fusarium* is found must be thoroughly disinfected.

The build-up of algae in the scales of grey or white tillandsias should

---

* There are now several proprietary preparations available which will assist the amateur grower in promoting bloom. *Ed.*

not be considered a disease. Too high a humidity level is the cause, for this algae build-up occurs even in nature, especially in the Peruvian desert tillandsias, whose leaves become grass green during the *garua* fogs. The build-up of algae in no way limits the growth of the tillandsias; it is only unsightly since the leaves lose their beautiful white or grey colour. The algae spread especially quickly during the winter months. If the tillandsias are placed outside in the summer, the algae quickly recedes since it cannot stand sunlight.

Insect Pests

These include a small gall gnat from the natural habitats of the bromeliads. It lays its eggs in the soft, succulent tips of young roots. The root tips swell into spheres and quickly begin to harden and become woody. The development of the insect takes place in the galls; such root galls are found on imported plants, especially on tillandsias from the dry, deciduous forests. The plants do not appear to suffer any significant damage from these gnats.

The important pests in cultivation are mealy bug, scale and root scale; seldom if ever are they noted in the natural habitat.

White mealy bugs prefer to nestle into inaccessible places in the axes of the foliage and bracts and protect themselves against moisture by surrounding themselves with a fine, white, mealy wax. They quickly multiply into a thick, wooly tuft and can cause considerable damage to the plant by sucking out the fluids. Insufficient air movement and too much dry heat favour the spread of mealy bugs. They are easily eliminated with various contact pesticides.

Scale is also a sucking insect, which moves freely about in its youth, but as an adult animal surrounds itself with a hard, brown or yellow shield, under which it also lays eggs and raises the young brood. Some scale seem to have specialized on certain bromeliads, e.g. the *Aechmea* scale, *Gymnaspis aechmeae* with a flat, bright armour, and the *Ananas* scale, *Diaspis bromeliae*. Another *Diaspis* species, *D. boisduvalii*, prefers to settle on green-leafed bromeliads, green tillandsias, vrieseas and guzmanias. Treatment can be a lengthy business but modern insecticides are available.

Root scale is by far the most dangerous form of scale because it lives beneath the soil, and its presence is not noted until the plant wilts. The bluish white pests resemble mealy bug and suck at the roots. They multiply especially quickly on plants that are kept very dry. Among bromeliads they are dangerous only to the terrestrial general such as *Hechtia*, *Cryptanthus*, *Fosterella* (*Lindmania*), *Vriesea* and *Guzmania*.

When the plants are grown in peat, these pests appear in the winter when the medium is kept too dry.

Also among the animal pests are slugs and snails, which can cause considerable chewing damage especially to seedlings. On older plants they frequently 'graze' the vital trichomes or devour the young tips of the inflorescence. Controlling these pests is possible by scattering snail powder around or picking them off during the evening and night-time hours.

All things considered, the number of pests attacking bromeliads is relatively few. The atmospheric tillandsias are usually free of pests.

5   Which Bromeliads to Collect*

First let us look at the selection of bromeliads that European flower shops and specialist nurserymen offer us at the present. Compared to the enormous number of bromeliads worthy of cultivating, the number available is quite small; some aechmeas, vrieseas, guzmanias, neoregelias, and cryptanthuses as well as a certain selection of hybrids. It is remarkable that from the genus *Tillandsia* the European market offers precisely those species that are least suited to home culture; *T. lindeni*, *T. cyanea*, *T. tricolor*, etc.

Many bromeliads can be grown in a small space. The green-leafed tillandsia species from the rain forests need warmth, humidity and shade while the atmospheric tillandsias, on the other hand, need aridity and sun. The latter require even less care than most cacti because they need neither a pot nor soil. You tie them to a piece of wood or bark and do not need to transplant them for 4 or 5 years. Then simply spray them morning and evening with pure rain water, to which you add a complete fertilizer at certain intervals. The atmospheric tillandsias also do not suffer from a lack of water for periods of up to a week. During a not too extensive vacation or holiday trip, you need not give them a thought. Also the atmospheric tillandsias are much less susceptible to pests. Atmospheric tillandsias can be taken outdoors in the summer in areas where rainfall is not too great. Except for spraying or dipping, you need not worry about them. In the autumn the plants are brought back to the greenhouse or the home and given the brightest spots.

The vegetative propagation of the atmospheric tillandsias is as easy as it is for the cacti. Of course, water plays an important role in the cultivation of this group of bromeliads. It must be completely neutral or slightly acid, since alkaline water encourages the growth of algae, thus causing the plants to die. Also they should not be grown too moist. A frequent alternation between humidity and slight drying is recommended and parallels the natural circumstances. One must exercise only a little patience because atmospheric tillandsias grow considerably more slowly than those of the green species. But the latter also require more care in comparison to the atmospherics. Just the question of light is crucial for

* Personal reflections and observations on tillandsias by Prof. Rauh. *Ed.*

their culture. Since the leaf blades lack a thick mantle of scales, they can stand no direct sunlight. For example, *Vriesea splendens* loses its beautiful leaf markings when it is exposed to too much light. The leaves even begin to yellow. Direct sunlight is recommended only during the dim months of November to February (in Europe) while the other months require shading.

The green-leafed Tillandsioideae can be taken outside only when they can be given favourable conditions, i.e. sufficient humidity, much shade, and an evenly moist, slow-drying potting medium because we know that *Vriesea*, *Guzmania* and most green-leafed tillandsias use their roots for the absorption of water. The tank bromeliads must always have water in the tanks. For this purpose use only clean rain water. When dirty water is used the tank type bromeliads usually react with 'heart rot'. Neither organic nor inorganic salts should be added to the tank water. On the other hand, so-called foliage fertilizing—spraying the leaves with a weak solution of nutriments—has been proved to be useful. For the remainder, the experiments on fertilizing the funnel-shaped bromeliads are still in their early stages. But we do know that the relatively slow growth of bromeliads is related to inherited factors and that exterior influences such as high temperature and humidity or the addition of fertilizer have only a limited influence.

In summary, the green-leafed *Tillandsioideae*, in comparison with the grey-leafed species, require different care.

In a greenhouse or glass case in the window both groups can be grown together. The green-leafed, tank type bromeliads can be planted in the ground in a central bed or on the benches. The grey-leafed tillandsias can then be hung on wires just under the glass of the greenhouse roof (Ill. 72). Here they will receive much light and at the same time will provide sufficient shade for the vase-shaped species. The large bromeliad collection of the Botanical Garden of the University of Heidelberg proves that such culture is easily possible (Ill. 72).

Mr. H. Pochert constructed a glass case for himself (Fig. 23). He says of it:

'To grow tillandsias in the house, I use a case which is situated on a window sill and occupies the middle wing of a three-part window in my living room.

The case consists of an angle-iron frame, which can be closed by two sliding glass doors on the room side. The top is closed with a glass pane. The two sides are glazed permanently. On the window side, the case is open. Here it is walled by the glass of the middle window.

I came to this conclusion because I consider the strong effect of the outdoor temperature to be valuable, especially in regard to the lower night-time temperatures. A double pane on the outside of the vitrine would provide too much insulation.

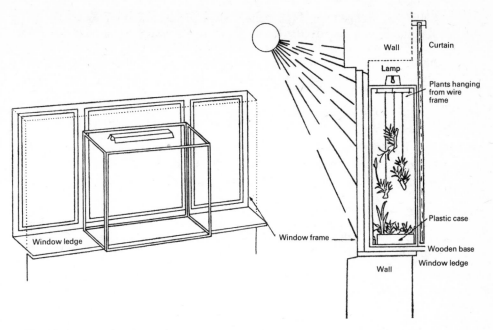

Fig. 23 Construction details for a home-made terrarium (after an original by H. Pochert).

Above the top glass there is a warm-tone fluorescent tube. In order to achieve humidity I have placed four square plastic pans in the bottom of the case. These are filled with damp leaf mould and are then planted.

The case is unheated. The temperature of the heated living room, even in winter, is quite sufficient. During extreme outdoor temperatures, I open the sliding glass doors on the room side. On very cold winter nights I use a low temperature heating cable, which raises the temperature in the lower third of the case by 3° (as protection for the plants in the plastic pans). When the room is heated in the winter and outside temperatures are around 0°C, the floor of the case is at 14–16°C. Just under the cover glass the temperature is about 24°C. I can achieve cooler night-time temperatures by drawing a simple linen curtain between the room and the case. It insulates the plants from the room heat. Then the temperatures at the bottom of the vitrine fall to 12°C, and the top falls to 15°C. When the outside temperatures permit, the two window panes at the left and the right of the case are opened. That gives the plants sufficient fresh air. The humidity conditions in the plant case are graduated.

In the winter, during the heating period, the humidity in the upper half of the case is at 20 to 30% day and night (immediately after dipping the plants it is somewhat higher for a while). Above the leaf mould pots the humidity is higher, about 40 to 50%. During dry cold spells in the winter it falls to 30% and sometimes even lower.

In the summer the humidity averages 50%. After dipping, it stays higher for a longer period of time than in the winter. In the spring, after evening watering, I can maintain a humidity of 70 to 80 % until morning because of the relatively cool nights, but to prevent rot I usually refrain from such practice.

During the cold winter weather I dip the plants every other day and let them dry quickly. In the summer they are dipped every day, seedlings twice per day. The relatively low humidity in the winter can be raised by using a vaporizer in the room.

The window is on the south side of the house. The plants in the window must be shaded from the sun's rays until May. From May to August the sun stands so high in the sky that the plants are not affected by its rays. Beginning in September, I must provide shade again. My shutters have movable slats for adjusting the amount of light. In the winter I give additional light by using the fluorescent tube.

I grow soft-leafed, green varieties and white ones with heavy scales. The white ones hang in the upper half of the case, the green ones in the lower half (Fig. 23). Here they benefit from the higher humidity. The white varieties in the upper part dry out faster. I have no losses from rot.'

Part II
Genera and Species

6 Classification of the Bromeliad Family

About 2000 species make up the family, based on authoritative studies by E. Morren, C. Mez and L.B. Smith, and are divided into three subfamilies:*

Sub-family Pitcairnioideae
Leaves entire or spiny, sometimes heterophyllous (Ill. 43); flowers large, conspicuous; ovary superior or half superior; fruits capsular; seeds winged (Fig. 17) or with comb-shaped or hood-shaped appendages; most are terrestrial.

Sub-family Tillandsioideae
Leaves entire; flowers small or conspicuous; fruits capsular; seeds plumose (with tufts of hair) (pappus, Fig. 18); epiphytes and terrestrials.

Sub-family Bromelioideae
Leaves seldom entire, usually spiny at the margins, spinous-dentate or serrate; flowers conspicuous; ovary usually inferior; fruits baccate (Ill. 63); seeds have no wing or pappus (Ill. 64); epiphytes or terrestrials.

Harms (1930) distinguishes one further sub-family Navioideae, with the single genus *Navia*. L.B. Smith places this genus under Pitcairnioideae, with *Navia* being the only genus among them with non-winged seeds.

In describing the Tillandsioideae we will follow L.B. Smith's classifications, which include the following genera: *Tillandsia, Vriesea, Guzmania, Mezobromelia, Catopsis* and *Glomeropitcirnia*. The traditionally separate genera, *Caraguata, Schlumbergeria, Sodiroa* and *Massangea*, are combined by Smith with the genus *Guzmania*; *Thecophyllum* and *Alcantarea* are included in the genus *Vriesea*; *Cipuropsis* is included under *Tillandsia*.

The delineation of the individual genera and even of the species is difficult and in no way can it be considered final since new discoveries continuously extend our knowledge.

* However, see page 82 for the Editor's note on the order of the subfamilies in this book.

Notes on Nomenclature

To facilitate international understanding, all plant names are derived from Greek or Latin. They refer either to an outstanding trait of the plant in question or to the names of persons who discovered the plant or who have acquired honours in the research of the plant group.

Since the time of the Swedish botanist Carl von Linné (Linnaeus) we have used a binary nomenclature so that every plant bears *two* names— the *genus* name and the *species* name. The abbreviated name often following the species name designates the *author* who was the first to describe the species officially.

As an example (from page 108) we give:

T. circinnata Schlechtend., 1844
(syn.: *T. paucifolia* Baker, 1878; *T. yucatana* Baker, 1887; *T. intermedia* Mez, 1903)

The explanation is:

Tillandsia is the genus name, given by Linné, to honour the Swedish botanist Tillands;

circinnata is the species name and comes from the Latin *circinnatus* and means rolled up, twisted, bent; this refers to a habit of the leaf blades.*

Schlechtend. (= Schlechtendahl) is the author, i.e. the botanist, who was the first to describe this plant.

1844 is the year in which the diagnosis (original description) was published.

Syn. stands for synonym and means a secondary name. It can happen that several scientists, unknown to each other, may write an original description of the same species under a different name. In 1878, Baker described this plant as *paucifolia*, originally described by Schlechtendahl as *T. circinnata*. In 1887 he changed it to *T. yucatana*, and in 1903 Mez described it as *T. intermedia*. If research proves that the same species was published under several names, then the earliest is the official name while the others are listed as synonyms.

Another example (from page 240):

Vriesea heterandra (André) L. B. Smith, 1951
(Syn.: *Tillandsia heterandra* André, 1888)

This means that in 1888 André described *Tillandsia heterandra*. Upon revising the genera *Tillandsia* and *Vriesea*, L. B. Smith determined that

* An explanation of the species name can be found in the description of each species.

Tillandsia heterandra is not a *Tillandsia* at all, but a *Vriesea*, and therefore changed names. Since he retained the species name *heterandra*, the original describer, André, appears in parentheses while the name *Tillandsia heterandra* is indicated as a synonym.

Also a book that is directed primarily at both the amateur and the serious collector should refer to the plant names accompanied by authors and dates of the original description as well as the most important synonyms in order to prevent a mass confusion of names.

The Order of Sub-families

Readers will note that the standard botanical arrangement of the sub-families is Pitcairnioideae, Tillandsioideae, and Bromelioideae.

This arrangement is *not* maintained in this book. The Editor has considered the matter in some depth and has decided that the translation into English should follow the arrangements contained in the original German two volumes of *Bromelien*; the first volume appeared in 1970 and covered the sub-family Tillandsioideae; the second volume was published in 1973 and dealt firstly with the sub-family Bromelioideae, followed by the sub-family Pitcairnioideae. These arrangements have, therefore, not been altered.

Ill. 1 Branches of the evergreen oaks of the cloud forest on Cerro San Felipe near Oaxaca (central Mexico), thickly covered with great funnels of *Tillandsia prodigiosa* Baker—'the Marvellous One'.

Ill. 2 In the cloud forest of Teotitlan near Tehuacan central Mexico, great rosettes of *T. bourgeaei* Baker indicate high humidity.

Ill. 3 *T. chaetophylla* Mez in thick nets on the cloud forest trees.

Ill. 4 *T. usneoides* hangs as long strands on the trees.

Ill. 5 Rosettes of *Tillandsia magnusiana* Wittm., like silver stars, at 2400 m in the oak forest near Cristobal de las Casas (Chiapas).

Ill. 7 One of the great tillandsias, with silver grey leaves, is *Tillandsia xerographica* Roh. from the dry forests of Mexico.

Ill. 6 Silver white rosettes of *Tillandsia matudai* L. B. Smith at 2400 m.

Ill. 8 *Tillandsia utriculata* L. grows in dry oak forests as well as on cacti.

Ill. 9 *Tillandsia makoyana* Baker on *Cephalocerus hoppenstedtii* near Tehuacan (central Mexico) at 2300 m.

Ill. 10 *Tillandsia albida* Mez et Purp. on *Marginatocereus* at 2200 m in the Barranco de los Venados near Pachuca central Mexico.

Ill. 11 *Tillandsia cacticola* L. B. Smith on *Armatocereus laetus* at 2100 m in the dry valley of Huancabamba northern Peru.

Ill. 12 Arms of *Armatocereus laetus* are completely covered with *Tillandsia recurvata* L. in the dry valley of Huancabamba, northern Peru, at 2100 m.

Ill. 13 Looking like a sea of white caps, great areas of *Tillandsia paleacea* Presl. in the Peruvian desert at Cajamarquilla near the capital Lima.

Ill. 14 *Tillandsia paleacea* Presl. The plants are arranged in long strands and have their growing tips pointed toward the sea as they die off from the landward side.

Ill. 15 Wide stretches of the Peruvian coastal desert are covered with *Tillandsia purpurea* Ruiz et Pav.

Ill. 16 A small clump of *Tillandsia purpurea* Ruiz et Pav. causes the formation of a sand dune.

Ill. 17 *Tillandsia latifolia* Meyen in the Peruvian desert near Chimbote.

Ill. 18 Large areas of *Tillandsia tectorum* Morren along with a *Haageocereus* in the rocky desert of the Casma Valley, central Peru, at about 800 m.

Ill. 19 In the dry forests of northern Peru, the tops of the deciduous trees are densely covered with the rosettes of *Tillandsia ebracteata* L. B. Smith.

Ill. 20 *Tillandsia ebracteata* L. B. Smith growing on *Espostoa lanata* in the valley of the Rio Saña at 1000 m in northern Peru.

Ill. 21 *Vriesea* (*Tillandsia*) *espinosae* L. B. Smith in the dry forest of the Piura valley at 800 m in northern Peru.

Ill. 22 *Tillandsia rauhii* L. B. Smith growing on steep cliffs of the Rio Saña valley at about 1000 m in northern Peru.

Ill. 23 *Tillandsia ferreyrae* L. B. Smith with an almost 2 m long, pendulous inflorescence on steep cliffs near Chachapoyas at 200 m northern Peru.

Ill. 24 *Tillandsia aff. fendleri* Griseb., one of the great tillandsias of Peru, forms mass stands on the steep banks of the Rio Huallaga near Huanuco central Peru.

Ill. 25 At 5–7 m *Puya raimondii* Harms is the largest bromeliad of the South American Andes; shown here at 4500 m in the Quebrada Pachacolo (Cord. blanca).

Ill. 26 *Tillandsia bryoides* Griseb. with its moss-like growth is one of the smallest Peruvian tillandsias; valley of the Rio Santa at 2000 m.

Ill. 27a The base of a blooming tillandsia with budding offshoots (Ek); b, c *Tillandsia tricholepsis* Baker; b blooming plant with inflorescence (J) and offshoots; c offshoot several years old with an acrotonal side shoot.

Ill. 28a–c *Neoregelia* sp. Mother rosette (R) with one offshoot and with two offshoots (b), which continue the propagation system sympodially; c sympodial–monochasial offshoot system with the growth periods from 1966 to 1968.

Ill. 29 *Tillandsia decomposita* Baker with rolled tips functioning as tendrils.

Ill. 30 *Tillandsia seleriana* Mez with pseudo-bulb. The hole (L) is the entrance for ants living in the plant.

Ill. 31 *Tillandsia ionantha* Planch.; an older, mounding specimen.

Ill. 32 *Tillandsia bulbosa* Hook. with pseudo-bulb.

Ill. 33 *Tillandsia butzii* Mez; a single pseudo-bulb; b same in longitudinal section.

Ill. 34 *Tillandsia disticha* H. B. K. Longitudinal section through the bulb-like base, which represents a transition from a hollow pseudo-bulb and a true bulb.

Ill. 35 *Tillandsia argentea* Griseb. (short, wide-leafed form) as a specimen of a true bulbous tillandsia.

Ill. 36 Longitudinal section through the bulb of *Tillandsia argentea* Griseb. (narrow, long-leafed form).

Ill. 37 *Abromeitiella brevifolia* (Griseb.) Castell. A mound about 1 m high.

Ill. 38 Section of a stand of *Tillandsia purpurea* Ruiz et Pav. Bromeliad leaf markings.

Ill. 39 *Cryptanthus zonatus* (Vis.) Beer.

Ill. 40 *Beoregelia carolinae* (Mez) L. B. Smith var. *tricolor* L. B. Smith.

Ill. 41 *Guzmania musaica* (Linden et André) Mez.

Ill. 42 *Vriesea gigantea* (Mart.) Mez.

Ill. 43 *Pitcairnia heterophylla* (Lindl.) Beer with thorny leaves and leafy foliage.

Ill. 44 *Tillandsia multicaulis* Steud. The base of a blooming plant with one normal pup (*left*) and one adventitious pup (*right*).

Ill. 45 *T. rauhii* L. B. Smith. Young plant with a tuft of adventitious pups.

Ill. 46 Longitudinal section and cross section of young shoots of *Puva raimondii* Harms (*top*) and of *P. roezlii* Morren (*bottom*), showing the course of the roots in the rind.

ak

Ill. 47 *Tillandsia complanata* Benth. Blooming plant with several inflorescences emerging from the plant axis.

Ill. 48 *Tillandsia latifolia* Meyen. Blooming plant. The rosette leaves merge into scape bracts becoming gradually smaller as they approach the tip. The end of the inflorescence has become a vegetative bud (ak).

Ill. 49 *Vriesea* x *morreniana* (*V. carinata* Wawra x *V. psittacina*) (Hook.) Lindl. Blooming plant showing scale-shaped scape bracts adjacent to the rosette leaves.

Ill. 50 Double racemes of *Fosterella penduliflora* Stapf.

Ill. 51 Double raceme of *Aechmea ferruginea* L. B. Smith.

Ill. 52 Double raceme of *Catopsis morreniana* Mez.

Ill. 53 Simple spike with a spiral arrangement of the flowers on *Guzmania calothyrsus* (Beer) Mez.

Ill. 54 Simple spike with the loose distichous arrangement of the flowers on *Tillandsia albida* Mez et Purp.

Ill. 55 Simple spike with imbricate arrangement of the flower bracts on *Vriesea incurvata* Gaud.

Ill. 56 Double spike inflorescence on *Tillandsia latifolia* Meyen var. divaricata (Benth.) Mez.

Vegetative propagation in tillandsias.
Ill. 57 *T. latifolia* Meyen var. *major* Mez. Old inflorescence (J), whose sterile tip has developed into a new plant (ak).

Ill. 58 *T.* aff. *denudata* André. Base of the bloom spike wth vegetative buds.

Ill. 59 *Tillandsia* sp. (Peru). Spent plant. Inflorescence scape (JS) with numerous axil buds; fading mother rosette (R).

Ill. 60 *T.* aff. *somnians* L. B. Smith. Blooming plant with emerging axil buds (ak).

Ill. 61 *Tillandsia utriculata* L. Bursting seed capsule.

Ill. 62 *Tillandsia fasciculata* Swartz. Seeds as they unfold their parachutes.

Ill. 63 *Aechmea chantinii* (Carr.) Baker. Inflorescence with berry fruits.

Ill. 64 *Billbergia farinosa* Koch. Longitudinal section and cross section of a berry.

Ill. 65 *Tillandsia caput-medusae* Morren. Fruiting plant scattering seed.

Ill. 66 *Tillandsia utriculata* L. The strewn seeds have germinated on the old inflorescence.

Ill. 67 Seedling cultures on *Thuja* branches with *Tillandsia streptophylla* (*left*).

Ill. 68 *Tillandsia rauhii*. Clumps of seedlings as they are often found in the wild.

Ill. 69 Epiphyte tree in the bromeliad nursery Gülz, Bad Vilbel, Germany.

Ill. 70 Transportable epiphyte tree showing various means of attaching the plants. (Further details in text.)

Ill. 71 Tillandsias in outdoor culture during the summer months.

Ill. 72 Inside the bromeliad house of the Botanical Gardens of the University of Heidelberg. Grey atmospheric tillandsias hang just beneath the glass; green bromeliads are planted in a side bed below.

Ill. 73 *Tillandsia adpressa* André var. *adpressa* in the cloud forest of Taulis at 2800 m; (*right*) inflorescence.

Ill. 74 *Tillandsia araujei* Mez; (*inset*) inflorescence.

Ill. 75 *Tillandsia aureobrunnea* Mez. Cajamarca, Peru, 2000 m.

Ill. 76 *Tillandsia bailevi* Rose. Mexico, Gulf of Veracruz.

Ill. 77 *Tillandsia bakeri* (Baker) L. B. Smith (central Peru, Cordillera blanca, Quebrada Yanganuco, 3000 m); (*inset*) inflorescence.

Ill. 78 *Tillandsia caulescens* Brongn. southern Peru, near Quillabamba, 2500 m.

Ill. 79 *Tillandsia balbisiana* Schult. Yucatan near Chichén Itzá, Mexico.

Ill. 80 *Tillandsia cauligera* Mez. Central Peru, near Tarma, 2000 m.

Ill. 81 *Tillandsia butzii* Mez.

Ill. 82 *Tillandsia caput-medusae* Morren.

Ill. 83 *Tillandsia circinnata* Schlechtend.

Ill. 84 *Tillandsia commixta* Mez. Peru, Chachapoyas, 2000 m.

Ill. 85 *Tillandsia cornuta* Mez et Sodiro.

Ill. 86 *Tillandsia monadelpha* (Morren) Baker.

Ill. 87 *Tillandsia narthecioides* Presl.

Ill. 88 *Tillandsia triglochinoides* Presl.

Ill. 89 *Tillandsia crispa* (Baker) Mez.

Ill. 90 *Tillandsia dasyliriifolia* Baker.

Ill. 91 *Tillandsia didisticha* (Morren) Baker (Paraguay).

Ill. 92 *Tillandsia disticha* H. B. K. (northern Peru).

Ill. 93 *Tillandsia caerulea* H. B. K.

Ill. 94 *Tillandsia calocephala* Wittm.

Ill. 95 *Tillandsia duratii* Viz.

Ill. 96 *Tillandsia dyeriana* Andre.

Ill. 97 *Tillandsia deppeana* Steud. var. *deppeana*, Chiapas, southern Mexico.

Ill. 98 *Tillandsia festucoides* Brongn. Teotitlan, Mexico.

Ill. 99 *Tillandsia filifolia* Cham. et Schlechtend. Mexico, near Fortin de las Flores, 1200 m.

Ill. 100 *Tillandsia floribunda* H.B.K. Huancabambatal, Nothern Peru.

Ill. 101 *Tillandsia flexuosa* SW.

Ill. 102 *Tillandsia funebris* Castell.

Ill. 103 *Tillandsia gardneri*. Terasopolis–Brazil.

Ill. 104 *Tillandsia grandis*. near Orizaba, Mexico, 1000 m.

Ill. 105 *Tillandsia guanacastensis* Standl. Costa Rica.

Ill. 106 *Tillandsia insularis* Mez. Galapagos Islands.

Ill. 107 *Tillandsia guatemalensis*. Mexico, Chiapas.

Ill. 108 *Tillandsia latifolia* Meyen var. *latifolia*. Coastal desert of Peru.

7 Sub-family Tillandsioideae

Key for Identification of the Genera

Identification Stage		Plant Characteristic	Species
1	a	Ovary half-superior; seeds with nearly equal length pappi on both ends	*Glomeropitcairnia*
	b	Ovary completely or mostly superior; seeds with pappus at the base or tip	see **2**
2	a	Pappus at the tip of the seed, folded in the capsule; flowers hermaphordite or unisexual	*Catopsis*
	b	Pappus at the base of the seed, straight in the capsule	see **3**
3	a	Petals separate or fused into a very short tube surpassed by the sepals	see **5**
	b	Sepals fused into a tube for at least $\frac{2}{3}$ of their length	see **4**
4	a	Petals without scales in their base	*Guzmania*
	b	Petals with scales at the inside of the base	*Mezobromelia*
5	a	Petals without a scale at their base	*Tillandsia*
	b	Petals with a scale at the inside of the base	*Vriesea*

Tillandsia L.

The genus *Tillandsia*, named after the Swedish botanist Tillands (1640 to 1693), was established in the year 1753 by Linné and with its more than 400 species is the largest genus in the family. For the fancier it is also the most interesting genus, since it contains many small species that can be grown in small greenhouses, flower windows or in terraniums.

Tillandsias occur throughout Central and South America and extend into the southern part of North America. They grow in tropical and subtropical regions, in cool mountain forests, in dry areas and in constantly moist rain forests.

The important recognition points for a tillandsia are:

1 Leaves never dentate or spinous, but always entire, often covered with a trichome (scale) mantle and therefore silver-grey or white in appearance;
2 Petals separate, with no scales at the base (contrasting to the closely related genus *Vriesea*, whose petals have scales at the base);
3 Ovary superior;
4 Fruit is a capsule;
5 Seeds have a pappus attached to the base;
6 Pappi lying straight in the capsule, not folded.

Tillandsia achyrostachys Morren, 1889 (Plate 20)

(*achyrostachys* with straw-like or chaff-like spikes; refers to the fact that the flower bracts after bloom take on a straw or chaff-like quality.)

Pl stemless, up to 40 cm high including inflorescence
Fo rosette, densely covered with grey scales
LS elongated oval, 4–6 cm long, 2–3·5 cm wide, grey or brownish scales. gradually blending into the leaf blade
LB upright or arched in the upper part, narrowly lanceolate, tapered, green, with dense grey scales, about 2·5 cm wide above the sheath, up to 20 cm long
Sc upright, naked, up to 15 cm long, 4–5 mm thick
SB tightly overlapped, the lower ones similar to the leaves, the upper ones almost surrounding the scape at their bases, thin, with more or less long, upright to arched tip, dense grey scales
I simple spiked, upright, laterally flattened, 10–20 cm long, 2 cm wide
FB upright, bipartite, closely overlapped, oval, pointed, up to 4 cm long, 2 cm wide, cutaneous, naked, with protruding veins, no keel, red
Fl upright, sessile, 4–5 cm long, green
Sa lanceolate, tapered, up to 2·6 cm long, thin-skinned, naked
Pe fused into a tube, 4–4·5 cm long, 6 mm wide
St protruding from the flower

Ha central and southern Mexico (around Puebla, Morelos, Oaxaca, vicinity of Acultzingo), epiphytic in cloud forests at 2,000 m.
Cu moderately moist, light shade. Since the plant does not get very large, it is appropriate for window greenhouses. Very attractive at bloom time.

VARIETY

var. *stenelopis* L. B. Smith, 1951 differs from the type (the 'main' species) by the presence of narrower flower bracts that do not completely surround the spike rhachis and are hardly 3 cm long.

Tillandsia acostae Mez and Tonduz., 1916 (Ills. 279–280)
(*acostae* named after plant collector, M. Acosta Solis)

Pl stemless, flowering to 25 cm high.
Fo numerous, forming a loose rosette from 30 cm in diameter
Ls oval, 3–4 cm long, 2–2·5 cm wide, dark brown scaled
LB narrow-triangular, long pointed, arched, with channelled bent upwards margins, about 15 cm long, 1 cm wide above the sheath, appressed grey scaled
Sc upright, shorter than the leaves, 10–12 cm long, 5 mm thick, bare, round, green
SB similar to the leaves, densely set with longer blade, appressed grey scaled
I simple or digitate, up to 3–4 spikes loosely compound, to 15 cm long and 10 cm wide; spikes upright or spreading

PB similar to the upper inflorescence bracts with shorter blade, much shorter than the spikes, about 2–3 cm long, reddish brown, appressed grey scaled
Sp 8–10 cm long, 2 cm wide, to 12 flowered
FB densely overlapping, somewhat inflated, keeled at the tip, naked, shiny red, 3 cm long, outspread 2·2 cm wide, beaked at the tip.
Se 1·5 cm long, the back one fused for 1 cm of its height, keeled, cutaneous, green, naked, 5 mm wide, pointed
Pe 3·5 cm long, formed into a tube, violet, shorter than the stamens
Ha Costa Rica (near San Ramon)
CU A decorative species.

Tillandsia adpressa André, 1888 (Ill. 73)
(*adpressa*(*us*) compact; because of the upright spikes pressed against the axis of the inflorescence)

Pl greatly variable in its shape, stemless, 20–70 cm high
Fo numerous, forming a slender, oval pseudo-bulb of 10–30 cm, thicker at its base
LS broadly elliptical to oval, dark brown, with a narrow cutaneous border, thickly covered with brown scales
LB narrow, triangular, upright or arched, dull green, pale scales, not over 2 m wide at the base; sometimes with

reddish-brown spots and with a thin, reddish-brown border

Sc upright or almost upright, with coarse, rust-coloured scales

SB narrowly oval, with a long tail, usually overlapping, with tightly clinging scales, the lower ones longer than the internodes, the upper ones shorter

I composed of 4 to 12 upright or flared spikes, upright or pendulous

PB broadly oval, with short tips or tails, as long as or shorter than the bipartite spikes of 8 to 12 flowers

FB broadly oval, pointed, usually shorter than the sepals, olive green, thickly covered with brown scales, up to 6 mm long

Fl sessile or with a short pedicel

Se asymmetrical, separate, with dense brown scales, up to 7 mm long, surpassed slightly by the yellow petals of 1 cm length and spread at their tips

St enveloped in the flower

CU a very decorative species, but like many other species of the subgenus *Pseudocatopsis* it is not easy to grow.

VARIETIES

T. adpressa var. *adpressa*
Syn *Catopsis schumanniana* Wittm., 1889; *T. schumanniana* Mez, 1896.
At the time of seed maturity the spikes are upright and pressed against the rhachis
Ha Colombia, Ecuador, northern Peru (Taulis), up to 2,800 m.
T. adpressa var. *tonduziana* (Mez) L. B. Smith, 1930
Syn *T. tonduziana* (Mez), 1901
At the time of seed maturity the spikes are flared or recurved
Ha Costa Rica, Panama, Colombia, Ecuador, Peru

Tillandsia aeranthos (Loisel.), L. B. Smith, 1943 (Plate 24)
(*aeranthos* air blooming; probably in reference to the fact that the plant frequently blooms without roots)
Syn *Pourretia aeranthos* Loisel., 1821; *T. dianthoidea* Rossi, 1825; *T. bicolor* Brongn., 1829; *T. microxiphion* Baker, 1893
T. aeranthos is closely related to *T. tenuifolia* (see page 181) and *T. bergeri* (see page 96), with which it can easily be confused.
This species is quite variable. Several varieties are distinguished which differ in the size of the inflorescence and the colour of the petals.

Pl rosette, but usually with a short stem, up to 25 cm high, growing in thick mounds

Fo numerous, spiral, more or less upright and very rigid

LS indistinct, gradually blending into the leaf blade, 2 cm wide, up to 1·5 cm long, with a thin skin, white

LB narrowly lanceolate, long-tapered, 8–14 cm long, more or less 1·5 cm wide at the base, green, both sides with dense grey scales, the edges curled

Sc slender, upright or pendulous, naked

SB overlapping, upright, green or red with green tip, partially covered with scales

I simple-spiked, open, multi-flowered

FB purplish-red, tip covered with scales, 1·7–2 cm long

Fl arranged in a spiral on the rhachis, up to 2·4 cm long, blue or grey blue (var. *griseus*)

Se shorter than the flower bracts, about 1·6 cm long, up to 4 mm wide, thin-skinned, white, the posterior se-pals are fused up to about 1·4 cm high

St enveloped in the flower, up to 1·5 cm long.

Ha *T. aeranthos* covers a wide range; it grows epiphytically on trees or on cliffs in Brazil, Uruguay, Argentina and Paraguay.

Cu an undemanding, hardy and re-sistant species, but also a grateful bloomer. It is especially good for the beginner. In the Mediterranean area *T. aeranthos* can be grown outdoors the year round.

Tillandsia albertiana F. Vervoorst, 1969 (Ill. 281 and Fig. 24a)
(*albertiana* named for Alberto Castellanos, Argentinian botantist)

Pl flowering to 29 cm high, with short, thin, divided stems, forming loose tufts.

Fo distichous (in two rows), spreading or upright (Fig. 43)

Ls 2–2·5 cm long, 1·5 cm wide, pale

Fig. 24a *Tillandsia albertiana* V. Vervoost; flowering plant.

brown scaled, nerved
Lb to 15 cm long, narrow awl-shaped, channelled upturned margins, 5 mm wide at the base, gradually lengthening and tapering to a sharp tip, underside clearly nerved towards the base, densely appressed grey scaled
Sc absent or 1–2 cm long, with 1 or 2 scale shaped scape bracts
I single flowered
Fb 2·5 cm long, 5–8 mm wide, short tipped, grey scaled

Se lanceolate, pointed, 2·5 cm long, free at the base, about 4 mm wide, green, reddish brown at the tip
Fl very large, outstanding, bright cinnabar-red, about 4 cm long, narrow tubed petals with flat flared elliptical plates
St and Pi enclosed within the flower
Fr almost cylindrical, 2 cm long
Ha Argentine (Prov. Salta, Dept. Canderlaria), on rocks

T. albertiana is one of the most beautiful plants to be discovered in recent years, and was found near a new road construction. It blooms freely in cultivation and tolerates very dry conditions and is recommended for those with limited space.

Tillandsia albida Mez et Purp., 1916 (Ill. 10, Ill. 54)

(*albida*(*us*) whitish, dull white; refers to the white colour of the leaves)
The species is variable; small, short-leafed and larger, long-leafed forms are recognized.

Pl forming a long stem, up to 80 cm long including the inflorescence
Fo numerous, spiral, with flared or slightly recurved blade
LS indistinct, 2–3 cm wide, 1–2 cm long, with dense, clinging, grey scales
LB narrowly lanceolate, long-tapered, 10–20 cm long, 2–2·5 cm wide above the sheath, both sides covered with dense, silvery scales
Sc slender, upright, up to 25 cm long, 3–4 mm thick, naked, red or green
SB overlapping, the lower ones similar to the foliage, with long, narrow blade, the upper ones lying close to the scape, scaled, green or reddish, with short blade
I simple-spiked, loosely bipartite; axis

zigzag-shaped, 6 to 10 flowers.
FB broadly oval, with blunt tip, about 2 cm long, up to 1·5 cm wide, green or reddish, few scales or bare
Se up to 2 cm long and 1 cm wide, the posterior sepals fused rather high, long oval, blunt at the tip, thin, green, naked
Pe white, upright, fused into a tube, up to 4 cm long and up to 1 cm wide
St projecting out of the flower
Ha Mexico. Epiphytic on cacti or on cliff walls
CU Like all white tillandsias, this species requires a bright, sunny location. Under cultivation it is usually rootless. A decorative and recommended species.

Tillandsia anceps Lodd., 1823 (Plate 28)

(*anceps* two-sided, two-edged; refers to the formation of the inflorescence)

Syn. *Phytarrhiza anceps* Morren, 1879; *Vriesea schlechtendalii* Wittm., 1889; *T. lineatifolia* Mez, 1896

Pl stemless, up to 30 cm high in bloom

Fo numerous, forming a compact rosette, as long or longer than the inflorescence

LS oval, striped with red, 2–3 cm wide, 4–5 cm long

LB longitudinally triangular, long-tapered, 7–12 mm wide, flared to recurved, green, reddish-brown stripes near the base, pliant

Sc upright, strong, short, usually covered by the bracts

SB tightly overlapping, oval, long-tapered

I simple sword-shaped, elliptical, much flattened, 10–15 cm long, about 5·5 cm wide, 10 to 20 flowers, naked

FB bipartite, overlapping, pointed, up to 4 cm long, much longer than the sepals, with a keel, green or pale pink with greenish edges

Se longitudinally lanceolate, pointed, up to 3 cm long, with a keel

Pe narrowly lanceolate to elliptical, pointed, blue, seldom white

St enveloped by the flower tube

Ha Central America, Trinidad and northern South America; epiphytic on trees. *T. anceps* in its non-blooming stage can easily be confused with *T. lindeni* (see page 148) and *T. cyanea* (see page 113). It differs from these, for it has much smaller flowers.

CU *T. anceps* has been in cultivation for a long time and is not difficult. It can be grown in a pot with a peat medium or epiphytically.

Tillandsia andicola Gill., 1878 (Fig. 24b)

(*andicola*(*us*) from the Latin *colere*, to live; living in the Andes)

Pl forming thick mounds or turf, with a stem 5–20 cm long

Fo numerous, bipartite arrangement, with flared blade

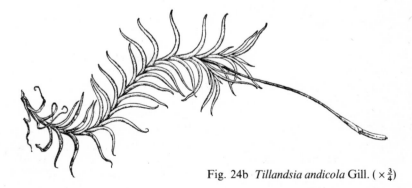

Fig. 24b *Tillandsia andicola* Gill. (× ¾)

LS oval, surrounding the stem, with grey scales, reddish-brown or green
LB round, awl-like, up to 5 cm long, reddish-brown or green, covered with dense, grey scales
Sc slender, upright, reddish-brown scales
I simple, a spike of 1 or 2 flowers
FB reddish-green, scaled

Fl very small, greenish-white
St very short, enveloped by the flower
Pi shorter than the ovary
Ha Argentina
CU Grown epiphytically with no difficulties. *T. andicola* is closely related to *T. capillaris* and can easily be confused with forms of this species.

Tillandsia andrieuxii (Mez) L. B. Smith, 1937 (Plate 16)
(*andrieuxii*, named after the collector Andrieux, active in Mexico)
Syn. *T. benthamiana* Klotzsch var. *andrieuxii* Mez, 1888

Pl usually growing in clumps, 10–15 cm high including inflorescence
Fo forming a loose, frequently leaning rosette
LS clearly delineated, broadly triangular, 2–3 cm long, 2–3 cm wide, thickly covered with silvery-grey scales
LB 10–12 cm long, up to 1 cm wide at the base, narrow, long-tapered, thickly covered with silver grey scales standing out from the surface
Sc 6–7 cm long, usually pendant
SB similar to the leaves, surrounding the scape with their sheaths, green, blade pink
I simple, pendulous

FB red, 3–4 cm long, 2 cm wide, thin, no keel on the back, rounded at the tips
Se approximately 2·5 cm long, up to 1 cm wide, thin, slightly protruding keel, pink toward the tip
Pe blue, up to 5 cm long, fused into a tube, shorter than the stamens
Ha Mexico; epiphytic on conifers and oaks at altitudes between 2,500 and 3,000 m.
CU Epiphytic, bright and sunny. Because of the altitude, not easy to cultivate. Decorative, beautiful and recommended species. Distinguished from *T. benthamiana* by the violet flowers.

Tillandsia araujei Mez, 1894 (Ill. 74)
(*araujei* named after the Brazilian river Araua)
T. araujei is closely related to *T. tenuifolia* (see page 181), especially to its variety *saxicola*.

Pl stemmed, 15–30 cm long; stem often horizontal or angled upward
Fo numerous, 5–6 cm long, closely arranged, spirally arranged, frequently turned to one side
LS triangular, surrounding the stem, up to 1·5 cm wide, 1 cm long
LB narrowly lineal to lanceolate, long-tapered, rigid, 5–6 mm wide at the

base, green with grey scales, the edges curved upward
Sc slender, longer than the leaves, 6–9 cm long, frequently grows horizontally
SB oval-lanceolate, long-tapered, overlapping, lying close to the scape
I 4–5 cm long, simple-spiked, 5 to 10 flowers

FB oval pointed, thin-skinned, naked, up to 2·2 cm long, much longer than the sepals

Fl upright, up to 2·2 cm long, arranged spirally on the rhachis, white

Se 1·3–1·5 cm long, lanceolate, long-tapered, naked, the posterior sepals fused approximately 4 mm high

St enveloped in the flower, 1·3 cm long

Pi 1·2 cm long

Ha Brazil. Epiphytic on trees and on rocks

CU Not fussy, easily grown and propagated species. Therefore especially recommended to the beginner

Tillandsia argentea Griseb., 1866 (Ills. 35–36)

(*argentea(us)* white as silver; refers to the silvery-white colour of the leaves)

T. argentea is quite variable. There are forms with short, sturdy leaves (Ill. 35) and some with longer, thin, almost thread-like leaves (Ill. 36).

Pl up to 25 cm high including the inflorescence; stem very short and not visible

Fo numerous, arranged spirally, forming a real 'onion' with its thick sheaths (Ill. 36)

LS almost triangular, with thick water tissue, 7–15 mm long, 6–7 mm wide

LB lineally awl-shaped, thread-like, long-tapered, 1–2 mm wide at the base, thickly covered with white, compact or flared scales

Sc upright to angled, longer than the leaves, 1 mm thick, red, naked

SB upright, partially surrounding the scape, oval, pointed or with a tail, thin, ribbed, with grey scales, green or red

I up to 7 cm long, simple-spiked, very open, up to 7 flowers, with a slender, zig-zag rhachis

FB up to 9 mm long, broadly eliptical, much shorter than the sepals, thin-skinned, red or green, with grey scales

Fl blue or red, almost perpendicular, up to 3 cm long, in bipartite arrangement

Se elliptical, blunt, thin-ribbed, green or red, up to 1·7 cm long, the posterior sepals slightly fused

St protruding from the flower

Ha Widely spread in Mexico, Guatemala, Cuba, Jamaica, at altitudes between 1,300 and 1,800 m.

CU Moderately humid, bright conditions. *T. argentea* is a graceful and attractive species; therefore very much favoured by bromeliad fanciers.

Tillandsia atroviridipetala Matuda, 1957 (Plate 22)

(*atroviridipetala(us)* *atro(ater)* black, dark; *viridis* green; *petalus* petaled; refers to the dark green colouration of the petals)

T. atroviridipetala is closely related to *T. mauryana* (see page 155) and *T. plumosa* (see page 163), but is easily distinguished from these.

Pl single or growing in groups, hanging downward, stemless, up to 7 cm high, quite variable in size

Fo numerous, forming a bulbous rosette with its thickened sheaths

LS clearly delineated, long-oval, up to 1·2 cm long, up to 7 mm wide, smooth at the base, covered with grey scales in the upper part

LB narrow, awl-shaped, pointed, perpendicular, flared to recurved, about 4·5 cm long, 3 mm wide at the base, thick grey scales, feathery

Sc perpendicular, barely developed, hidden in the rosette, 1–2 cm long

SB similar to the leaves, arranged spirally and surrounding the scape

I hardly protruding from the leaves, composed of 2 to 6 single flowered spikes

FB up to 1·5 cm long, as long as the sepals, green, red tips, heavily covered with grey scales

Se up to 1·5 cm long, thin-skinned, red-tipped, free, covered with grey scales

Pe green, narrow ribbon-shaped, up to 2 cm long and 2·5 mm wide

St enveloped by the petals, up to 1·5 cm long

Ha Mexico, at 2,400 m above the city of Cuernavaca and near Tehuacan. At Cuernavaca *T. atroviridipetala* usually grows in a hanging position, preferably on *Commiphora* trees (family Burseraceae), which are among the characteristic trees of the pedrigalles (lava flows).

CU Sunny and bright, little watering. Especially suitable for growers with little room; pretty, small species. When mounting these plants in cultivation, the hanging position should be observed.

Tillandsia aureobrunnea Mez, 1906 (Ill. 75)

(*aureobrunnea*(*us*) *aureus* golden yellow; *brunneus* deep brown; refers to the yellow-brown coloration of the petals)

Pl stemless or with a short stem, up to 40 cm high including the inflorescence, forming thick clumps

Fo pliant, thick and coarse, covered with almost frost-like scales, up to 30 cm long and 2·5 cm wide

LS hardly delineated from the blade

LB narrowly triangular, long-tapered, with edges bent sharply upward

Sc slender, shorter than the leaves, heavily scaled

SB upright, similar to the leaves, with a long, heavily scaled blade and edges curved upward

I simple or composed of few spikes

PB elliptical, pointed, heavily scaled, green, shorter than the loose spikes of 7 flowers, spikes up to 6 cm long

FB elliptical, pointed, much shorter than the sepals, no keel on the back, heavily scaled, green

Fl up to 2·5 cm long, short-stemmed, strongly fragrant; the 6 mm long, open plates of the petals are yellow and spotted with dark brown, or they are pale yellow

Se separate, elliptical, pointed, up to 1·8 cm long, scaled

St enveloped in the flower

Ha northern and central Peru (vicinity of Cajamarca, Huanuco and Lima) between 2,000 and 2,500 m; epiphytic on trees, but also frequently on cliffs

CU Bright, sunny, little watering. Decorative, small species.

Tillandsia baileyi Rose, 1903 (Ill. 76)
(*baileyi* named for the American L. H. Bailey, the publisher of many botanical and horticultural works)

Pl stemless, 20–40 cm high, forming thick clumps
Fo as long as or longer than the inflorescence, tightly pressed, covered with grey scales
LS spoon-shaped, outside clearly nerved, forming a 2–5 cm long, hollow pseudo-bulb
LB lineally awl-shaped, long-tapered, frequently twisted, up to 5 mm wide at the base, about 30 cm long, its edges curled into a gutter shape
Sc perpendicular or angled upward, 2 mm thick, covered with grey scales
SB similar to the leaves, the lower ones with a long, narrow blade, the upper ones with a short tip, upright, surrounding the scape
I simple or composed of 2 to 4 spikes
PB much shorter than the spikes, which are lineal, approximately 4 cm long, 1 cm wide and flattened
FB bipartite, closely arranged, oval, pointed, about 2 cm long, 8 mm wide, no keel, covered with grey scales, greenish-red
Se lanceolate, 1·6 cm long, scaled, the posterior sepals fused to about 3 mm, green to red
Pe blue, up to 3·5 cm long, fused into a tube
St protruding from the flower
Ha southern U.S.A., Mexico and Guatemala, at altitudes between 900 and 1,200 m.
CU bright, airy location, little watering; epiphytic. An especially desirable species but very variable in regard to the size of the pseudo-bulb and the length of the inflorescence.

Tillandsia bakeri L. B. Smith, 1931 (Ill. 77)
(*bakeri* named after the English botanist J. G. Baker 1834–1920)
Syn. *Catopsis flexuosa* Baker, 1887; *T. flexuosa* (Baker) Mez, 1896, non SW. 1788

Pl stemless, over 1 m high including inflorescence
Fo numerous, forming a funnel rosette of 40–50 cm diameter
LS wide oval, 8–10 cm long, 7–8 cm wide, brown, red violet at the upper end
LB up to 30 cm long, 4–5 cm wide above the sheath, triangular, long-tapered, their tips recurved, pliable, green, both sides heavily covered with scales
Sc up to 40 cm long, 5 mm thick, upright, round, heavily grey scaled, green, partly reddish-brown
SB the basal scape bracts similar to the rosette leaves, the upper ones approximately 1·5 times as long as the internodes, narrow, lanceolate, long-tapered, upright, close to the scape, their tips somewhat curved outward, thickly covered with grey scales
I 50–55 cm long and 14 cm wide, spikes in a loose panicle with 15 to 20 panicle branches; the axis is green or reddish-brown, with brown scales, round, straight
PB similar to the scape bracts, shorter than the panicle branches, somewhat

longer than their stemmed, flowerless segments; side branches flared to slightly recurved, 5–9 cm long, somewhat flexible toward the tip, with 2 to 3 multi-curved, laterally flattened, scaled spikes

FB shorter than the sepals, about 5 mm long, about 4 mm wide, triangular, heavily covered with grey scales, green, sometimes with a reddish-brown edge

Fl loose, bipartite, connected directly to the rhachis, flared, up to 7 mm long

Se asymmetrical, 4–5 mm long, separate, heavily covered with grey scales, green or reddish-brown

Pe about 5 mm long, 2·5 mm wide, thin-skinned, greenish white

St about 13 mm long

Ha Peru and Bolivia; epiphytic on trees.

CU Suitable only for larger greenhouses.

Tillandsia balbisiana Schult., 1830 (Ill. 79)

(*balbisiana* named after the Italian botanist G. Balbis 1765–1831)
Syn. *Platystachys digitata* Beer, 1857; *T. urbaniana* Wittm', 1889; *T. cubensis* Gandoger, 1920

Pl stemless, up to 65 cm high, very variable

Fo numerous, up to 60 cm long, heavily covered with short grey scales

LS upper side pale rust brown, underside green with grey scales, spoon-shaped, forming an overall hollow, egg-shaped or elongated onion up to 12 cm long; the exterior leaf sheaths forming scoop-shaped rudimentary leaves with no blade

LB spread and recurved to pendulous, narrowly lanceolate, 1·5 cm wide at the base, thread-shaped, long tapered, twisted, their edges slightly bent upward

Sc upright or angling upward, slender and naked

SB overlapping, surrounding the stalk with its sheath, the blades narrow, thread-like, pointed and recurved

I compound, seldom simple

PB shorter than the spikes, similar to the scape bracts

Sp sessile, lineal, flattened, up to 12 cm long, 1 cm wide

FB overlapping, oval 1·5–2·2 cm long, surpassing the sepals, usually naked and with no central nerve

Se 1·5 cm long, posteriorly slightly fused, green, naked

Pe blue, 3–4 cm long, perpendicular, blunt, fused into a tube

St protruding from the flower

Ha Widespread, southern U.S.A., through Central America to Venezuela, at altitudes between 1,000 and 1,500 m and just above sea level in Florida

CU Epiphytic, bright and sunny, little watering. A recommended species.

Tillandsia bandensis Baker, 1887 (Fig. 25)

(*bandensis* named after the Uruguayan mountain range Banda Oriental)
Syn. *T. quadriflora* Baker, 1889; *T. recurvata* var. *majuscula* Mez, 1894

Pl small, with a short stem, growing in mounds, up to 15 cm high including inflorescence

Fo distichous (in two rows), with

arched, open blade

LS thin-skinned, surrounding the stem, 1·2 cm long, up to 7 mm wide

LB awl-shaped, long-tapered, forming a gutter, up to 9 cm long, 4 mm wide at the base, average width 1 mm, green, heavily covered with grey, feathery scales

Sc upright, slender, few scales to naked, 6–7 cm long, 1 mm thick

SB similar to the leaves in the lower part of the scape, longer than the stalk segments, in the upper part shorter, lying close to the scape, grey-scaled or naked

I simple-spiked, open, 2 to 6 flowers, up to 4 cm long

FB long-oval, pointed, thin-skinned, naked, up to 1 cm long, as long as the sepals, sepals separate, 9 mm long, 3 mm wide

Pe blue, 1·5 cm long, fused into a tube at the base, 1·5 mm wide, the upper part shaped into a broad plate 5 mm wide and 8 mm long

St enveloped in the flower, somewhat longer than the 1 mm pistil

Ha Brazil, Paraguay, Uruguay and Argentina. Epiphytic on trees.

CU The cultivation of this miniature tillandsia entails no problems. When not in bloom it is easily confused with *T. capillaris* (see page 104) and *T. recurvata* (see page 171).

Fig. 25 *Tillandsia bandensis* Baker (× $\frac{3}{4}$).

Tillandsia benthamiana (Beer) Klotzsch, 1857 (Plate 17)
(*benthamiana* named after the English botanist G. Bentham 1800–1884)
Syn. *Anoplophytum benthamianum* Beer, 1857; *T. vestita* Benth., 1840, non alior, 1831; *Anoplophytum vestitum* Beer, 1889; *T. hartwegiana* Morren, 1889

Pl medium size, stemless

Fo numerous, forming a thick rosette sometimes bent to the side, depending on its location

LS not clearly delineated

LB 3 cm wide at the base, average width up to 2 cm, triangular, long-tapered, both sides closely covered with silver grey scales

Sc up to 12 cm long, pendulous

SB similar to the leaves, with a long, narrow blade

I simple, lined, up to 8 cm long and 3 cm thick

FB close, pale pink, 5–7 cm long, thin-skinned, broadly elliptical, drawn out into a long point, scaled

Fl tubular, up to 7 cm long, greenish-white

Pe up to 4 cm long, fused slightly at

equal heights at the base, lanceolate, long-tapered, thin-skinned, with a keel
St protruding from the flower tube
Ha Mexico, at altitudes between 2,500 and 3,000 m growing on conifers. Very closely related to the Mexican species *T. macdougallii* (see page 152) and *T.*
andrieuxii (see page 90), with which it is found.
CU Bright and sunny; little watering. Like all other white tillandsias, a very beautiful, but somewhat difficult species.

Tillandsia bergeri Mez, 1916 (Fig. 26)

(*bergeri* named after the well-known succulent researcher A. Berger 1871–1931)
Syn. *Anoplophytum strictum* var. *krameri* André, 1888, non Mez, 1894

Pl stemmed, up to 18 cm long, growing in thick clumps
Fo numerous, thick, arranged spirally, with an open to upright blade
LS not clearly delineated, thin-skinned, white, gradually melding into the blade, about 2 cm wide and up to 1·5 cm long
SB narrowly lineal, long-tapered, about 1·5 cm wide at the base, up to 10 cm long, grey green scales, edge distinctly bent upward
Sc slender, upright, as long as the leaves or longer
SB tightly overlapping, almost surrounding the stalk with their sheaths, short blades, upright, scaled
I simple-spiked, 7 to 12 flowers
FB green, grey-scaled, longer than the sepals
Fl arranged in a spiral on the rhachis
Se naked, up to 1·6 cm long, the posterior ones about 6 mm high, fused
Pe blue, white at the base, flared, pink as they fade, about 3 cm long
St enveloped in the flower
Ha Argentina.
CU The plant needs little care; it is just as tough as *T. aeranthos* and therefore recommended to the beginner but in its non-blooming stage can easily be confused with *T. aeranthos* (see page 86) and *T. tenuifolia* (see page 181).

Fig. 26 *Tillandsia bergeri* Mez (× ½).

Tillandsia biflora Ruiz et Pav., 1802 (Plate 21, Fig. 27)
(*biflora*(*us*) double-flowered; refers to the usually double-flowered spikes)
Syn. *Diaphoranthema biflora* Beer, 1857; *T. tetrantha* sensu Griseb., 1865, non Ruiz et Pav., 1802; *T. grisebachiana* Baker, 1888; *T. augustae-regiae* Mez, 1901

Pl medium size, stemless, up to 30 cm high including inflorescence
Fo numerous, forming a funnel rosette of 25–30 cm diameter
LS clearly delineated, long-oval, 6–7 cm long, 4–5 cm wide, sometimes spotted reddish-brown or striped, thin
LB narrowly lanceolate, pliable, 10–15 cm long, 2–2·5 cm wide at the base, green, partially dotted red, naked, frequently curled back at the tip
Sc upright or slightly arched, up to 10 cm long, round, naked, spotted red brown at the nodes
SB tightly overlapped, similar to the leaves, much longer than the stalk segments, the sheaths circling the scape; blades upright with curled tips, green and pliable
I open panicle, with 8 to 12 spikes, 6–9 cm long in all, 6–7 cm wide
PB similar to the scape bracts, spread, the lower ones longer than the spikes, the upper ones shorter, surrounding the rhachis with the base, reddish-brown to violet, tips rolled back
Sp 2–3 cm long, stem approximately 1 cm long, 1 to 3 flowers, but usually 2 flowers with the third rudimentary (Fig. 27)
FB shorter than the sepals, 1 cm long, 7 mm wide, thin-skinned, keel, naked,

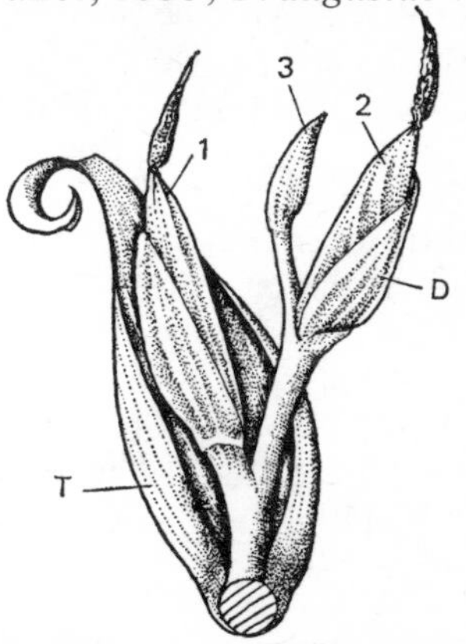

Fig. 27 *Tillandsia biflora* Ruiz et Pav. Single bloom spike with two fertile flowers (1 and 2) and one stunted flower (3). T primary bract; D flower bract (larger than life-size).

violet
Se 1·5 cm long, naked, the posterior ones fused approximately 5 mm high, violet
Pe blue or purple
St enveloped in the flower tube
Ha Colombia Costa Rica and from Venezuela to Peru and Bolivia, epiphytic on trees, at altitudes from 1,000 to 3,000 m.
CU Epiphytic or in a pot. Like all green-leafed tillandsias must be grown rather damp and shady. Decorative, pretty and small species. Very variable in regard to the leaf colour.

Tillandsia bourgeaei Baker, 1887 (Ill. 2, Plate 13)
(*bourgeaei* named after the collector Bourgeau)
Syn. *T. violacea* Baker, 1889

Pl stemless, 50–60 cm high including inflorescence

Fo numerous, upright or flared, forming a rosette of 40–50 cm diameter

LS not clearly delineated, about 8–9 cm long, 6 cm wide, both sides brown-scaled

LB lanceolate, narrowing into a thread-like tip, up to 40 cm long, 5·5 cm wide above the sheath, both sides heavily grey-scaled

Sc strong, upright, about 20 cm long, angular, up to 1·5 cm thick, grey-scaled

SB tightly overlapping, covering the internodes, the basal ones similar to the leaves with flared to recurved blades

I about 25 cm long, up to 8 cm wide, composed of about 20 spikes arranged in a loose spiral, each 6–7 cm long and 2 cm wide, short-stemmed, almost upright, flat

PB heavily grey-scaled, the sheaths shorter than the spikes, the blades always bent back and lacking on the upper primary bracts

FB closely distichous, firm, slightly nerved on the back, about 3·5 cm long and up to 2 cm wide, green or pink, heavily grey-scaled, sometimes completely bare, the two basal flower bracts without flowers

Fl green, tubular, upright, up to 5·5 cm long

Se thin-skinned, grey-scaled, about 2·5 cm long, about 6 mm wide, separate, the posterior ones fused about 6 mm high, approximately as long as the flower bracts

St surpassing the petals by 1 cm; filaments twisted at the base, otherwise straight

Ha Mexico; epiphytic on trees, primarily oaks, at 2,000 m frequently found with *Tillandsia prodigiosa*.

CU Bright, sunny, little watering. A very beautiful and decorative species at bloom time.

Tillandsia brachycaulos Schlechtend., 1844 (Plate 2)

(*brachycaulos*(*on*) short-stalked; refers to the short scape)
Syn. *T. cryptantha* Baker, 1888; *T. bradeana* Mez et Tonduz, 1916; *T. flammea* Mez, 1935

Pl medium large, up to 20 cm high, but greatly variable

Fo numerous, tight rosette, occasionally bent to one side, up to 26 cm long, longer than the inflorescence, sparse, short scales, green, colouring intensively red in strong sun at bloom time

Sc short, upright, naked

SB similar to the leaves, upright to open, green, at bloom time bright red

I drawn together into a head, appearing to be simple since the small spikes produce only 1 or 2 flowers

PB similar to the scape bracts, hiding the spikes

FB lanceolate, thin, as long as the sepals; the sepals are elliptical, 1·2–2 cm long, the posterior ones slightly fused

Pe blue, 5–7 cm long, upright, fused into a tube from which the stamens protrude

Ha Southern Mexico and Central America; epiphytic in forests between 900 and 2,000 m.

CU Bright and sunny, moderately humid, epiphytic. In a sunny location the upper rosette leaves colour up brilliant red and make a splendid show for several months.

Tillandsia bryoides Griseb., 1878 (Ill. 26 and Fig. 28)
(*bryoides* moss-like; reminiscent of the moss *Bryum*)
Syn. *T. coarctata* Gill., 1878, non Willd., 1830
This miniature tillandsia reminiscent of a large moss or of club moss can easily be confused with the similar *T. pedicellata*, but the latter has blue violet flowers (see page 162)

Pl of a moss-like form; many small, simple or branched stems up to 5 cm long gather into thick mounds
Fo leaves 4–9 mm long, arranged very closely and in a spiral, upright, stiff, completely surrounding the axis
LS oval to almost round, dry-skinned, with 3 ribs
LB almost triangular, pointed, 2 mm wide, thickly grey-scaled, overlapping
Sc very short or lacking, after bloom sometimes elongated to 3 cm, upright, naked
SB usually lacking
I a single-flowered spike giving the impression of a simple flower
FB up to 7 mm long, triangular, oval, thin, single-ribbed, usually naked
Fl yellow, small, insignificant
Se narrowly elliptical, 5–9 mm long, 3 ribbed
St short, enveloped in the flower

Fig. 28 *Tillandsia bryoides* Griseb. with flower and seed capsules (larger than life-size).

Ha Peru, Argentina, Bolivia; epiphytic on trees and on cliffs.
CU Not very difficult in cultivation. This bizarre species is especially recommended to fanciers with small greenhouses or terramums.

Tillandsia bulbosa Hook., 1826 (Ill. 32)
(*bulbosa(us)* bulbous; refers to the bulbous swelling of the base of the plant)
Syn. *T. bulbosa* var. *brasilensis* Schult. fil., 1830; *T. bulbosa* var. *picta* Hook., 1847; *Pourretia hanisiana* Morren, 1847; *T. inanis* Lindl., 1850; *T. erythraea* Lindl., 1850; *T. pumila* Lindl. et Paxt., 1850; *Platystachys inanis* (Lindl.) Beer, 1857; *Platystachys bulbosa* Beer, 1857; *Platystachys erythraea* (Lindl.) Beer, 1857

Pl 7–30 cm high, greatly variable in size, usually growing in thick clumps
Fo few in number, covered with close grey scales
LS up to 2·5 cm long in all, forming a hollow, egg-shaped pseudo-bulb, thin red edge, tightly appressed grey scales
LB up to 30 cm long, 2–7 mm wide, folded into a gutter, sometimes red longitudinal lines on the edges

Sc short, upright, horizontal or pendulous, few scales
SB leaf-like, with elongated blades longer than the inflorescence; the upper ones often having red and grey scales
I simple or composed of few spikes, red or green
PB egg-shaped, pointed, usually shorter than the spikes, seldom longer
Sp upright or flared, lanceolate, pointed, flattened, 2–6 cm long, 2 to 8 flowers
FB overlapping, pointed oval, central nerve, 1·5 cm long, longer than the sepals, heavily scaled, green or red
Se long lanceolate, thin-skinned, naked, green 1·2 cm long, the posterior ones slightly fused.
Pe blue, lineal, forming a tube 3–4 cm long from which the stamens protrude, so possible confusion with *T. baileyi* when not in bloom.
Ha From southern Mexico and the West Indies to Colombia and Brazil; epiphytic on trees, from the coast up to 700 m.
CU Moderately humid, bright; decorative, recommended species.

Tillandsia butzii Mez, 1935 (Ills. 33 and 81)
Syn. *T. variegata* Schlecht., 1844, non Vell., 1825

Pl stemless, with inflorescence and leaves up to 30 cm long, longer-leafed forms up to 50 cm, growing in thick clumps since the plants that have already bloomed live on for a long time
Fo sparse, up to 50 cm long, green, spotted with reddish-brown, grey-scaled
LS scoop-shaped, together forming a hollow pseudo-bulb up to 4 cm long and 3 cm thick (Ill. 33), on the underside (outside) scaled with numerous red brown, frequently joined spots, upper side with red brown scales
LB rolled into a gutter, therefore awl-shaped or thread-like, long-tapered, frequently twisted, the edges fringed at the beginning, 3–4 mm wide at the base
Sc depending on the location of the bulb perpendicular or pointed downward, grey-scaled, much shorter than the leaves
SB similar to the leaves, surrounding the scape with their sheaths
I compound, seldom simple
PB similar to the upper scape bracts, but with sheaths much shorter than the lineal, flattened spikes of 4–8 cm length and about 1 cm width
FB oval-lanceolate, pointed, overlapped, 2–2·8 cm long, greenish-red, with brownish-grey scales
Se narrowly elliptical, blunt, 1·2–1·5 cm long, green, naked, the posterior ones fused 4–5 mm high
Pe blue, forming a perpendicular tube 3–3·5 cm long and surpassed by the stamens. In regard to the formation of the pseudo-bulb and the length of the leaves, *T. butzii* is just as variable as *T. bulbosa*
Ha Throughout all of Central America to southern Mexico; epiphytic on trees in the mountain forests at altitudes between 1,300 and 2,300 m. In the cloud forest of Teotitlan at 2,200 m a form of *T. butzii* was found, which differs from the normal forms by the possession of a very slender, hardly formed pseudo-bulb and pale red-violet flowers (collection number Rauh, 15486)

CU Moderately moist, somewhat shady. Especially suitable for mount-ing on epiphyte trees. Decorative, and a favoured species.

Tillandsia cacticola L. B. Smith, 1954 (Ill. 11 and Plate 15)

(*cacticola*(*us*) dwelling on cacti. The name is misleading since *T. cacticola* grows more commonly on trees than on cacti)

In its non-blooming stage it can be confused with *T. purpurea* (see page 169) and *T. straminea* (see page 176).

Pl stemless, rosette form, up to 60 cm high including inflorescence

Fo pliant, open to recurved, up to 25 cm long, thick silver-grey scales and therefore of a white appearance

LS not clearly delineated

LB lineal-triangular, long-tapered, 2–3 cm wide at the base, with curled edges

Sc thin, about 4 mm thick, upright or angling upward, green, scaled to naked

SB grey-scaled, the basal ones leaf-like, with long upright blade surpassing the stalk segments

I compound, with 3–7 open, wide elliptical, greatly flattened spikes 3–5 cm long and 2–3 cm wide, 5 to 10 flowers

PB elliptical, pointed, as long as or shorter than the spikes, cutaneous, scaled, lavender colour

FB close, almost perpendicular, much longer than the internodes of the spike rhachis, up to 2 cm long, thin, distinct keel on the back, barely scaled, greenish-white at first, intensive lavender colour at bloom time

Fl very aromatic; petals cream colour or pale yellow with blue, recurved tips, after peak bloom reddish-lavender

Sc lanceolate, pointed, cutaneous, up to 1·6 cm long

St enveloped in the flower

Ha Northern Peru (around Huancabamba and Chachapoyas), epiphytic on cacti and on acacias, between 1,800 and 2,000 m.

CU Bright, sunny, little watering. Also appropriate for flower windows with a southern exposure. A splendid species whose inflorescence lasts several months and is sold in the market places near its habitat as 'chapullas'.

Tillandsia caerulea H. B. K., 1816 (Ill. 93)

(*caerulea*(*us*) sky blue)

Syn. *T. squamulosa* Willd., 1830; *Diaphoranthema squamulosa* Beer, 1857

Pl small, with 4–6 cm long, thin stems gathering into thick mounds

Fo spiral, with open to recurved blades

LS oval, about 1 cm long, surrounding the stem, gradually merging into the blade

LB 10–15 cm long, 3 mm wide at the base, narrow awl shape, merging into a thread-like tip, thick silver-grey scales

Sc upright, thin, 1–2 mm thick, scaled

SB lying close to the scape, perpendicular, longer than the stalk segments, scaled

I simple-spiked, open 4–7 cm long

FB elliptical, pointed, about 1·5 cm long, as long as or somewhat longer

than the sepals, spread, scaled, with protruding nerves
Fl brilliant blue with a white throat; petals about 2 cm long, with lineal unguis (claw-shaped base) and spread, almost rhombic plate up to 7 mm wide
Se lanceolate, pointed, thin, naked, up to 1·3 cm long
St enveloped by the flower
Ha Peru, Ecuador. Epiphytic in dry forests or on rocks.
CU Bright and sunny. A very recommended species because of its small size.

Tillandsia califani Rauh, 1971 (Ill. 282 and 283)
(*califani* after Prof. Dr L. Califano, Italian bromeliad fancier)
Can be confused with *T. achyrostachys* (Plate 20).

Pl stemless, flowering to 80 cm high
Fo about 15, forming a slender, upright funnelled rosette
LS not clearly distinguished, to 6 cm long and 4 cm wide, densely brown scaled
LB narrow-triangular, about 30 cm long, 3·5 cm wide above the sheath, with slightly upward curved margins ending in a long curved back point, grey green, frequently somewhat violet, densely appressed scaled.
Sc upright, sturdy 20–30 cm long, shorter than the leaves
SB upright, the lower ones with shorter lineal blade, the upper ones broadoval, spine tipped, pale carmine-red, longer than the internodes
I upright, simple sword shaped, to 50 cm long, 2–3 cm wide
FB densely overlapping, short tipped, not keeled, pale carmine-red with whitish appressed scales, clearly nerved.
Se free, thin cutaneous, to 3 cm long, 9 mm wide, lanceolate, white to pale green
Pe forming a narrow tube, to 6 cm long, 6 mm wide, dark violet
St and **Pi** projecting from the flower; anthers pale-yellow green; pistil and stigma violet
Ha Mexico, near Teotitlan (Tehuacan) predominately on columnar cereus.

Tillandsia calocephala Wittm., 1916 (Ill. 94)
(*calocephala*(*us*) *calos* beautiful; *cephalus* head)
In the non-blooming stage, can be confused with *T. cauligera* (see page 106).

Pl stemmed, 20–30 cm long
Fo in a tight spiral, frequently bent to one side, bent upward, 6–8 cm long
LS not clearly delineated, about 2–2·5 cm wide, upper side with thick brown scales
LB narrowly triangular, long-tapered, 1·5–2 cm wide at the base, with curled edges, thick grey-brown scales
Sc very short, 4–5 cm long, upright or angled upward, enveloped by the leaves
SB similar to the leaves
I compound, compact head, 2·5–3 cm long
PB wide oval, pointed awl shape,

surpassing the basal spikes, the upper ones united with the spikes

Sp very close, compacted into a head, about 1·5 cm long, laterally flattened, 2 to 3 flowers

FB oval lanceolate, pointed, about 1 cm long, somewhat shorter than the sepals, keel on the back, scaled, red

Fl about 1·7 cm long, red

Se lanceolate, keel, naked, about 1·1 cm long, cutaneous

St enveloped by the flower

Ha Southern Peru (Cuzco) and Bolivia, epiphytic on cliffs at altitudes between 2,500 and 3,500 m.

CU Easily grown; an interesting species.

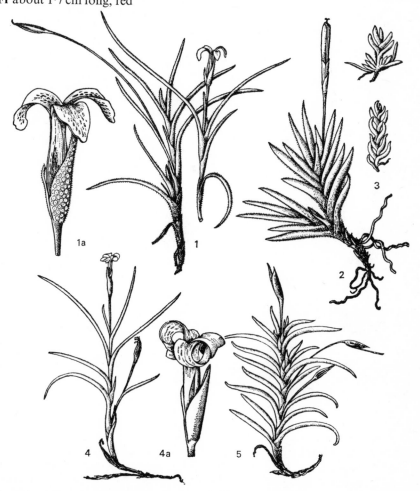

Fig. 29 *Tillandsia capillaris* Ruiz et Pav. Examples of growth forms which are sometimes described as separate species: 1 *T. capillaris* (typical form); 2 *T. lanuginosa* Gill.; 3 *T. pusilla* Gill.; 4 *T. dependens* Hieron.; 5 *T. hieronymi* Mez. All plants are life-size; flowers are enlarged.

Tillandsia capillaris Ruiz et Pav., 1802 (Fig. 29)

(*capillaris* as fine as hair; refers to the very thin leaves, perhaps also to the hair-like scape)

Syn. *T. virescens* Ruiz et Pav., 1802; *T. propinqua* Gay, 1853; *T. lanuginosa* Gill., 1878; *Diaphoranthema capillaris* Beer, 1857; *Diaphoranthema virescens* Beer, 1857; *T. pusilla* Gill., 1878; *T. incana* Gill., 1878; *T. cordobensis* Hieron., 1885; *T. stolpi* Phil., 1895; *T. hieronymi* Mez, 1896; *T. dependens* Hieron., 1896; *T. williamsii* Rusby, 1910.

Very variable, easily confused with *T. recurvata* (see page 171) and *T. andicola* (see page 89).

Pl miniature species, with short, simple or branched stems up to 16 cm long combining into thick clumps or mounds
Fo distichous, seldom spiral, 1–4 (6) cm long, thick grey scales
LS barely delineated, elliptical, cutaneous
LB lineal, up to 2 mm wide, up to 4 cm long with recurved or upright blade
Sc up to 8 cm long, because of longer axis growth it appears to grow laterally, slender, naked or slightly scaled toward the tip, without scape bracts
I a 1 to 2 flowered spike
FB oval, cutaneous, 3 or more nerves, heavily scaled to naked, as long as or longer than the 8 mm long sepals
Fl very small, greenish-yellow to brown; petals recurved at the tips
Ha Mexico to Peru, Bolivia, Argentina, Chile, from the coast to the high Andes, epiphytic on cacti, on trees and bushes, sometimes covering them in thick coats
CU Fast growing, easy to propagate, but a small and insignificant species.

Tillandsia capitata Griseb., 1866 (Ill. 281)

(*capitata(us)* headshaped: so named because of the glomerate inflorescence.

Pl stemless, with inflorescence about 40 cm high
Fo numerous, forming a funnel shaped rosette from 30–40 cm in diameter
LS longish-oval, 9–11 cm long, 5–6 cm wide, dark brown scaled
LB narrow-triangular, long pointed, 30–35 cm long, 2·3 cm wide above the sheath, appressed grey scaled, recurved
Sc upright, 15–20 cm long, 6–10 mm thick, soon becoming bare
SB similar to the leaves, densely overlapping, about as long as the leaves, the lower ones green, the upper ones reddish, grey scaled
I densely (glomerate) upright, consisting of several 2 flowered spikes 7–8 cm long
PB similar to the upper scape bracts, but smaller, covering the spikes, much longer than these, bright red or yellow, grey scaled
FB oval, pointed, keeled, as long as the sepals, thin cutaneous, densely grey scaled

Se lanceolate, pointed, 25–27 mm long, keeled, bare, thin cutaneous, white
Pe to 6 cm long, forming an upright tube, dark violet, lighter at the tip, white at the base

St projecting beyond the flower
Ha Mexico, Honduras, Guatemala
CU A good species, remaining small. The primary bracts become bright red at blooming time making this a really decorative species.

VARIETY *guzmanioides* L. B. Smith, 1939 is distinguished from the type in the bare sepals and the arched flower stem.

Tillandsia caput-medusae Morren, 1880 (Ill. 82)

(*caput-medusae* head of Medusa; refers to the shape of the plant)
Very variable in size; in the non-blooming stage can be confused with *T. pruinosa* (see page 167), *T. circinnata* (see page 108) and *T. seleriana* (see page 173). The hollow pseudo-bulbs in the natural habitat are inhabited by ants.
Syn. *T. langlassei* Poiss. et Menet, 1908

Pl 15–40 cm high, very variable in size, stemless, growing in thick clumps and resembling a Medusa head
LS wide-oval to elliptical, spoon-shaped, forming in all a 2–5 cm, hollow pseudo-bulb inhabited by ants
LB longer than the inflorescence, lineal awl-shaped, gutter-like, up to 1·5 cm wide at the base, long-tapered in the upper part, frequently twisted and rolled, closely scaled
Sc depending on the plant's growth direction upright, angular or pendulous, slender, naked
SB tightly overlapping, similar to the leaves and likewise grey-scaled
I simple, but usually composed of 2 to 6 upright or flared, lineal-lanceolate spikes up to 18 cm long and with 6 to

12 flowers
FB overlapping, oval-lanceolate, rounded on the back, up to 2 cm long, ribbed, naked, red or green, sterile at the base of the spikes, i.e. without flowers
Fl blue, perpendicular, tubular, 3–4 cm long
Se long-blunt, ribbed, naked, green
St protruding from the flower
Ha Mexico, Guatemala, El Salvador, Honduras and Costa Rica; epiphytic on trees, from the coast (Mexico) up to 2,500 m.
CU Bright and sunny, moderately humid. Very decorative much desired by fanciers, best suited for epiphyte trees.

Tillandsia carlsoniae L. B. Smith, 1959 (Plate 25)

(*carlsoniae* named after Mrs. Carlson)

Pl stemless, up to 25 cm high in bloom
Fo numerous, spread to recurved, forming a thick rosette

LS wide to elliptical, dark brown, up to 12 cm long, 6–7 cm wide
LB narrowly triangular, long-tapered,

30–35 cm long, 3–4 cm wide over the sheath, densely grey-scaled, pliant, with curled edges
Sc very short, sometimes covered by the leaf sheaths
SB tightly overlapping, similar to the foliage but smaller
I composed of 5 to 6 spikes
PB widely oval, short-tipped, shorter than the upright spikes that are elliptically pointed, pushed to the side, about 10 cm long and with 8 to 10 flowers
FB elliptical, pointed, up to 5 cm long, 2 cm wide, no central nerve, densely grey-scaled, the lower flower bracts are without flowers, pink
Fl short-stemmed, blue, upright, up to 6 cm long
Se lanceolate, pointed, 3·5–4 cm long, scaled, 1 cm wide at the base, separate
St protruding from the flower
Ha Mexico (Province of Chiapas, near Comitan de las Casas); epiphytic on oaks and conifers, from 2,500–3,000 m.
CU Sunny and bright. Very beautiful, very decorative species at bloom time.

Tillandsia caulescens Brongn., 1889 (Ill. 78)
(*caulescens* forming a stalk or stem)

Pl stemmed, up to 45 cm long; stem often branched
Fo numerous, compact, spiral, 10–15 cm long, rigid, almost perpendicular
LS narrowly triangular, gradually merging into the blade
SB lineally triangular, long-tapered, 5–8 mm wide at the base, distinctly curled edges
Sc upright to angling upward, naked, up to 15 cm long and up to 3 mm thick
SB tightly overlapping, the lower ones similar to the leaves, the upper ones elliptical, ribbed, red, scaled
I simple-spiked; spikes lineally lanceolate, pointed, flattened, 5–7 cm long, 1–1·2 cm wide, about 14 flowers
FB upright, tightly distichous, lanceolate-oval, pointed, no keel, 1·5–2 cm long, longer than the sepals, slightly ribbed, naked, red
Fl white, 2·5 cm long; petals with recurved tips
Se lanceolate, pointed, naked, 1·4–1·7 cm long, separate
St enveloped in the flower, about 2 cm long
Ha Bolivia and Peru (around Cuzco); epiphytic on trees and on cliffs between 1,300 and 3,300 m. Easily confused with *T. tenuifolia* (see page 181) in its non-blooming stage
CU A very hardy and not delicate species, therefore especially recommended to the beginner.

Tillandsia cauligera Mez, 1906 (Ill. 80)
(*cauligera(us)* bearing a stalk or stem)
Somewhat variable in regard to leaf length and leaf width. In its non-blooming stage can easily be confused with *T. calocephala* (see page 102).

Pl low lying, creeping or hanging from cliff walls, with a stem 0·5–1·5 m long, covered by the old leaves
Fo tightly spiral, up to 20 cm long

LS not clearly delineated, oval, 2–4 cm
wide and 2–3 cm long
LB up to 15 cm long, upright or recurved, triangularly lanceolate, somewhat long tapered, with curled edges,
grey-scaled
Sc upright, about 15 cm long, 4 mm
thick, naked
I simple-spiked (seldom with a second
spike), 6–10 cm long, about 2 cm wide,
10 to 14 flowers, lanceolate to lineal,
pointed, flattened
FB red, oval, up to 2·5 cm long, ribbed,
longer than the sepals
Fl blue, tubular, up to 3·7 cm long
St about 1 cm long, enveloped by the
flower tube
Ha Peru (near Cajamarca, Tarma,
Junin, Cuzco), scattered, but usually
forming large stands; terrestrial or on
cliffs at altitudes of 2,500–3,000 m.
CU Bright and sunny, little watering.
Can be grown epiphytically (hanging).
Because of its silver grey to white
colour it is a very decorative species.

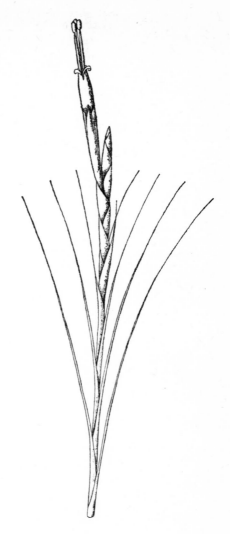

Fig. 30 *Tillandsia chaetophylla* Mez; inflorescence (life-size).

Tillandsia chaetophylla Mez, 1896 (Ill. 3, Fig. 30)

(*chaetophylla chaetos* bristly; *phyllos* leafed)

With its grass-like leaves, this plant resembles a clump of grass more than
a tillandsia. In its non-blooming stage it can easily be confused with the
similar *T. linearis* (see page 150) and *T. remota* (see page 171).

According to Smith, 'as far as known' the stamens are enveloped in the
flower tube. But Mez places *T. chaetophylla* in the subgenus *Platystachys*,
in which he includes all species whose stamens protrude from the flower
tube. The plants in the Heidelberg Botanical Garden collected from the

cloud forest of Teotitlan (Puebla Province, central Mexico) at 2,000 m (collection number Rauh 15487, 1966) show stamens clearly protruding from the flower (see Fig. 30).

Syn. *T. subulata* Morren, 1889, non Vell., 1825

Pl 20–40 cm high including inflorescence, forming thick turf

Fo upright, numerous, in bushy rosettes with dark brown-scaled sheaths and thin, thread-like, green blades that are dark brown at the base, otherwise grey-scaled

Sc slender, 15–20 cm long, 2 mm thick, upright or angling upward

SB similar to the leaves, tightly overlapping, surrounding the scape with the sheaths; blades thread-like upright, pale red

I simple-spiked, lanceolate-pointed, distichous, 4–8 cm long, 3 to 8 flowers

FB overlapping, lanceolate-oval, pointed, 2·5–3·5 cm long and about 1 cm wide, keeled near the tip, scaled, thin-skinned, nerved, red

Se lanceolate, long-tapered, up to 3 cm long, 5 mm wide, the posterior ones fused about 9 mm high, naked, shorter than the flower bracts

Pe 5–7 cm long, pale blue, tubular, with tips somewhat flared

St 4 cm long, enveloped in the flower

Ha Mexico (Veracruz, Puebla, Oaxaca); epiphytic, primarily on oaks

CU Moderately cool and damp. Fast growing and recommended species.

Tillandsia circinnata Schlechtend., 1844 (Ill. 83)

(*circin(n)ata(us)* rolled snail-like; because of the frequently curled leaf blades)

Can easily be confused with the smaller forms of *T. caput-medusae* (see page 105).

Syn. *T. paucifolia* Baker, 1878; *T. yucatana* Baker, 1887; *T. intermedia* Mez, 1903

Pl usually stemless, 10–45 cm high, sometimes because of vegetative reproduction it forms a pseudo-stem of several meters length in the area of the inflorescence

Fo not numerous, grey-scaled

LS forming an egg-shaped to ellipsoid, hollow pseudo-bulb 5–15 cm long

LB up to 20 cm long, with distinctly rolled edges, therefore awl-like, frequently twisted, curled or snail-shaped, grey-scaled

Sc short, surpassed by the foliage

SB similar to the foliage, heavily scaled

I simple or composed of few spikes, surpassed by the scape bracts

PB green or pink, heavily scaled, with their blades surpassing the perpendicular or almost perpendicular, lineally lanceolate, pointed spikes, which are up to 12 cm long and have 2 to 10 flowers

FB green or pink, distichous, overlapping, elliptical, pointed, no distinct keel on the back, 2–3 cm long, ribbed, heavily scaled

Fl blue, up to 4 cm long, tubular, perpendicular

Se lanceolate-long, pointed, up to 2 cm long, naked or slightly scaled, ribbed, the posterior ones fused about 4 mm high
St protruding from the flower
Ha Widely spread from southern Florida, the Bahamas, Cuba, Mexico over Central America to Colombia. Epiphytic on trees and bushes from sea level to about 1,000 m.
CU Moderately humid, bright. Like all other bulbous species highly recommended for epiphyte trees.

Tillandsia commixta Mez, 1919 (Ill. 84)
(*commixta(us)* mixed)
Syn. *T. parviflora* auct., non Ruiz et Pav., 1802; *T. parvifolia* Baker, 1887, non Ruiz et Pav.
T. commixta is related to *T. weberbaueri*, *T. parviflora* and *T. pallidoflavens*. It differs from *T. weberbaueri* (Ill. 142), described on page 191, by its scape bracts, which are shorter than the stalk segments.

Pl stemless, up to 55 cm high
Fo forming a swollen, slender, upright rosette
LS clearly delineated, wide-oval, 10–12 cm long, 10 cm wide, spoon-shaped, upper side dark reddish-brown, underside light brown-scaled
LB up to 30 cm long, 3–4 cm wide at the base, average width 1·8 cm, narrowly triangular, long-tapered, the wavy edges distinctly curled, green, spotted reddish-brown, pale, short scales
Sc slender, upright
SB not numerous, lying close to the scape, upright, shorter than the stalk segments, scaled
I loose panicle, compound, about 20 cm long about 10 cm wide, the basal branches also composed of 2 spikes
PB about 9 mm long, surrounding the base of the joint
Sp short-stemmed, loose, about 18 flowers, lineal, 4 cm long, 9 mm wide, with a winged rhachis
FB 2 mm long, oval, blunt, slightly scaled
Fl pale green, distichous, 4 mm long
Se asymmetrical, broadly elliptical, blunt, 3·5 mm long
St enveloped in the flower
Ha Peru, Amazon district (near Chachapoyas), about 2,000 m high
CU Cool and humid. Culture difficult, since the plant rarely pups; very decorative at bloom time.

Tillandsia complanata Benth., 1846 (Ill. 47)
(*complanata(us)* flattened; because of the flattened spikes and scape. The synonym *T. axillaris* refers to the fact that the inflorescences emerge from the axes)
Syn. *T. axillaris* Griseb., 1864
T. complanata and *T. multicaulis* (see page 157) are the only tillandsias with several inflorescences in the axes of the foliage and therefore easily recognizable. As a result of the wide range it is very variable.

Pl stemless

Fo upright or spreading, forming a funnel rosette up to 30 cm diameter and up to 50 cm high

LS clearly delineated, elliptical, 4–6 cm wide, 6–8 cm long, green, underside has a brown ring at the base

LB very variable; ribbon-shaped lanceolate, long or short-pointed or blunt, up to 30 cm long, 3–5 cm wide at the base, light green to dark green, sometimes dark red or waxy-frosted or red-fringed

Sc numerous, in the axes of the middle rosette leaves, pendulous, naked, flattened, 10–20 cm long

SB narrowly lanceolate, perpendicular, surrounding the scape, naked, shorter than the internodes

I simple-spiked, lanceolate or lineal, pointed, flattened, 4 to 24 close flowers, 2–8 cm long, 1·5–2 cm wide

FB distichous, overlapping, elliptical, blunt, 1·5–2 cm long, rounded on the back, longer or as long as the sepals, green, carmine or cinnabar red

Fl pink, carmine red or violet, up to 2·5 cm long

Se lanceolate, pointed, 1–1·5 cm long, the posterior ones fused more than half way

St somewhat shorter than the flowers

Ha From the Antilles and Costa Rica to Peru and Bolivia, epiphytic on trees, at altitudes of 1,000–3,000 m.

CU Cool and damp. Very decorative at bloom time, but not easy to grow. Since the plant sends out no offshoots, it is usually lost after a few years. Propagation by seed is difficult.

Tillandsia concolor L. B. Smith, 1960 (Plate 32)

(*concolor* all one colour; probably because of the uniformly green, shiny flower bracts)

In its habitat very variable and related to *T. fasciculata* (see page 125) through transitory stages; perhaps only a variety thereof.

Pl stemless, 20–25 cm high in bloom

Fo forming an upright to spread, stiff rosette

LS not clearly delineated, light brown, merging into the leaf blade

LB up to 25 cm long, up to 1·7 cm wide at the base, narrowly triangular, long-tapered with curled edges, thick, short, grey scales

Sc short, strong, upright, up to 6 cm long, covered by the leaves

SB similar to the leaves, but smaller, close, upright, surrounding the scape

Sc simple-spiked or composed of 3 to 4 spikes

PB similar to the scape bracts

Sp upright, lineal-lanceolate, pointed, 9–13 cm long, up to 3 cm wide, laterally compressed, multi-flowered

FB distichous, tightly overlapping, oval, pointed, 3·5–4 cm long, 2–2·5 cm wide, keeled, the lower flower bracts are smaller and without flowers, green or bright red

Se lanceolate, pointed, thin-skinned, 2·5–3·5 cm long, the posterior ones fused 1·5–2·5 cm high

Pe upright, tubular, 6 cm long, lilac violet

St protruding from the flower

Ha Scattered but frequently in spots in central and southern Mexico, epiphytic, from the coast up to about 300 m.

CU Robust, easily grown, especially recommended to the beginner.

Tillandsia cornuta Mez et Sodiro, 1905 (Ill. 85)
(*cornuta(us)* horned; because of the horn-like flower bracts that stick out from the rhachis)
Can be confused with *T. monadelpha* (see page 157), *T. narthecioides* (see page 159) and *T. triglochinoides* (see page 184). All four species are therefore depicted on one page (Ill. 85–88).

Pl stemless, 40 cm high in bloom
Fo forming a rosette, lineal, about 20 cm long, 1·5 cm wide, close, scaled, green, sometimes with reddish-brown streaks, sheaths not clearly delineated
Sc upright, stout, usually shorter than the leaves, naked
SB longer than the stalk segments, the basal ones with long, thread-like recurved blades, the upper ones upright, short-pointed, almost shiny
I simple-spiked, lineal, up to 17 flowers, 10–12 cm long, 2·8 cm wide open
FB distichous, almost perpendicular or horizontal, widely oval, thorn-tipped, 1·8–2·2 cm long, rigid, smooth or slightly ribbed near the tip, green, surrounding the flower, no keel
Fl white, about 3 cm long, standing away from the rhachis but not spread
Se blunt, rigid, smooth, naked, 1·8 cm long, the posterior ones slightly fused, green
St enveloped in the flower, 9 mm long
Ha Ecuador, epiphytic at about 600 m.
CU Shady, humid. Can be grown in a pot in pure peat or on an epiphyte tree. Very interesting at bloom time.

Tillandsia crispa (Baker) Mez, 1896 (Ill. 89)
(*crispa(us)* curly; Because of the curly, wavy leaves)
Syn. *Guzmania crispa* Baker, 1887; *T. undulifolia* Mez, 1896; *T. plicatifolia* Ule, 1907

Pl stemless, 10–30 cm high including inflorescence
Fo in a thick, upright rosette with a pseudo-bulb base
LS large, wide elliptical, somewhat spoon-shaped, brown-scaled, 6–8 cm long, 3–4 cm wide
LB very narrowly lanceolate, long-tapered, 1 cm wide at the base, up to 30 cm long; leaf edges curled and very wavy, a trait that is partially lost in cultivation
Sc upright or slightly curved, slender, 10–15 cm long, 3 mm thick, round, green to reddish-brown, greyish-brown-scaled
SB overlapping, longer than the stalk segments, upright, lying close to the scape, reddish-brown, grey-scaled, 3–4 cm long
I 10–15 cm long, simple or composed of a few spikes
PB similar to the scape bracts, somewhat longer than the stemmed section of the upright, elongated spikes, which are 3–8 cm long, 1·5–2·5 cm wide and have 6 to 36 flowers
FB distichous, overlapping, 9–15 mm long, longer than the sepals, slightly scaled, somewhat inflated, yellowish
Fl sessile, white, surpassing the sepals
Se broadly elliptical, asymmetrical,

firm, slightly scaled or naked
St enveloped in the flower
Ha Panama, Colombia, Ecuador, Peru

CU Interesting, small, but somewhat sensitive species. The waves on the leaves disappear when culture is too humid.

Tillandsia crocata (Morren) Baker, 1887 (Fig. 31a)
(*crocata*(*us*) saffron-like; because of the saffron yellow flowers)
Syn. *Phytarrhiza crocata* Morren, 1880; *T. mandonii* Morren, 1896

Pl short-stemmed, up to 20 cm long
Fo distichous, spread, up to 15 cm long
LS triangularly oval, 1·5–2 cm long, 1 cm wide
LB underside round, upper side gutter-like, awl-shaped, long-tapered, 3 mm wide at the base, heavily grey-scaled; scales eccentric with 1·1–1·5 mm long, very narrow, dangling wings
Sc slender, 10 cm long, 1 mm thick, bent at the base, grey-scaled
SB 0 to 2, lying close to the scape

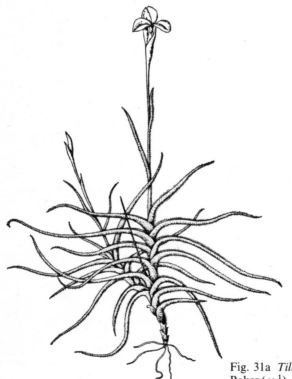

Fig. 31a *Tillandsia crocata* (Morren) Baker ($\times \frac{1}{2}$).

I simple, distichous, 3 to 4 flowers, compact
FB up to 1·3 cm long, green, heavily grey-scaled, the lower ones longer than the sepals, the upper ones shorter
Se 1 cm long, slightly scaled, green
Pe yellow, 2 cm long with spread plates
St enveloped in the flower, 6 mm long
Ha Argentina, Brazil
CU Bright, sunny, dry. Very pretty, white, miniature tillandsia with bright yellow flowers.

Tillandsia cryptopoda L. B. Smith, 1951 (Ill. 285 and 286)

(*cryptopoda(us)* hidden foot; because of the short, inflorescence scape hidden in the rosette.)
Syn. *T. miniatispica* Rohweder, 1953

Pl stemless, flowering to 25 cm high
Fo a dense, rosette, upright, to 18 cm high and 10 cm in diameter, the base sometimes swollen
LS distinct, broad-oval, about 6·5 cm long, 4 cm wide, thin, underside leather brown, upper side violet brown scaled towards the apex
LB short-triangular, 2 cm wide above the sheath, gradually narrowing towards the tip, more or less 10 cm long, the upper third recurved, underside densely appressed grey scaled, upper side lighter and less thickly scaled
FB to 5 cm long, cinnabar red, narrow-triangular pointed, 1·5 cm wide, longer than the sepals, scales very dispersed, dry nerved, not keeled at the back
Se the back ones keeled, fused for 1–2 mm of the height; the front ones free, narrow-triangular, pointed, to 3·5 cm long, upper third dry nerved
Pe to 5 cm long, narrow-tongue shaped (ligulate)
St and Pi projecting
Ha Honduras and Salvador, epiphytic
CU Very fine, remaining small but somewhat difficult in cultivation.

Tillandsia cyanea Linden, 1867 (Plate 29)

(*cyanea(us)* dark, corn flower blue; refers to the corn flower blue coloration of the attractive flowers)
Syn. *T. lindeni* Morren, 1869; *Vriesea lindeni* Lem., 1869; *T. morreniana* Regel, 1870; *T. coerulea* Linden, 1870; *Wallisia lindeni* Morren, 1870; *T. lindeni vera* Dombrain, 1872; *T. lindeni* var. *genuina* Morren, 1879; *Phytarrhiza lindeni* var. *violacea* Hort., 1886; *T. lindeniana* sensu Mez, 1896, non Regel 1896; *T. lindeni superba rosea* Dauthenay, 1898; *T. lindeni vera superba* Duval, 1901.

Pl stemless, 25–30 cm high including inflorescence
Fo numerous, forming a tight rosette of about 40 cm diameter, at first almost perpendicular, later open
LS elliptical, clearly delineated, up to 6 cm long, 3–4 cm wide, brownish, red stripes

LB narrowly triangular, long tapered, up to 35 cm long, 1–2 cm wide, dark green, red-striped near the base, slightly scaled, edges curled
Sc upright, short, up to 15 cm high, 5 mm thick, round, green
SB close, overlapping, the lower ones similar to the leaves, the upper ones elliptically pointed, upright, green, slightly scaled
I simple-spiked, sword-shaped, elliptical, blunt or widely pointed, up to 16 cm long, 7 cm wide, up to 20 flowers
FB distichous, tightly overlapping, keeled, elliptical, pointed, 4–5 cm long, longer than the sepals, pink or red
Fl large, attractive, uniformly dark blue, no white eye in the throat (differing from the similar *T. lindeni*, see page 148). Petals with spread plates 2–3 cm long and 2·3 cm wide
Se 3–3·4 cm long, separate, narrowly lanceolate, pointed, naked
St about 1·8 cm long, enveloped by the perianth tube
Ha Ecuador; epiphytic on trees between 600 and 1,000 m.
CU Half shade, moderately humid, in a pot or on an epiphyte tree. Favoured species long in collections. Usually only select forms in cultivation.

VARIETIES

Of *T. cyanea* we distinguish the var. *tricolor* (André) L. B. Smith (syn. *T. lindeni* Morren var. *tricolor* André). This one has a long, wide, many-flowered sword, and the blue petals have a white spot at the base.
All other varieites of the short-scaped species described as *T. lindeni* Morren, according to L. B. Smith (1951), are only stouter cultivars and therefore need no mention.
T. cyanea, with which L. B. Smith also includes *T. morreniana* Regel, in its non-blooming stage can easily be confused with *T. anceps* (see page 89) and *T. lindeni* (see page 148). Only at bloom time are the differences clear. *T. anceps* has small pale blue flowers; *T. lindeni* has large, deep blue flowers with a white eye in the centre. It can be distinguished from *T. cyanea* var. *tricolor* by the elongated scapes. In *T. cyanea* they are short and frequently completely hidden in the leaf rosette.

Tillandsia dasyliriifolia Baker, 1887 (Ill. 90, Fig. 31b)
(*dasyliriifolia(us)* has leaves like the Mexican plant *Dasylirion*)
Syn. *T. drepanoclada* Baker, 1889; *T. pulvinata* Morren, 1889

Pl stemless, 50 cm–1·70 m high during bloom
Fo 70–90 cm long, numerous, forming a tight rosette
LS large, elliptical, dark brown
LB lineally triangular, pointed, up to 6 cm wide, firm, tight, short scales
Sc upright, stout, usually longer than the leaves
SB upright, oval or lanceolate, pointed
I open panicle
PB oval pointed, shorter than the sterile, basal segment of the spikes, climbing these almost upright or arched, up to 45 cm long with naked,

slightly zigzag rhachis
FB widely oval, surrounding the rhachis, 1·6–2·5 cm long, blunt, open, distichous
Se narrowly elliptical or inverse egg-shaped, blunt, up to 2·5 cm long
Pe upright, blunt, 3–4 cm long, white or greenish
St protruding from the flower
Ha Southern Mexico to Colombia. Epiphytic on rocks, but also terrestrial in low places (up to 150 m).
CU Robust, not difficult in cultivation. Suitable only for larger greenhouses. In its non-bloom stage can easily be confused with *T. utriculata* (see page 186).

Fig. 31b Segments of the inflorescences of *Tillandsia dasyliriifolia* (1), *T. makoyana* Baker (2) and *T. utriculata* L. (3) (all × ½).

1   2   3

Tillandsia decomposita Baker, 1881* (Ill. 29, Fig. 32)

(*decomposita(us)* break down, take apart; refers to the multiple-branched inflorescences with the segmented inflorescences spreading out from the main axis)
Can easily be confused with *T. duratii* (see page 120).
Syn. *T. weddellii* Baker, 1889; *T. tomentosa* N. E. Brown, 1894

Pl stemmed; stems sometimes over 1 m long and usually rootless, hooking into branches with its spirally rolled leaf tips
Fo spiral, usually recurved after developing
LS wide-oval, surrounding the stem, 3–4 cm wide at the base, 3–4 cm long, upper side naked, lower side scaled
LB narrow, lineal-lanceolate, awl-shaped, long-tapered, up to 25 cm long, 1–2 cm wide at the base, thick, firm, edges greatly curled (this feature is often lost in cultivation), white-scaled
Sc upright, stout

* Recently named *T. duratii* Visiani var. *saxatilis* (Hassler) L. B. Smith comb. nov.

SB overlapping, upright
I compound-spiked, up to 40 cm long and 10 cm wide
Sp lineal, pointed, up to 12 flowers, up to 7 cm long and 7 mm wide, standing out from the main axis
FB tightly overlapping, up to 1·2 cm long, naked, as long as the sepals or a little longer
Fl upright, up to 2·5 cm long, blue; petals spread, up to 1·5 cm wide
Se the posterior ones fused
St enveloped in the flower
Ha Argentina, epiphytic
CU Bright and sunny. Little watering. A very remarkable and interesting species because of the curling of the leaves, which is lost under humid cultivation.

Fig. 32 *Tillandsia decomposita* Baker (1) and *T. duratii* Vis. (2) (both × ½).

Tillandsia deppeana Steud., 1841 (Ill. 97)
(*deppeana* named after the botanist Deppe)

Pl stemless, up to 2 m high, very variable
Fo 60–100 cm long, forming a large funnel rosette

LS almost elliptical, not distinct, 10–30 cm long, heavily brown-scaled
LB ribbon-shaped to lineal, long-tapered, up to 8 cm wide, flat, nearly or

completely naked
Sc upright, sturdy
SB similar to the leaves, tightly over-lapping, blades recurved
I compound, seldom simple or double
PB similar to the upper scape bracts, much shorter than the long, distichous spikes, but longer than the flower bracts
Sp pointed, laterally flattened, close flowers (6 to many), often spread or recurved, more or less stemmed, with reduced, sterile bracts at the base
FB inverse egg-shaped, keeled near the tip, firm, thorned, naked, approximately as long as the sepals
Fl short-stemmed
Se long-lanceolate, 2·5–4·5 cm long, central nerve, equally high, almost separate
Pe blue.
Ha Colombia, Peru
CU Shady, humid; suitable only for larger greenhouses.

VARIETIES:
T. deppeana var. *deppeana*
Spikes upright, long-stemmed
Syn. *T. paniculata* Schlecht. et Cham., 1831, non L., 1762
Ha Mexico, West Indies
T. deppeana var. *clavigera* (Mez) L. B. Smith, 1956
Spikes pendulous, slightly flattened, the sterile base usually elongated
Syn. *T. clavigera* Mez, 1896

Tillandsia diaguitensis Cast., 1929 (Fig. 33)
(*diaguitensis* named for the Indian landscape Diaguitas, the location of the type)

Pl stemmed, up to 60 cm long
Fo spiral, with upright or spread to recurved blade
LS 1·3–1·8 cm wide, 1–2·5 cm long, surrounding the stem
LB up to 8 cm long, 6 mm wide at the base, narrowly lanceolate, awl-shaped, heavily grey-scaled, lower side ribbed, with curled edges
Sc up to 8 cm long
SB yellowish, upright, overlapping, scaled at the tip, multi-nerved
I simple-spiked, usually 5-flowered, 4·5–8 cm long, 1·5 cm wide
FB up to 3 cm long, boat-shaped, surpassing the sepals, naked, thorn-tipped, edges thin, papery
Fl stemmed up to 3 mm long, white, aromatic
Se naked, long-tapered, up to 3·2 cm long
Pe up to 7 cm long; claw-like petal lineal, plate spade-shaped, open, finely dentate on the edges
St about 5 cm long, enveloped in the flower
Ha Argentina, epiphytic
CU Bright, sunny. Robust. Recommended for room culture.

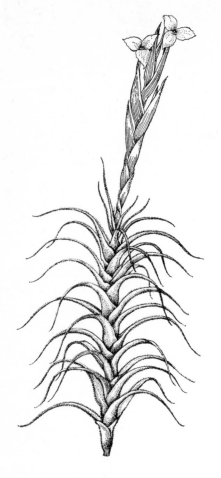

Fig. 33 *Tillandsia diaguitensis* Castell. (× ½).

Tillandsia didisticha (Morren) Baker, 1888 (Ill. 91)
(*didisticha*(*us*) doubly two-part; refers to the distichous arrangement of the actually two-part spikes)
Syn. *Anoplophytum didistichum* Morren, 1881; *T. oranensis* Baker, 1889; *T. crassifolia* Baker, 1889; *T. goyazensis* Mez, 1901; *Guzmania complanata* Wittm., 1916
In the non-blooming stage possibly confused with *T. vernicosa* (see page 187) or *T. ixioides* (see page 143).

Pl stemless, up to 30 cm high in bloom
Fo numerous, forming a compact, spreading rosette

LS 1·5–2 cm wide, up to 2 cm long
LB narrowly lanceolate, long-tapered, 15–20 cm long, 1–1·5 cm wide at the

base, the edges curved upward, rigid, heavily grey-scaled
Sc upright, angling upward or pendulous, 10–12 cm long
SB tightly overlapping, broadly triangular, pointed, grey-scaled, upright, lying close to the scape
I composed of doubly distichous spikes, in all, up to 12 cm long and 7·5 cm wide
Sp distichous, up to 4·5 cm long, 9 mm wide, laterally compressed, narrowly lanceolate, pointed, 8 to 16 flowers

FB tightly overlapping, the lower 3 to 4 flower bracts are sterile, up to 1·1 cm long, red, grey-scaled
Fl white, upright, up to 1·6 cm long
Se narrowly lanceolate, long-tapered, 1·2 cm long, scaled, the posterior ones fused 3 mm high
St enveloped in the flower, 1·3 cm long
Ha From Brazil to Paraguay, Bolivia and Argentina; usually epiphytic
CU Moderately humid and sunny; very decorative at bloom time.

Tillandsia disticha H. B. K., 1816 (Ill. 92)
(*disticha(us)* two-part; refers to the two-part nature of the spikes)
Syn. *T. cinerascens* Willd., 1830; *Platystachys cinerascens* Beer, 1857

Pl stemless, 16–30 cm high, forming groups
Fo up to 30 cm long
LS broadly oval, spoon-shaped, in all forming a spherical to egg-shaped pseudo-bulb 2–4 cm in diameter, upper side heavily brown-scaled; edges partially trimmed in violet brown and equipped with spreading, coarse, grey scales; the exterior sheaths of the bulb are without a blade or have a short blade
LB awl-shaped, long-tapered, 3 mm wide with curled edges, silver grey-scaled
Sc upright or angling upward, slender
SB tightly overlapping, similar to the leaves
I composed of few spikes, seldom simple
PB lanceolate, long-tapered, much shorter than the spikes

Sp almost sessile to short-stemmed, lineal, 4–6 cm, seldom up to 14 cm long, 8 mm wide, laterally flattened
FB distichous, triangular, pointed, 6–10 mm long, longer than the sepals, keeled, ribbed, almost naked or scaled
Se pointed, keeled, naked, 8 mm long, the posterior ones fused about 2 mm high
Pe small, yellow, 1·5 cm long; petals narrowly lanceolate with flared tips
St enveloped, 1·2 cm long
Ha Colombia, Ecuador, Peru; epiphytic in dry forests, frequently at altitudes between 500 and 800 m, usually accompanied by *Vriesea espinosae* (see page 234).
CU Moderately humid, sunny. A pretty, small, bulbous species. Especially suited for small greenhouses and flower windows.

Tillandsia dura Baker, 1889 (Plate 26)

(*dura*(*us*) hard; because of the hard leaves)
Syn. *T. linearis* Wawra, 1880, non alior

Pl stemmed, up to 40 cm long
Fo up to 20 cm long, almost upright, arranged spirally on the stem, sometimes turned to one side, rigid and hard
LS not distinct, thin, papery, about 1·4 cm wide
LB narrowly lanceolate, long-tapered, 1·1 cm wide above the sheath, with curled edges, heavily grey-scaled, dark brown near the base
Sc up to 13 cm long, naked
SB tightly overlapping, the lower ones similar to the leaves, the upper ones lanceolate pointed, 2 cm long, up to 1 cm wide

I simple-spiked, distichous, up to 13 cm long, 1·4 cm wide, up to 20 flowers
FB tightly overlapping, oval, long-tapered, thorny tip, up to 1·7 cm long, red, grey-scaled
Se shorter than the flower bracts, longitudinally lanceolate, rounded, pointed, naked, 8 mm long
Pe blue, 1·7 cm long
St not protruding beyond the flower
Ha Brazil; usually epiphytic
CU Robust, recommended species for flower windows and small greenhouses.

Tillandsia duratii Vis., 1840 (Ill. 95, Fig. 32)

(*duratii* named for the Italian Durat, who was the first to cultivate this plant in Europe)
Syn. *Phytarrhiza duratii* Vis., 1855; *Anoplophytum duratii* Beer, 1857; *T. floribunda* Durat, 1840, non alior; *T. circinalis* Griseb., 1874; *Phytarrhiza circinalis* Morren., 1889; *T. revoluta* Burb., 1889; *T. gigantea* Ruchinger, 1876, non alior
In its non-blooming stage can easily be confused with *T. decomposita* (see note on page 115).

Pl stemmed, up to 30 cm long
Fo about 20 cm long, upright depending on the age, spread, recurved
LS surrounding the stalk, up to 5 cm long, 3–4 cm wide, clearly delineated
LB 2 cm wide, lineal, long-tapered, almost succulent, the edges greatly curled, densely covered with silvery white scales, the tips rolled like a clock spring
Sc strong, upright, 15–20 cm long
SB tightly overlapping, upright, heavily scaled, surrounding the scape

I composed of several short spikes, up to 30 cm long, more or less pyramidal
PB similar to the scape bracts, shorter than the upright, distichous spikes, which have 4 to 6 flowers and are laterally flattened
FB elliptical, up to 1·4 cm long, heavily scaled
Fl blue, upright, up to 3 cm long, petals with spread plates
S approximately as long as the flower bracts, elliptical, naked, fused toward the rear

St enveloped
Ha Argentina, Uruguay and Bolivia;

usually epiphytic
CU like *T. decomposita* (see page 115)

Tillandsia dyeriana André, 1888 (Ill. 96)

(*dyeriana(us)* named for the English botanist Dyer, 1843–1929)
Very similar to it is *T. hamaleana* (see page 137), which, however, has blue flowers.

Pl stemless, 30–35 cm high including inflorescence
Fo forming a funnel rosette 30–35 cm in diameter
LS clearly delineated, broadly oval, 8–9 cm long, 6–7 cm wide, green, spotted irregularly brown
LB tongue-shaped, 4–5 cm wide, 15–18 cm long, lanceolate-pointed, green, irregularly spotted with reddish-brown, short, grey scales on the underside
Sc slender, upright, 17–20 cm long, 4 mm thick, round, green, verrucose
SB slightly longer than the stalk segments, 4 cm long, 1 cm wide, upright, scaled, green with reddish-brown spots
I open, compound, with 3 to 5 horizontal spikes, in all about 10 cm long, about 10 cm wide
PB similar to the scape bracts, but drying already before bloom, shorter

than the spread spikes, which have stems 1 cm long and are about 5 cm long, 3–4 cm wide
FB loosely distichous, flared, triangular, 2 cm long, 1·5 cm wide, green, reddish-brown spots, grey-scaled, drying before bloom, slightly verrucose
Fl very fragrant
Se slightly longer than the flower bracts, about 2 cm long, about 9 mm wide, separate, green or reddish-brown
Pe white, 2·8 cm long, the claw about 4 mm wide, plate spread to recurved, up to 9 mm wide
St 1 cm long, enveloped in the lower part of the tube
Ha Ecuador, epiphytic
CU epiphytic or in a pot, semi-shade. A very beautiful, decorative species with large flowers smelling strongly of carnations.

Tillandsia ebracteata L. B. Smith (Ill. 19 and 20)

(*ebracteata(us)* without flower bracts; because of the very small, hardly visible flower bracts)

Pl stemless, 50–70 cm high including the inflorescence
Fo forming a spreading, flat rosette 30–40 cm in diameter
LS not distinct
SB lineally triangular, long-tapered, 4–5 cm wide at the base, up to 60 cm long, lower side with longitudinal ribs, both sides with thick, short, grey

scales, the edges somewhat curled, dark green in cultivation, yellow green in the natural habitat
Sc upright, strong 6 mm thick, round, naked, up to 20 cm long
SB similar to the leaves, much longer than the stalk segments, surrounding the scape with the sheaths, the blades upright or flared, short, grey scales

I up to 30 cm long, up to 10 cm wide, upright, much branched into a panicle of spikes

PB flared, lineally, lanceolate, green, grey-scaled, the lower ones longer than the branches, the upper ones shorter

Panicle branches of the first order: with 3 to 5 short-stemmed, distichous, open spikes; spikes are 2–3 cm long, up to 2 cm wide

FB very small, about 4 mm long, 4 mm wide at the base, green

Se about 5 mm long, 2 mm wide, separate, the posterior ones briefly fused, green, naked

Pe white, 6 mm long, with flared tips

St enveloped in the flower, 3 mm long

Ha Southern Ecuador and northern Peru, frequently in dry forests between 700 and 1,000 m, epiphytic on cacti and trees, often completely enveloping their crowns (Ill. 19)

CU In a bright, sunny location the rosette leaves will keep the prominent yellow green coloration; in shady culture it is lost. A pretty species worthy of cultivation.

Tillandsia ehrenbergiana Klotsch, 1857 ex Baker (Ill. 287)
(*ehrenbergiana* named for botanist C. A. Ehrenberg
Syn. *Platystachys ehrenbergii* Beer, 1857; *T. ehrenbergiana* Hemsl.
T. ehrenbergiana is near to *T. lepidosepala* (Ill. 111) differing from it by its more distinctive form and the longer inflorescence scape.

Pl short stemmed, densely tufted, flowering to 20 cm high

Fo a dense, short, outspread or upright rosette

LS distinct, broad-elliptical, 1–5 cm long, to 1·3 cm wide, upperside white, bare

LB 10–15 cm long, 5 mm wide above the sheath, awl-shaped (subulate) long threadlike apex, thick, with curled upward margins, outspread or recurved, densely covered with grey, spreading large scales

Sc upright, 4–8 cm long, 1 mm thick, round, bare

SB similar to the leaves, lanceolate, long pointed, much longer than the internodes, densely grey scaled, rose-coloured upper part

I upright, simple, swordshaped, elliptical, 3–4 cm long, 1–1·5 cm wide, densely flowered 3 to 8 flowers

FB arranged in two rows (distichous), densely overlapping, lanceolate, pointed, 2·3 cm long, outspread for 1 cm, thin, densely grey scaled, rose coloured, keeled and nerved towards the tip, longer than the sepals

Se lanceolate, long pointed, free, 1·6 cm long, flared for 5 mm, the back ones keeled, thin cutaneous, grey scaled, weakly nerved

Pe 3 cm long, 3 mm wide, rounded and somewhat flared at the tip; emerald-green

St and Pi enclosed in the flower; pistil 2 cm long

Ha Mexico

CU A nice, small, free flowering Tillandsia, requiring little water in cultivation and a light sunny position.

Tillandsia esseriana Rauh and L. B. Smith, 1970 (Ill. 288)
(*esseriana* named for Dr. G. Esser, a Heidelberg Botanist)
The plant has flowers very similar to the large, blue-violet flowers of T. lindenii. It is worthy of cultivation. It leads an extraordinary hard life, requiring very little water, yet in much sun. Cactus culture should suit it well.

Pl with inflorescence to 30 cm tall, stemless or short stemmed, forming groups
Fo few forming a star shaped, upright rosette
LS not clearly defined from the blade, 4–5 cm long, 3–4 cm wide, brown scaled
LB narrow-lanceolate, long pointed, stiff, to 30 cm long, 3 cm wide at the base; channelled curled upwards margins, reddish-brown to reddish-violet, appressed grey scaled, underside finely grooved longitudinally
Sc 15–18 cm long, 4 mm thick, round, green, bare
SB upright, dense, longer than the internodes (stem segments) the lower ones similar to the leaves, the upper ones grey appressed scaled, blade-less
I simple or compound in 2–3 spikes, about 15 cm long
Sp upright or spreading 2 cm wide
FB densely overlapping, arranged in two rows, upright, 4·4 cm long, 1·5 cm wide, green, weakly scaled, much longer than the sepals
Se 2·8 cm long, 8 mm wide, thin cutaneous, green, bare, free
Pe 6·5 cm long, clawed; claw 4·3 cm long, 4 mm wide, white; the plate 2·2 cm long, 2 cm wide, bright violet
St and **Pi** enclosed within the flower
Ha Paraguay, on steep rocks of the Sierra Guazu and Sierra Cova (Amambay), in company with Pilocereus.*

Tillandsia exserta Fernald, 1895 (Fig. 34)
(*exserta(us)* protruding; probably refers to the stamens protruding from the bloom)
Syn. *T. cinerea* Mez, 1896, non Raf., 1840

Pl stemless, 20–70 cm high including inflorescence
Fo numerous, in a tight rosette
LS 2–3 cm long, triangularly oval, 1·5–2·5 cm wide
LB hard, up to 30 cm long, greatly recurved, lineal, pointed, curled edges at the base, 1·5 cm wide, heavily grey-scaled
Sc upright, up to 25 cm long, about 3 mm thick, naked
SB the lower ones leaf-like, the upper ones elliptical, pointed, thick, short scales, green to pink
I simple or composed of few spikes
PB red, similar to the upper scape bracts, but much shorter than the upright or slightly flared spikes, which

* The genus *Pilo* has been discontinued and broken down into several separate genera. *Ed.*

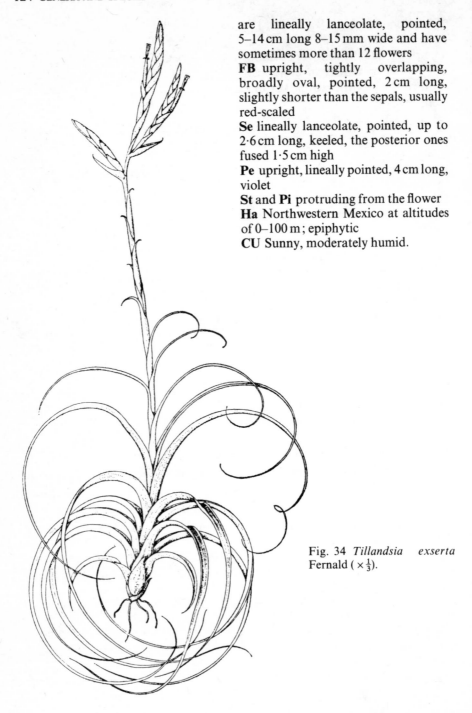

are lineally lanceolate, pointed, 5–14 cm long 8–15 mm wide and have sometimes more than 12 flowers

FB upright, tightly overlapping, broadly oval, pointed, 2 cm long, slightly shorter than the sepals, usually red-scaled

Se lineally lanceolate, pointed, up to 2·6 cm long, keeled, the posterior ones fused 1·5 cm high

Pe upright, lineally pointed, 4 cm long, violet

St and **Pi** protruding from the flower

Ha Northwestern Mexico at altitudes of 0–100 m; epiphytic

CU Sunny, moderately humid.

Fig. 34 *Tillandsia exserta* Fernald (× ⅓).

Tillandsia fasciculata SW., 1788 (Plate 34)
(*fasciculata*(*us*) gathered into a bundle; probably because of the bundle of bloom spikes; possibly also refers to the bushy appearance of the plant)
Syn. *T. compressa* Bert., 1830; *T. setacea* Hook., 1833; non SW., 1788; *Vriesea glaucophylla* Hook., 1848; *T. pungens* Mez, 1896
This species is very variable in regard to size, coloration and number of bloom spikes. There are all sorts of transitions between *T. concolor* (see page 110) and *T. fasciculata*.

Pl stemless, 20–100 cm high, forming rather large clumps
Fo numerous, in a thick, rigid, spread to perpendicular rosette
LS large, oval, rust-brown
LB 2–4 cm wide at the base, scaled, triangularly lanceolate, long-tapered, rigid
Sc upright, strong
SB overlapping, the lower ones resembling the leaves
I simple-spiked, but usually composed of few to many spikes in a bundle
PB widely oval, long-tapered, shorter than the sessile to long-stemmed, upright or flared spikes, which are usually over 10 cm long, greatly flattened and sterile at the base

FB broad, pointed, 2–4 cm long, smooth to heavily ribbed, naked or almost naked, usually vivid red to reddish-yellow
Se usually shorter than the flower bracts
Pe 6 cm long, upright, white at the base, blue violet in the upper portion
St surpassing the petals
Ha From Florida and Mexico to Colombia and Peru; usually epiphytic in dry forests at altitudes of 600–1,900 m.
CU Robust, easily grown. Splendid at bloom time. Bloom spikes and their colouration last a long time. Suitable only for larger greenhouses.

Tillandsia ferreyrae L. B. Smith, 1963 (Ill. 23)
(*ferreyrae* named for the Peruvian botanist Prof. Dr. R. Ferreyra)

Pl stemless, 1·5–3 m high including the inflorescence
Fo numerous, forming a large funnel rosette 1–2 m in diameter
LS long-elliptical, brown-scaled on the inner side
LB lanceolate, long-tapered, 30–60 cm long, up to 20 cm wide above the sheath, grey green, waxy frosted, with a reddish tip, both sides with short grey scales
Sc 50 cm–1 m long, angular, greyish-red, waxy frosted, arched
I loosely compound, with stemmed,

multi-flowered, pendulous, flat spikes, that are 10–20 cm long, up to 3 cm wide
PB about 3·5 cm long, much shorter than the sterile base of the spikes
FB distichous, tightly overlapping, rounded at the back, as long as the sepals, about 3·5 cm long, 1·7 cm wide, grey waxy frosted with violet tip
Fl almost sessile, about 4 cm long, upright, dark purple
Se long-oval, rounded
St protruding from the flower along with the pistil
Ha *T. ferreyrae*, like *T. rauhii*, is one of

the large Peruvian tillandsias with arched inflorescences, which are very decorative at bloom time

CU suited for large greenhouses

Tillandsia festucoides Brongn., 1896 (Ill. 98)

(*festucoides* Festuca-like; refers to the growth pattern, which is reminiscent of tufts of the grass *Festuca*)

Syn. *T. caricifolia* Morren, 1896

In its non-blooming stage can easily be confused with *T. remota* (see page 171) and *T. chaetophylla* (see page 107).

Pl stemless, up to 55 cm high, growing in thick, grass-like bundles

Fo numerous, grass-like, green, only slightly scaled

LS narrowly triangular, up to 2 cm long, up to 1 cm wide

LB lineally awl-like to thread-like, 30–40 cm long, the edges curled into a gutter, 5 mm wide above the sheath

Sc slender, upright or angling upward, 15–20 cm long

SB close, the lower ones resembling the leaves

I up to 17 cm long with about 10 almost perpendicular to flared spikes, in cultivation sometimes simple

PB resembling the scape bracts, their sheaths shorter than the spikes, the blades thread-shaped and partially surpassing the lineal spikes, which are 3–9 cm long and have 6 to 10 flowers

FB lanceolate-oval, 1·7–2·2 cm long, keeled, green or red

Se 1·7 cm long, the posterior ones fused

Pe blue, fused into a tube 3 cm long

St surpassing the flower

Ha Southern Mexico and Central America; epiphytic on trees at altitudes of 60–600 m.

CU Easy, moderately humid and semi-shady. An interesting plant reminiscent of a tuft of grass or sedge.

Tillandsia filifolia Cham. et Schlecht., 1831 (Ill. 99)

(*filum* thread; *folium* leaf i.e. thread leafed; because of the thin, thread-like leaf blades)

Syn. *Platystachys filifolia* Beer, 1857; *T. staticiflora* Morren, 1871

Pl stemless, forming a bulb, up to 30 cm high, growing in rather large clumps

Fo numerous, sticking out in all directions

LS triangular, 1 cm wide, 2 cm long, brown, grey-scaled

SB upright to almost upright and arched, spread, thread thin, about 1 mm wide, up to 30 cm long, green, grey-scaled

Sc upright to angling upward, slender, 1 mm thick

I about 15 cm long, loosely compound, perpendicular or pendulous

PB very short, 8–10 mm long

Sp 5 to 10, open, 8 to 16 flowers, 5–8 cm long, with zigzag axis

FB loosely distichous, oval, 7–8 mm long, thin, nerved

Se oval, 8 mm long, 3 mm wide, as long as the flower bracts
Pe 1 cm long, pale blue, spread in the upper third
St hardly surpassing the petals
Ha Southern Mexico, Guatemala, Honduras and Costa Rica; epiphytic between 100 and 1,300 m.

CU Moderately humid; the plant makes a very graceful appearance with its thin, thread-like leaves and the bulb-like formation of the basal leaves. It is primarily suitable for small epiphyte trees and should be in any collection.

Tillandsia flabellata Baker, 1887 (Plate 35)

(*flabellata fan-like*; because of the fan-like arrangement of the long, narrow bloom spikes)

Pl stemless, 20–50 cm high
Fo numerous, rosette-shaped, upright to spread, 30–50 cm long, green or red
LS wide-oval, 10 cm long, up to 5 cm wide, grey-scaled, gradually merging into the blade
LB lineally lanceolate, long-tapered, about 4 cm wide above the sheath, short scales
Sc very short and hidden by the leaves or elongated, upright and surpassing the leaves
SB close, overlapping, very large, similar to the leaves
I loose, with 3 to 10 upright or flared spikes up to 35 cm long and arranged fan-like or hand-shaped
PB very large in the lower part of the inflorescence, similar to the scape bracts, but shorter than the spikes, short scales, red or green
Sp narrow, rounded, or laterally flat-tened, long-stemmed, 15–30 cm long, up to 1·5 cm wide
FB distichous, overlapping, lan-ceolate, pointed, keeled, 2·5–4 cm long, 1 cm wide, thin, vivid red, at first scaled, later naked
Se thin-skinned, up to 1·5 cm long, the posterior ones fused 3 mm high, other-wise separate
Pe fused into a tube, up to 4 cm long, blue
St surpassing the petals
Ha Mexico, Guatemala, El Salvador, from 100 m up into the cloud forest region (1,300 m).
CU Moderately humid. Few prob-lems. At bloom time *T. flabellata* is an extremely decorative plant with its brilliant red, long, narrow spikes. It is, however, very variable in regard to the colouration of the leaves, the length of the inflorescence and the spikes.

Tillandsia flexuosa SW 1788 (Ill. 101, Fig. 35)

(*flexuosa* arched, bent, twisted)
Syn. *T. tenuifolia* sensu Jacq., 1763, non L., 1762; *T. flexuosa* var. *fasciata* Lindl., 1823; *T. aloifolia* Hook., 1826; *T. patens* Willd., 1830; *Vriesea aloefolia* (Hook.) Beer, 1857; *Vriesea tenuifolia* Beer, 1857; *Plastystachys patens* K. Koch, 1873; *T. flexuosa* var. *vivipara* André, 1889

Pl stemless, 30–150 cm high including inflorescence
Fo 10 to 20, forming a long, hollow, upright, sometimes spirally twisted

pseudo-bulb
LS large, oval, merging into the blade, dark chestnut brown on the inside, green on the outside, heavily grey-scaled, often equipped with lateral banding; the outer sheaths are scoop-shaped, smaller and without a blade
LB firm to hard, narrow ribbon shape and spirally twisted, 10–20 cm long, 1·5 cm wide at the base, both sides heavily grey-scaled, often laterally banded, edges curled into a gutter
Sc upright, slender, naked, 2–4 mm thick
SB the lower ones similar to the leaves, the upper ones shorter than the stalk segments, upright, scaled
I simple or open compound, with a slightly zigzag rhachis
PB similar to the upper scape bracts, much shorter than the lower sterile part of the open, flared spikes
FB open distichous, flared, 2–3 cm long, as long as or shorter than the sepals, elliptical, pointed, not keeled, ribbed, scaled to naked
Pe red
St surpassing the petals. Quite variable in regard to the twisting and the marking of the leaves
Ha Southern Florida, West Indies, Panama, Venezuela, Colombia,

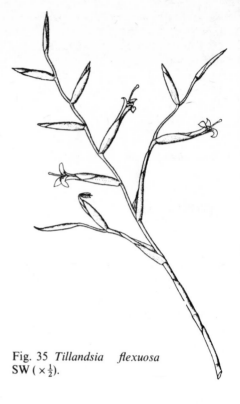

Fig. 35 *Tillandsia flexuosa* SW (× ½).

Guyana, from sea level to 600 m.
CU Moderately humid, sunny. Because of the leaf twisting a quite remarkable species.

Tillandsia floribunda H. B. K., 1816 (Ill. 100)
(*floribunda* greatly flowered)
Fine for small greenhouses. *T. floribunda* in its non-blooming stage can easily be confused with the variable *T. juncea* (see page 144). At bloom time the two species are easily distinguished by the fact that *T. juncea* stamens protrude far out of the flower tube, while in *T. floribunda* they are enveloped in the flower.
Syn. *Platystachys floribunda* Beer, 1857; *T. coarctata* Willd., 1830

Pl grows turf-like, 20–30 cm high
Fo numerous, arranged in tufts, heavily grey-scaled, surpassing the inflorescence

LS almost triangular, up to 2 cm wide at the base, 4 cm long, merging into the blade, both sides dark brown-scaled
LB fine awl-shaped, stiff, average width 4 mm, 35–45 cm long, gutter-like
Sc upright, 20–25 cm long, 4 mm in diameter, round, naked, green
SB similar to the leaves, upright, tightly overlapping, longer than the stalk segments, surrounding the scape with the sheaths; blade long, fine awl-shaped, heavily grey-scaled
I simple or composed of 5 to 8 close, almost finger-shaped, upright to almost upright spikes
PB similar to the scape bracts, much

shorter than the spikes, which are 7–8 cm long, lanceolate or lineally lanceolate and have 10 to 15 flowers
FB distichous, tightly overlapping, oval, 1–1·5 cm long, naked, smooth, shiny, distinct central nerve, as long as the sepals
Fl almost sessile, 2 cm long, tubular, dark blue violet
Se elliptical, pointed, central nerve, naked, 1·1 cm long
St enveloped in the flower
Ha Ecuador, Peru (Huancabam-batal); epiphytic.
CU Moderately humid, sunny.

Tillandsia fraseri Baker, 1889 (Ill. 290 and 291)
(*fraseri* named after the collector L. Fraser)
Syn. *T. erectiflora* André, 1888 non *T. erectiflora* Bak., 1887
T. fraseri is a 'green' small flowered Tillandsia in the Pseudocatopsis group. Its charm lies in the bright red flower-bracts and sepals which contrast with the white flowers at flowering time.

Pl stemless, flowering from 1·50 m to 3 m high
Fo several, forming a flat funnel shaped rosette from 60–80 cm in diameter
LS longish-oval, 16–18 cm long, 7–9 cm wide, dark brown
LB tongue-shaped, 6–8 cm wide, to 50 cm long, lanceolate, recurved at the apex, green or red, somewhat scaled
Sc upright, sturdy, round, bare, green 8–10 mm thick, 1–2·5 m long
SB longish-oval, pointed, longer than the internodes, upright, covering the scape
I upright, loose-pyramidal, 30–40 cm long, 20–30 cm wide, much branched
PB lanceolate, long pointed, shorter than the branches, the lower ones 4–5 cm long, green or red

Sp lineal, pointed, to 11 cm long, about 1 cm wide, short stalked, loose 20–30 flowered, with knee-jointed (geniculate), bare, red rachis
FB at first branched, but soon becoming partially so for the upper third, triangular, 1 cm long, weakly keeled at the tip, naked, red, nerved
Fl at first loosely branched, later becoming partially so for the upper third, short stalked, white
Se 6 mm long, longish, pointed, keeled, bare, red
Pe 2 mm longer than the sepals with flared tip
St and Pi enclosed within the flower
Fr a 2–3 cm long cylindrical capsule
Ha Colombia, Ecuador.

Tillandsia friesii Mez 1906 (Ill. 289)
(*friesii* after a botanist named R. E. Fries)
A fine, small tillandsia requiring extremely dry conditions (xerophytic), a light and full sun position and very little water.

Pl forming a short stem to 15 cm long
Fo spirally arranged, dense
LS not clearly defined from the blade, about 2–3 cm wide
LB arching, triangular, long pointed, 2 cm wide above the sheath, 10–11 cm long, dense grey scaled
Sc upright, short, covered by the leaves and not extending beyond them, about 8–10 cm long, 3 mm thick
SB densely enclosing the scape, longer than the internodes, the basal ones similar to the leaves, the upper ones bladeless, bare, green, upright.

I simple, sword shaped, 3–6 flowered to 5 cm long and 8 mm wide
FB to 2·5 cm long, brownish-green, bare, not keeled, upright, arranged in two rows, overlapping, oval-lanceolate
Fl upright, sessile, to 4 cm long
Se 1·7 cm long, the back ones keeled, thin cutaneous, free
Pl 3·5 cm long, flared or curled back, 6 mm wide, violet-blue or reddish
St and **Pi** enclosed within the flower
Ha Argentina, Bolivia.

Tillandsia funckiana Baker, 1889 (Plate 27)
(*funckiana* named after the collector Funck)
T. andreana Morren, 1888 (named for the bromeliad researcher André) is similar and also comes from Colombia. For a long time L. B. Smith placed it with *T. funckiana*, but today it is recognized as a separate species. *T. andreana* differs from *T. funckiana* in its tight rosette growth (Fig. 36), i.e. the axis in *T. funckiana* is elongated but in *T. andreana* it is short. Both inflorescences are single-flowered and have a large, red flower. It would be better to list *T. funckiana* as a variety of *T. andreana*.

Pl with elongated axis
Fo numerous, spiral, upright or turned to the side depending on the plant growth habit, up to 5 cm long, firm
LS clearly delineated, triangularly oval, about 5 mm long, grey-scaled
LB lineal, long, thin, pointed, 1–2 mm wide at the base, reddish-brown or green, heavily grey or brown-scaled
Sc lacking
I simple-spiked with 1, seldom 2 flowers

FB lanceolate-long, pointed, thin-skinned, naked, half as long as the sepals
Se elliptically egg-shaped, blunt, longer than 1·5 cm, naked.
Pe large, brilliant red, up to 4·4 cm long, recurved at the tip
St protruding from the flower
Ha Colombia
CU Moderately humid. A pretty, small species with large, bright red flowers.

Ill. 109 *Tillandsia juncea* Ruiz et Pav.; as an enlarged inflorescence (northern Peru, valley of the Rio Saña, 800 m).

Ill. 110 *Tillandsia kirchhoffiana* Wittm.

Ill. 111 *Tillandsia lepidosepala* L. B. Smith. Central Mexico.

Ill. 112 *Tillandsia loliacea* Mart. Small form with short inflorescences, from Paraguay.

Ill. 113 *Tillandsia lorentziana* Griseb. Paraguay.

Ill. 114 *Tillandsia oroyensis* Mez. Near Oroya, Central Peru, 3800 m.

Ill. 115 *Tillandsia paleacea*.

Ill. 116 *Tillandsia polystachya* L. Chiapas, Comitan, southern Mexico.

Ill. 117 *Tillandsia pendulispica* Mez. near Ayabaca, northern Peru, 2000 m; close up of inflorescence (*right*).

Ill. 118 *Tillandsia pohliana* Mez. Quillabamba, southern Peru.

Ill. 119 *Tillandsia peiranoi* Castell.

Ill. 120 *Tillandsia prodigiosa* (Lem.) Baker. S e g m e n t   o f inflorescence.

Ill. 121 *Tillandsia subulifera* Mez.

Ill. 122 *Tillandsia pueblensis* L. B. Smith.

Ill. 123 *Tillandsia schiedeana* Steud.

Ill. 124 *Tillandsia rectangula* Baker.

Ill. 125 *Tillandsia remota* Wittm.

Ill. 126 *Tillandsia schreiteri* Lillo et Castel.

Ill. 127 *Tillandsia sphaerocephala* Baker.

Ill. 128 *Tillandsia streptophylla* Scheidw.

Ill. 129 *Tillandsia tectorum* Morren.

Ill. 130 *Tillandsia setacea* SW.

Ill. 131 *Tillandsia streptocarpa*; (*inset*) simple-spiked inflorescence of a very small plant only 15 cm high.

Ill. 132 *Tillandsia tetrantha* Ruiz et Pav. var. *tetrantha* (Peru, Canchaque).

Ill. 133 *Tillandsia tetrantha* Ruiz et Pav. var. *aurantiaca* (Griseb.) L. B. Smith (northern Peru, Ayabaca).

Ill. 134 *Tillandsia stra-minea* H. B. K.

Ill. 135 *Tillandsia tri-cholepis* Baker.

Ill. 136 *Tillandsia utri-culata* L.

Ill. 137 *Tillandsia triticea* Burchell.

Ill. 138 *Tillandsia valenzuelana* A. Rich.

Ill. 139 *Tillandsia vernicosa* Baker.

Ill. 140 *Tillandsia vicentina* Standl.

Ill. 141 *Tillandsia walteri* Mez. Tarma, central Peru, at 2000 m.

Ill. 142 *Tillandsia weberbaueri* Mez. Near Nazareth, Peru (Amazon basin).

Ill. 143 *Tillandsia violacea* (Beer) Baker. Valley of the Rio Y, near Oaxaca, Mexico.

Ill. 144 *Tillandsia xiphioides* Kerr; (*inset*) inflorescence.

Ill. 145 *Catopsis brevifolia* Mez et Wercklé.

Ill. 146 *Catopsis hahnii* Baker.

Ill. 147 *Catopsis sessiliflora* Ruiz et Pav.

Ill. 148 *Catopsis wangerinii* Mez et Werckle.

Ill. 149 *Guzmania dissitiflora* (André) L. B. Smith.

Ill. 150 *Guzmania erythrolepis* Brongn.

Ill. 151 *Guzmania mucronata* (Griseb.) Mez.

Ill. 153 *Guzmania variegate* L. B. Smith.

Ill. 152 *Guzmania lindenii* (André) Mez.

Ill. 154 *Guzmania vittata* (Mart.) Mez.

Ill. 155 *Vriesea espinosae* L. B. Smith. Valley of Ayabaca, northern Peru.

Ill. 157 *Vriesea chrysostachys* Morren.

Ill. 156 *Vriesea espinosae* (L. B. Smith) L. B. Smith; blooming plant.

Ill. 158 *Vriesea cereicola* L. B. Smith. Valley of the Rio Saña, northern Peru.

Ill. 159 *Vriesea erythrodactylon* Morren.

Ill. 161 *Vriesea fenestralis* Lind. et André.

Ill. 160 *Vriesea friburgensis* Mez var. *friburgensis*.

Ill. 162 *Vriesea gigantea* Gaud.

Ill. 163 *Vriesea appenii* Rauh. (holotype). Valley of Olmos, northern Peru, near the pass Abra Porculla, 2000 m.

Ill. 164 *Vriesea heterandra* (Andre) L. B. Smith (Colombia).

Ill. 165 *Vriesea hieroglyphica* (Carr.) Morren.

Ill. 166 *Vriesea malzinei* Morren.

Ill. 167 *Vriesea olmosana* L. B. Smith. (clonotype) valley of Olmos, northern Peru.

Ill. 168 *Vriesea patula* (Mez) L. B. Smith. Between Oxapampa and San Ramon, Peru.

Ill. 169 *Vriesea platzmannii* Morren; plant and inflorescence.

Ill. 170 *Vriesea rodigasiana* Morren.

Ill. 171 *Vriesea racinae* L. B. Smith.

Ill. 172 *Vriesea vanhyningii* L. B. Smith.

Ill. 173 *Vriesea tillandsioides* L. B. Smith. Valley of Canchaque, northern Peru.

Ill. 174 *Vriesea saundersii* (Carr.) Moren.

Ill. 175 *Acanthstachys strobilacea*. Klotsch shown fruiting top) and flowering (bottom).

Ill. 176 *Aechmea angustifolia* Poepp. and Endl.

Ill. 177 *Aechmea calyculata* (Morren) Baker.

Ill. 178 *Aechmea bambusiodes* L. B. Smith; (*inset*) close-up of inflorescence.

Ill. 179 *Aechmea bromeliifolia* (Rudge) Baker.

Ill. 180 *Aechmea candida* Morren.

Ill. 182 *Aechmea caudata* Lindn.

Ill. 181 *Aechmea chantinii* (Carr.) Baker.

Ill. 183 *Aechmea coelestis* (C. Koch) Morren.

Il. 184 *Aechmea filicaulis* (Griseb.) Mez.

Ill. 185 *Aechmea distichantha* Lem. var. *glaziovi* (Baker) L. B. Smith.

Ill. 186 *Aechmea fasciata* (Lindl.) Baker.

Ill. 187 *Aechmea fosteriana* L. B. Smith.

Ill. 189 and 190 *Aechmea germinyana* (Carr) Baker (*right*) and *Aechmea veitchii* Baker (*left*).

Ill. 188 *Aechmea gamosepala* Wittm.

Ill. 191 *Aechmea gracilis* Lindm.

Ill. 192 *Aechmea luddemanniana* (C. Koch) Brongn.

Ill. 193 *Aechmea mariae-reginae* Wendl.

Ill. 194 *Aechmea mexicana* Baker.

Ill. 195 *Aechmea penduliflora* Andre.

Ill. 196 *Aechmea pineliana* (Brongn.) Baker.

Ill. 197 *Aechmea pubescens* Baker.

Ill. 198 *Aechmea tillandsioides* (Mart.) Baker.

Ill. 199 *Aechmea triangularis* L. B. Smith.

Ill. 200 *Aechmea warasii* E. Pereira.

Ill. 202 *Aechmea × Bert* M. B. Foster.

Ill. 201 *Aechmea weberbaueri* Harms.

Ill. 203 *Billbergia pyramidalis* (Sims). Lindl. var. *concolor* L. B. Smith.

Ill. 204 *Ananas comosus* (L.) Merill, young inflorescence.

Ill. 205 *Ananas comosus* (L.) Merill, flowering.

Ill. 206 *Araeococcus flagellifolius* Harms.

Ill. 207 *Billbergia chlorantha* L. B. Smith.

Ill. 208 *Billbergia distachia* (Vell.) Mez var. *straussiana* Wittm.

Ill. 209 *Billbergia iridifolia* (Nees and Mart.) Lindl.

Ill. 210 *Billbergia morelii* Brongn.

Ill. 211 *Billbergia aff. rosea* Beer.

Ill. 212 *Billbergia sanderiana*. Morren.

Ill. 214 *Bromelia urbaniana* (Mez) L. B. Smith.

Ill. 213 *Billbergia viridiflora* H. Wendle.

Ill. 215 *Bromelia urbaniana* (Mez). L. B. Smith; inflorescence.

Ill. 216 *Canistrum aurantiacum* Morren; habit.

Ill. 217 *Canistrum aurantiacum* Morren; inflorescence.

Ill. 218 *Canistrum fosterianum* L. B. Smith.

Ill. 219 *Hohenbergia augusta* (Vell.) Morren.

Ill. 220 *Cryptanthus beucheri* Morren.

Ill. 221 *Cryptanthus bivattatus* (Hook) Regal.

Ill. 222 *Cryptanthus lacerdae* Antoine.

Ill. 223 *Cryptanthus zonatus* (Vis.) Beer var. *fuscus* Mez.

Ill. 224 *Neoregelia concentrica* (Vell.) L. B. Smith. Longitudinal section through rosette and inflorescence.

Ill. 225 *Neoregelia pauciflora* L. B. Smith.

Ill. 226 *Neoregelia ampullacea* (Morren) L. B. Smith.

Ill. 227 *Neoregelia cyanea* (Beer) L. B. Smith.

Ill. 228 *Neoregelia cruenta* (R. Graham) L. B. Smith.

Ill. 229 *Neoregelia laevis* (Mez). L. B. Smith.

Ill. 230 *Neoregelia myrmecophila* (Ule) L. B. Smith.

Ill. 231 *Neoregelia pendula* L. B. Smith. var. *brevifolia* L. B. Smith.

Ill. 232 *Neoregelia wurdachii* (R. Graham) L. B. Smith.

Ill. 233 *Nidularium purpureum* Beer.

Ill. 234 *Orthophytum foliosum* L. B. Smith; habit.

Ill. 235 *Orthophytum foliosum* L. B. Smith; inflorescence.

Ill. 236 *Portea leptantha* Harms.

Ill. 238 *Quesnelia lateralis* Wawra showing terminal flower growth (Jt), lateral growth with inflorescence (Jl).

Ill. 237 *Portea petropolitana* (Wawra) Mez.

Ill. 239 *Quesnelia liboniana* (De Jonghe) Mez.

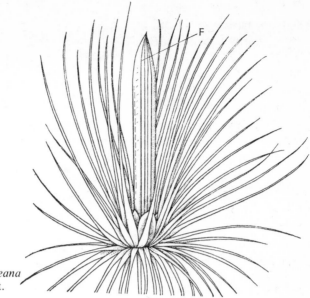

Fig. 36 *Tillandsia andreana* Morren (life-size); F fruit.

Tillandsia funebris Cast., 1933 (Ill. 102)

(*funebris* mourning; probably refers to the dingy yellow brown petals)
T. funebris is very closely related to *T. rectangula* (see page 170) and is linked with it through transitional varieties.

Pl short, stemmed, 8–10 cm high including the inflorescence
Fo loose, spiral, spread, heavily grey-scaled
LS 4–5 mm long, 4 mm wide, surrounding the stem
LB recurved, awl-shaped or thread-like, long-tapered, up to 4 cm long, 3–4 mm wide above the sheath, with distinctly curled edges
Sc 4 cm long, thin, upright, naked, greenish-brown
SB usually lacking, seldom one in the upper part of the scape; it will be about 1 cm long, naked, soon withers
I a 1 to 2 flowered spike

FB 8–9 mm long, naked, brownish-skinned, soon withers, shorter than the sepals, which are 1 cm long, 2 mm wide, thin-skinned, naked, brownish, separate
Pe 1·4 cm long; claw 2 mm wide, plate flared to recurved, 5 mm wide, dingy yellow brown
St enveloped in the tube, about 7 mm long
Ha Dry areas of Bolivia, Paraguay and Argentina
CU Warm, dry and sunny. Because of its small size it is especially suitable for the smallest greenhouses and terraniums.

Tillandsia gardneri Lindl., 1842 (Ill. 103, Plate 36)
(*gardneri* after the collector Gardner)
Syn. *T. fluminensis* Mez, 1894; *T. regnellii* Mez, 1894; *T. cambuquirensis* A. Silveira, 1931; *T. venusta* A. Silveira, 1931
Very variable in regard to the length of the scape and the length of the spikes. In its non-blooming stage it can be confused with *T. geminiflora* var. *incana* (see page 134).

Pl stemless, 15–25 cm high
Fo numerous, forming a compact, white rosette, often turned to one side
LS indistinct, merging into the blade
LB narrowly triangular, long-tapered, up to 25 cm long, up to 2·5 cm wide at the base, heavily silver grey-scaled, flared to recurved
Sc almost upright or pendulous, slender, up to 15 cm long, thickly scaled
SB similar to the leaves, spiral, much longer than the stalk segments, heavily scaled
Pe fused into a tube, with spread tips, pink, up to 1·8 cm long
I compound, overall shape elliptical or spherical up to 6 cm long, 3–4 cm wide
PB similar to the scape bracts, the lower ones longer than the spikes, the upper ones shorter, heavily scaled, green to pink
Sp 4 to 12, spiral, usually 3 to 4 (7) flowered, laterally compressed, oval to lineally lanceolate
FB distichous, oval, pointed, keel at the tip, longer than the sepals, up to 2 cm long, heavily scaled, green to pink
Se almost oval, pointed or blunt, keel, about 1 cm long, separate
St enveloped in the flower, 1·3 cm long
Ha Venezuela, Trinidad, Colombia and Brazil; epiphytic at altitudes around 700–1,300 m.
CU Bright and sunny. Very pretty species worthy of cultivation and one whose silver grey rosettes can be the foliar charm of any collection.

Tillandsia gayi Baker, 1889 (Fig. 37)
(*gayi* after the botanist Gay)

Pl stemless, up to 40 cm high including inflorescence
Fo rosette-shaped, up to 25 cm long, heavily covered with silver grey scales
LS not clearly delineated, narrowly oval
LB narrowly lanceolate, 2 cm wide at the base, long-tapered
Sc upright, naked, round, green, 15–20 cm long, about 3 mm in diameter
SB similar to the leaves, surrounding the scape with the sheath; blades of the lower scape bracts are narrow, long-tapered, upright, heavily grey-scaled, longer than the stalk segments, green or red
I compound, 5 to 10 upright to spreading spikes, 15–18 cm long, 5–6 cm wide
PB similar to the upper scape bracts, shorter than the spikes, slightly scaled, red or green
Sp 8 to 10 flowers, lanceolate, red, 5–6 cm long, 7–8 mm wide
FB overlapping, distichous, oval-elliptical, 1·4–1·6 cm long, no keel,

naked, distinctly nerved
Se lanceolate, blunt, 1·4 cm long, green, naked, posterior ones fused 3 mm high
Pe pink violet, 2 cm long, tips flared and somewhat recurved
St and **Pi** enveloped in the upper part of the bloom
Ha Peru
CU Sunny, little watering. With its white-scaled leaves it is a small pretty species.

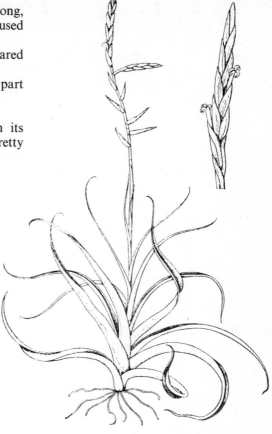

Fig. 37 *Tillandsia gayi* Baker (× $\frac{1}{3}$).

Tillandsia geminiflora Brongn., 1829 (Plate 37)

(*geminiflora* paired, double flowered; frequently there are only 2 flowers in each spike)
Syn. *Anoplophytum geminiflorum* Morren, 1880; *T. rubida* Lindl., 1842; *Anoplophytum rubidum* Beer, 1857, 1857; *T. coccinea* Platzm., 1857; *Anoplophytum paniculatum* Morren, 1889

Pl stemless, 15–20 cm high including inflorescence
Fo numerous, forming a compact rosette often turned to one side
LS indistinct, gradually merging into the blade
LB 10–15 cm long, up to 2·5 cm wide at the base, triangularly lanceolate, long-tapered, edges somewhat curled, heavily covered with short grey scales, green, pliant
Sc slender, angling upward to pen-

dulous, 6–9 cm long, 4 mm thick
SB similar to the leaves, loosely spiral, much longer than the stalk segments, short scales; sheaths usually pink or green, the thread-like blades green
I 8–10 cm long, compound
PB similar to the scape bracts, the lower ones much longer than the spikes, the upper ones shorter, short scales, reddish-brown
Sp 10 to 15 spiral, 3–4 cm long, almost upright, stemmed, seated on a wide base, open, 3 to 4 flowers
FB distichous, 1·2 cm long, 7 mm wide, oval-pointed, scaled, shorter than the sepals, brownish-red or pink

Se 1·2 cm long, lanceolate, long-tapered, stiff, the posterior ones fused 2–3 mm high
Pe forming a tube up to 1·7 cm long; the tips flared, rounded, reddish-violet
St enveloped in the flower, 1·1 cm long
Pi 1 cm long, several times as long as the ovary
Ha Brazil, Paraguay, Uruguay and Argentina
CU A very pretty, easily grown species, which shows a certain variability in regard to the formation of the inflorescence: they are sometimes drawn together into a head, sometimes elongated.

VARIETY

T. geminiflora Brongn. var. *incana* (Wawra) Mez, 1894
Syn. *T. incana* Wawra, 1880; *T. caldasiana* Baker, 1889; *Anoplophytum rollissonii* Morren, 1878
This variety differs from the type by its silver grey, thick, somewhat separate, coarse scales and somewhat thicker leaves.

Tillandsia gilliesii Baker, 1878 (Fig. 11, see page 33)
(*gilliesii* after the botanist John Gillies (1747–1836), who travelled throughout South America)
Syn. *T. compressa* Gill., 1878, non Bertero, 1830

Pl short-stemmed, simple or branched, growing in thick clumps, up to 18 cm high including inflorescence
Fo distichous
LS surrounding the stem, 1 cm long, 1·2 cm wide
LB flared to recurved, up to 8 cm long, 8–9 mm wide at the base, succulent, laterally compressed, upper side gutter-like, long-tapered, thick silver grey-scaled
Sc slender, upright, up to 7 cm long, 1 mm thick, slightly scaled
SB lacking
I simple-spiked, 1 to 13 flowers

FB distichous, upright, oval-lanceolate, pointed, 1·5–2 cm long, up to 1 cm wide, silver grey-scaled, as long as or somewhat shorter than the sepals, which are long-lanceolate, 15–16 mm long, scaled and all somewhat fused with each other
Pe somewhat surpassing the sepals, inclining into a tube, 3 mm wide, yellow
St enveloped in the bloom, 6 mm long
Pi shorter than the ovary
Ha Peru, Argentina and Bolivia; usually growing on rocks, at altitudes around 2,500 m.

CU Dry, bright, sunny. *T. gilliesii* belongs to the truly distichous miniature tillandsias, which, because of their location requirements, can easily be grown along with cacti. Closely related to *T. myosura* (see page 158).

Tillandsia grandis Schlecht., 1844 (Ill. 104)

(*grandis* large; refers to the size of the plant)
Syn. *Platystachys viridiflora* Beer, 1857; *T. macropetala* Wawra, 1887; *T. viridiflora* Baker, 1888; *T. orizabensis* Baker, 1888; *T. longiflora* Sesse et Moc, 1894; *T. virginalis* sensu Wittm., 1895, non Morren, 1880

Pl stemless, 50–200 cm including the inflorescence
Fo numerous, bluish-frosted, forming a large, almost upright or spread rosette
LS distinct, oval, brown-scaled
LB 30–100 cm long, 3–7 cm wide, ribbon-shaped, pointed, upper side naked, underside slightly scaled, green
Sc upright, strong, usually longer than the leaves
SB tightly overlapping, similar to the leaves
I simple or composed of few spikes, bluish-frosted
PB very short, not longer than the flower bracts
Sp open distichous, a stemmed, sterile base in the lower part, up to 30 cm long, 6–7 cm wide
FB 3–5 cm long, 1·5–2 cm wide, broadly oval or elliptical, somewhat shorter than the sepals
Fl very large, separate at bloom time with strong stems up to 1 cm long, green or greenish-white
Se elliptical, blunt, almost separate, 3–4·5 cm long, no keel
Pe upright at bloom time, later pendulous, lineal, blunt, 8–12 cm long
St far surpassing the flower, later pendulous
Ha Southern Mexico, Guatemala and Honduras; epiphytic, usually on rocks at altitudes between 800 and 1,200 m.
CU Like all cistern bromeliads, *T. grandis* requires rather high humidity and a shady location. It develops into its full beauty only when standing alone, and is therefore recommended only for the owners of larger greenhouses.

Tillandsia guanacastensis Standl., 1927 (Ill. 105)

Pl stemless, 15–20 cm high including the inflorescence
Fo in an upright rosette with a loose, pseudo-bulb base
LS clearly defined, almost round, spoon-shaped, 5–6 cm long, 4–5 cm wide, dark brown
LB triangularly narrow, long-tapered, 1·5 cm wide above the sheath, 10–14 cm long, edges curled and somewhat wavy, green with dark lavender spots, close grey scales
Sc upright, covered by the leaves
SB oval elliptical, pointed, approximately as long as the stalk segments, pale-scaled
I as long as the leaves or somewhat longer, loosely compound, with 2 to 9 spikes
FB similar to the scape bracts, not longer than the flowerless lower part of the upright to almost upright spikes,

which are 3–6 cm long and have 12 to 22 closely arranged flowers
FB 4 mm long, shorter than the sepals, oval, blunt, thin, no keel, pale-scaled, green
Fl tightly distichous, sessile, almost perpendicular; petals yellow about 6 mm long, with spread tips
Se almost oval, asymmetrical, 5–6 mm long, slightly scaled
St enveloped in the flower

Ha Costa Rica and Panama
CU *T. guanacastensis* is a representative of the subgenus *Pseudocatopsis*. They are all distinguished by the presence of small blooms and asymmetrical sepals. Since they come from the greater altitudes, primarily the cloud forests, they are somewhat difficult to grow. They require a cool location and high humidity. They are best grown epiphytically.

Tillandsia guatemalensis L. B. Smith, 1949 (Ill. 107)
(*guatemalensis* coming from Guatemala)
T. guatemalensis is closely related to *T. leiboldiana* (see page 147), with which it is joined by many transitory forms.

Pl up to 1 m high including the inflorescence; very variable in regard to size and form
Fo numerous, forming a large, green or red (in cultivation green) funnel rosette 50 cm in diameter
LS long-oval, up to 13 cm long, up to 7 cm wide, green, upper side partially brown-scaled
LB up to 30 cm long, up to 5 cm wide at the base, ribbon-shaped, long-tapered, almost naked
Sc upright, naked, angular, up to 1 cm thick
SB tightly overlapping, similar to the rosette leaves, with flared red or green blades
I upright, open, panicle-spiked, overall up to 80 cm long and 20 cm wide
PB those of the first order panicle branches leaf-like, the basal ones longer than the branches, lanceolate to broadly oval with long-pointed, green or red blades; those of the spikes shorter, about 2 cm long, red or green

Sp lineally lanceolate, pointed, flattened, 5 to 6 flowers, 2–4 cm long, 1 cm wide; in the upper part of the inflorescence the spikes can be up to 12 cm long and have up to 25 flowers
FB distichous, tightly overlapping, 10–14 mm long, elliptically oval, 7 mm wide, slight keel on the back, naked, ribbed, somewhat longer than the lanceolate-oval, thin-skinned sepals, of which the posterior ones are slightly fused
Fl short-stemmed, blue, about 2 cm long, the lineal, blunt petals are fused into a tube
St enveloped by the tube
Ha Widely spread in southern Mexico (Chiapas), Guatemala, Honduras, El Salvador and Costa Rica
CU Shady, cool and damp. Like the other large cistern bromeliads, suitable only for rather large greenhouses. Seems to produce few pups in cultivation; after bloom the plants usually perish.

Tillandsia gymnobotrya Baker, 1887 (Ill. 291)
(*gymnos* naked; *botrys* racemose, panicle, bunch, cluster; because of the bare, bunch like inflorescence)
Syn. *T. tridactylites* Morr., 1889; *T. liebmanniana* Morr., 1889

Pl stemless, flowering to 50 cm high, forming several offshoots before flowering
Fo numerous, forming a rosette 30–40 cm in diameter
LS longish-oval, 7–8 cm long, 4 cm wide, both sides brown scaled
LB supple, narrow-triangular, long threadlike apex, 20–25 cm long, 2·5 cm above the sheath, dull-green, both sides appressed grey scaled
Sc upright, to 25 cm long, 7 mm thick, round, bare, green
SB densely overlapping, much longer than the internodes, fully covering the scape, similar to the leaves but with longer blade, appressed grey scaled
I loose, double-compound, upright, 30 cm long, to 15 cm wide with red, angular naked axis
PB the upper ones similar to the inflorescence bracts, green, appressed

grey scaled, the upper ones reddish and shorter; the lower ones with the blades however longer than the horizontal 2 to 3 compound spiked 8–9 cm long branches
Sp 5–6 cm long, 8 mm wide, sword shaped
FB densely overlapping, branched, to 12 mm long, outspread, 9 mm wide, keeled only near the apex, naked, smooth, red, thin
Fl upright, to 2·2 cm long
Se 1·4 cm long, outspread 6 mm, green, with lilac-coloured tip, cutaneous, bare, free
Pe 2·1 cm long, 4 mm wide, rounded but short pointed tip
St and **Pi** enclosed in the flower
Ha Mexico (near Orizaba)
CU Very fine when in flower, decorative and not becoming too large.

Tillandsia hamaleana Morren, 1869 (Plate 31)
Syn. *Wallisia hamaleana* Morren, 1870; *Phytarrhiza hamaleana* Morren, 1870; *T. commelyna* Morren, 1870; *T. platypetala* Baker, 1888.

Pl stemless, 30–40 cm high including the inflorescence
Fo soft, forming a funnel rosette 30–35 cm in diameter
LS long-oval, 12 cm long, 8–9 cm wide, pale green, short scales, gradually merging with the blade
LB ribbon-shaped, 20–30 cm long, 5·5 cm wide above the sheath, average width 4–5 cm, lanceolate-pointed with recurved tip, green, both sides with short grey scales
Sc slender, upright, 10–15 cm long,

round, 3 mm thick, green to slightly reddish, slightly short-scaled
SB surrounding the scape with the sheaths and with short blade, upright, the lower ones shorter than the stem segments, the upper ones somewhat longer, reddish-brown, slightly grey-scaled, distinctly ribbed lengthwise
I compound, usually with 3 flared spikes, which are 5–6 cm long, short-stemmed and have 6 to 8 flowers
PB shorter than the spikes, 2 cm long, reddish-brown, grey-scaled, soon

withering

Fl up to 3 cm in diameter, very aromatic, dark blue with white throat, fading to light blue

FB distichous, distinctly keeled, reddish-brown at the tip, grey-scaled, 1·8–2 cm long, 1·5 cm wide, thin

Sc 2 cm long, 8 mm wide, slightly asymmetrical, thin-skinned at the edge, wider at one edge than the other, green in the lower part, reddish-brown toward the tip, naked

Pe claw 1·8 cm long, 3 mm wide, narrowly lineal, white; plate spread, 1·6 cm wide, 1·7 cm long, rhombic, dark blue

St 1 cm long, longer than the pistil, which is 2·5 mm long, deeply enveloped in the tube

Ha Ecuador and northern Peru (Ayabaca); epiphytic at 2,600 m.

CU Moderately damp, semi-shady (like *T. lindeni*, see page 148). At bloom time a very pretty species, which is very closely related to the Ecuadorian *T. dyeriana* (see page 121)

Tillandsia heterophylla E. Morren, 1873 (Ill. 295 and 296)
(*heterophylla*(*us*) different leaves)
Syn. *T. virginalis* Morren (non Wittm.), 1880

Pl stemless, flowering to 1 m high

Fo forming a wide outspread funnel shaped rosette from 50–60 cm in diameter

LS broad-oval, 10–14 cm long, 7–8 cm wide, naked, pale green

LB tongue-shaped, long pointed, 30–35 cm long, 4–5 cm wide above the sheath, bare, green, partly becoming grey frosted

Sc to 60 cm long, 1 cm thick, upright, round, bare, green

SB upright, dense, longer than the internodes, enclosing the scape with the sheaths, green, becoming grey frosted, with short, spreading, green blades, occasionally with lilac coloured margins

I simple-swordshaped, 20–23 cm long, 4 cm wide

FB in two rows (bipartite) densely overlapping, 5–6 cm long, pointed, keeled, green, grey frosted, firm

Se free, 5 cm long, flared for 1·5 cm, pale green, bare, the back ones keeled, shorter than the flower bracts

Pe to 9 cm long, strap shaped, white, rounded at the tip

St and **Pi** enclosed within the flower; filament 6·7 cm long; white, spirally twisted; anthers 1·2 cm long; pistil 6·5 cm long

Ha Mexico (near Orizata). Frequently found growing on *Taxodium mucronatum* trees.

Tillandsia hildae Rauh, 1972 (Ill. 293 and 294)
(*hildae* after Hilda Rauh who found the first flowering plant)
T. hildae is one of the best Peruvian epiphytic Tillandsias, resembling an Aechmea more than a Tillandsia. It is similar to the Mexican *T. leucolepis* MacDougall from which it differs by the form of its inflorescence.

Pl flowering to 2 m high, stemless or with short thick stem, with many

offshoots forming at the base

Fo numerous, rigidly upright, very

firm and compact, forming a rosette to 90 cm high and 90 cm in diameter

LS not distinct, broad-oval, to 12 cm wide and 20 cm long, dark-chestnut brown

LB to 70 cm long, 8–10 cm wide gradually narrowing to a fine, spined tip, with upward curved margins above the sheath, underside fine and narrow grooved, light to dark brown or dark brown with strong white scaled cross-banding, upper side grey green and less clearly banded

Sc to 1 m long, 3 cm thick at the base, with dark-black violet axis

SB densely overlapping, longer than the 5 cm long internodes, the basal ones almost leaf like with upright densely scaled blades lying against the scape

I very erect, twice to three times forked pyramid, to 80 cm long, to 90 cm wide at the base, with thin, bare, weakly knee-jointed (genuflect) axis; branches very thin; the basal ones divided to 45 cm long, horizontally positioned on the axis, the upper ones spreading and mostly single

PB dark black-violet, longer than the sterile, leaf covered section of the branch arrangement of the first order (see page 43) broad-triangular, pointed, to 5 cm long, with cutaneous margins, with scattered scales

FB loosely bipartite, longer than the internodes and the sepals, lying upright against the axis, thin margins covering the base, firm, almost bare, dark violet, the border green edged, when young indistinct when dry strongly nerved, not keeled, lanceolate-oval, pointed, to 2·5 cm long and 1 cm wide

Fl upright-angled, much exceeding the flower bracts, more or less 4·5 cm long and frequently with weak crooked tube, sometimes one sided after flowering

Se shorter than the flower bracts, 2 cm long, 7 mm wide, longish-oval short pointed tip, green and violet tipped, not keeled, only the back ones weakly keeled at the base; all sepals fused for about 2 mm of their height

Pe violet, tongue shaped with reflexed bent tip, to 4·5 cm long and 4 mm wide

St and **Pi** much exerted from the flower

Ha frequently on steep sides of rock walls in the dry valleys of the Rio Chamaya in North Peru, between 800 to 1,100 m.

CU *T. hildae* is dry growing and protected from dessication in nature by its stout thick leaves, its moisture absorbing bands of scales and its water holding capacity between the leaves. Cactus cultivation suits it well.

Tillandsia ignesiae Mez, 1903 (Ill. 297 and 298)
(*ignesiae* after the Mexican mountain Monte de Santa Ignes)
Syn. *T. lecomtei* Poiss et Menet.
Unless in flower, this plant cannot be distinguished from *T. plumosa* (page 163). The difference is that in *T. gnesiae* the inflorescence is simple and not compound as in *T. plumosa*.

Pl stemless, flowering to 10–17 cm high

Fo numerous, recurved, outspread or upright, not longer than the influorescence scape

LS roundish, 8–10 mm wide, thickened forming as a whole, a ball-like bulb, about 3 cm in diameter

LB lineal-awl shaped, long pointed, 8–9 cm long, 4 mm wide above the sheath, dense white, almost feathery scaled

Sc slender, upright to pendulous, about 10 cm long, 2 mm thick, round, green, naked

SB similar to the leaves, much longer than the internodes, enclosing the scape with their rose-coloured sheaths; narrow blades, long pointed, densely set with scales, frequently on one side

I forming a simple, sword shaped, 5–6 cm long, 1 cm wide, pendulous axis

FB densely overlapping, in two rows, longer than the sepals, sharply keeled, 1·9 cm long, flaired 1 cm, triangular-oval, long pointed, rose coloured, grey scaled, nerved towards the tip.

FB sessile, 2·5 cm long

Se lanceolate, pointed, 1·2 cm long, keeled, thin, whitish-rose, white wooly scaled especially along the keel; the back ones fused for 1 mm of their height

Pe 2·3 cm long, 3 mm wide with rounded slightly flared tip, pale green

St and **Pi** enclosed in the flower

Ha Southern Mexico, epiphytic on oak trees, at 1,500 m high

CU light and sunny, little water.

Tillandsia imperialis Morren, 1889 (Plate 14)

(*imperialis* imperial; with its brilliant red, fat spike, this tillandsia is a truly imperial among the bromeliads)

Syn. *Guzmania imperialis* Rowzl, 1889; *T. strobilantha* Baker, 1888, non Poir.

Pl stemless, up to 50 cm with inflorescence

Fo numerous, forming a large cistern rosette up to 50 cm in diameter

LS broadly oval

LB ribbon-like, pointed, the tips often recurved, up to 40 cm long, up to 6 cm wide above the sheath, crisp green, shiny, naked or sparsely short-scaled

Sc in the native habitat sometimes reddish, short, strong, hidden by leaves and scape bracts

SB similar to the leaves, tightly overlapping, covering the scape with their wide sheaths; the upper sheaths shiny red, the spread, leaf-like blades green

I compound, forming a thick cone up to 18 cm long and up to 10 cm thick

PB upright, thick, overlapping, broadly oval, long-tapered, covering the spikes, brilliant red, with green, flared,

recurved tips

Sp up to 4 flowers, laterally flattened, almost sessile, up to 6 cm long, up to 2·5 cm wide, elliptical, almost completely hidden by the primary bracts

FB up to 3·5 cm long, tightly overlapping

Se separate, up to 2·2 cm long

Pe forming an upright, blue tube 5·3 cm long

St protruding from the flower

Ha Central and southern Mexico, in cloud forests between 2,400 and 3,000 m; epiphytic on oaks and conifers

CU Damp, cool, half shady. In spite of the formation of numerous pups in the natural habitat, *T. imperialis* is difficult to cultivate. With its cone-shaped, brilliant red inflorescence it makes a splendid sight at bloom time. In

Mexico it is sold in the market places at Christmas time and serves as a Christmas decoration.

Tillandsia insularis Mez, 1869 (Ill. 106)
(*insularis* living on an island; refers to the habitat)

Pl stemless, 40–50 cm high including the inflorescence
Fo few, forming a funnel rosette
LB ribbon-shaped, with broadly blunt tips, 3·5–4·5 cm wide, 20–25 cm long, green
Sc upright, about 30–35 cm long, round, 4–5 mm thick, green
SB upright, longer than the stalk segments, green, 2 cm wide, 4–5 cm long, grey-scaled, triangularly lanceolate
I 20 cm long, 4–5 cm wide, composed of 8 to 12 spikes arranged loosely distichous
PB 1·5–2·5 cm long, green, grey-scaled, much shorter than the short-stemmed spikes, which are 4–6 cm long, 1·5 cm wide
FB loosely overlapping at bloom time, 7–8 mm long, 7 mm wide at the base, broadly triangular, green, as long as or slightly longer than the sepals, which are 6 mm long, 3 mm wide, thin-skinned
Pe white, insignificant, 8 mm long, only slightly surpassing the sepals
St 3 mm long, enveloped in the flower
Ha Galapagos Islands
CU Moderately damp, semi-shady. Rarely seen in collections.

Tillandsia intumescens L. B. Smith, 1955 (Plate 5)
(*intumescens* puffed; refers to the puffed, spoon-shaped primary bracts and flower bracts)

Pl stemless, 50–60 cm high with inflorescence
Fo numerous, soft, recurved, forming a rosette 30–35 cm in diameter
LS large, 14–18 cm long, 8–9 cm wide, brownish-violet on the inside, dull grey green on the outside, heavily scaled
LB 35–38 cm long, 5–6 cm wide at the base (average width 3·5 cm), lanceolate, long-tapered, upper and underside dark green to reddish, thick, short, grey scales
Sc strong, upright, about 30 cm long, 1 cm thick, dull green, reddish in the upper part, round, finely furrowed
SB similar to the leaves, soft, recurved with a spreading, dish-like base; the lower ones dull greenish-red, the upper ones bright rose red or pale green, heavily grey-scaled, much longer than the stalk segments; the basal ones 28–32 cm long, the upper ones 15–18 cm long, surrounding the scape with the sheath-like base
I composed of 3 to 8 loose, upright spikes arranged in a whorl, in all about 18–22 cm long, 12–15 cm wide
PB similar to the scape bracts, bright rose red or pale green, longer than the spikes, with a wide, spoon-shaped base and long, recurved, thread-like tips, heavily grey-scaled
Sp 9–11 cm long, short-stemmed, 3 cm wide, almost round-oval, somewhat flattened, 5 to 10 flowers
FB distichous, loose, overlapping, dull

green with grey wax covering, spotted with glands, broadly oval, puffed, 3–4 cm wide, 3 cm long, keeled, short-tipped
Se not surpassing the flower bracts, 2·5 cm long, 1·5 cm wide, broadly oval, lanceolate, pointed, thin-skinned, greenish colour showing through
Pe yellow, 3·5 cm long, 9 mm wide, fused into a tube, tips slightly flared
St surpassing the petals, filaments

4·3 cm long, straight
Ha Mexico. This species was discovered in 1952 by Matuda in conifers and oaks (1,800 m) near Ixtapan, later also found by Rauh on the Pacific side between Cilpancingo and Acapulco (800 m)
CU Moderately damp, semi-shady. Easily grown species with a splendid bloom, especially suitable for epiphyte trees.

Tillandsia ionantha Planch., 1855 (Plate 1, Ill. 31)
(*ionantha* violet-flowered; refers to the intensive violet flower colour)
Syn. *T. scapus* Hook., 1871; *T. rubentifolia* Poisson et Menet, 1908; *T. erubescens* sensu Mez, 1935, non Schlecht., 1844.

Pl stemless, usually growing in thick clumps or mounds (Ill. 31)
Fo numerous, in a tight bulb-like rosette
LS long-oval, 2–2·5 cm long, 1–1·5 cm wide, light brown-scaled
LB awl-shaped, long-tapered, 3–4 cm long, 5 mm wide above the sheath, upright, slightly spread in the upper third, heavily covered with grey scales, at bloom time the inner rosette leaves are brilliant red
Sc lacking or very short
I sessile, simple-spiked
FB long-oval, pointed, 2 cm long, up to 6 mm wide, thin-skinned, white, slightly scaled, longer than the sepals

Se 1·5 cm long, 3 mm wide, thin-skinned, naked, lanceolate-pointed, the posterior ones fused 2 mm
Pe united into a tube, 4·5 cm long, white in the lower part, violet blue at the tips
St surpassing the petals
Ha Mexico (from the Atlantic and the Pacific coasts to about 1,300 m), Guatemala, Nicaragua
CU Easy, moderately damp, bright and sunny, can be grown outside in summer. The leaf tips become bright red at bloom time. A favourite species sought by tillandsia fanciers. Because of the small size, suitable for terrariums.

VARIETIES
T. ionantha exists in many forms in regard to size, shape and the formation of the inflorescence. There are forms with almost spherical rosettes the size of a hazel nut and sold under the name 'Hazel Nut', some with large rosettes and upright leaves, and forms with spread, elongated leaves. Of all these we list only the two following forms, since they are considered actual varieties.
T. ionantha var. *scaposa* L. B. Smith, 1941, differs from the type by a very loose foliage as well as a short scape and a branched inflorescence.

T. ionantha var. *van hyningii* M. B. Foster, differs from the type by the formation of elongated axes; therefore it forms a loose turf and open, silver grey rosettes. This variety grows only on steep walls of the Rio Gravilja canyon (Chiapas, Mexico).

Tillandsia ionochroma André et Mez, 1896 (Plate 38)

(*ionochroma*(*us*) violet-coloured; because of the bright, violet blue flowers)
Syn. *Caraguata violacea* André, 1888

Pl stemless, up to 40 cm high
Fo numerous, in a tight rosette, naked, hardly scaled, green to grey green or made wine red by the sun
LS narrowly oval
LB tongue-shaped, up to 5 cm wide, up to 40 cm long, pointed
Sc slender, naked, approximately as long as the leaves, pale reddish
SB tightly overlapping, the basal ones similar to the leaves, with recurved blade, either reddish at the base and green at the tip or all bright red
I compound, cylindrical, pendulous, about 25 cm long
PB spiral, as long as the spikes or surpassing them, bright cinnabar red when young, the basal ones somewhat separate, broadly oval, with a long, recurved, often greenish tip; the upper ones arranged tightly, broadly oval, dish-shaped, pointed
Sp covered by the primary bracts, distichous, broadly oval, about 3 cm long, flattened, about 4 flowered (the upper spikes may have as few as one flower)
FB elliptical, blunt, scarcely or not keeled on the back, the basal ones distinctly shorter than the sepals
Fl short-stemmed, bright blue violet, the petals about 2 cm long, with recurved tips
Se long, blunt, keeled, clearly nerved, about 13 mm long
St and **Pi** enveloped in the flower
Ha Peru (Valley of Canchaque; Collection number Rauh 20092, Huanuco, Cuzco), Ecuador; epiphytic in mountain forests from 2,000–2,800 m.
CU Well worth cultivating and at bloom time is a splendid plant with its bright cinnabar red, cylindrical, pendulous inflorescences. In intensive sunlight foliage turns burgundy red.

Tillandsia ixioides Griseb., 1879 (Plate 33)

(*ixioides* (Greek) refers to the clump-shaped growth of the plant)
Syn. *Anoplophytum luteum* Morren, 1889; *T. lutea* Baker, 1889; *T. canescens* Hort., 1889, non SW., 1788

Pl short-stemmed, 15–20 cm high
Fo spiral, rosette type, upright to spread, rigid
LS 1·5 cm long, up to 2 cm wide, naked, surrounding the stalk
LB 10–13 cm long, 1·5 cm wide above the sheath, lanceolate, awl-shaped, pointed, sharp, the edges curled into a gutter, thick grey scales

Sc upright to pendulous, up to 10 cm long, 2 mm thick, naked

SB dense, overlapping, longer than the stalk segments, upright, surrounding the scape with the sheath, thick silver grey scales

I simple-spiked, 5–6 cm long, 2·5 cm wide

FB oval-lanceolate, pointed, as long as the sepals or a little shorter, silver grey scales, ribbed lengthwise

Se separate, up to 2 cm long, naked, thin-skinned, the side ones somewhat winged

Fl spiral, about 1 cm long, short-stemmed, golden yellow

St enveloped in the upper part of the tube which is about 3·5 cm long, up to 2·5 cm long

Ha Uruguay, Argentina, Paraguay, Bolivia; epiphytic on trees, usually growing in clumps

CU Bright, sunny. A recommended, small, easily grown species, which in its non-blooming stage can be confused with *T. didisticha* (see page 118).

VARIETY

T. ixioides var. *occidentalis* Castell., 1933 is larger than the type; it is up to 30 cm high, and the leaves reach up to 18 cm.

Tillandsia juncea (Ruiz et Pav.) Poir., 1817 (Ill. 109)
(*juncea(us)* rush-like; refers to the rush-like appearance)
Syn. *Bonapartea juncea* Ruiz et Pav., 1802; *T. quadrangularis* Mart., 1843; *T. juncifolia* Regel, 1874; *T. setacea* sensu Baker, 1887, non SW., 1797

Pl stemless, growing in tufts, up to 50 cm high; the pups beginning with a short stolon segment

Fo numerous, upright, forming a thick, busy rosette

LS clearly delineated, broadly triangular, dark brown-scaled

LB narrow, grass-like or rush-like, hard, long-tapered, dark brown at the base, otherwise silver grey-scaled, the edges curled into a gutter

Sc upright or arched, strong, approximately as long as the leaves

SB similar to the leaves, upright, surrounding the scape with the sheath, overlapping, heavily scaled, green or pink, the blades narrow, thread-like long-tapered, scaled, green

I composed of a few spikes, overall egg-shaped, up to 7 cm long, up to 6 cm wide

PB surpassing the lower spikes with long, thread-like blade, the upper ones shorter than the spikes, pink, heavily grey-scaled

Sp almost upright to flared, distichous, few flowers, elliptical to lanceolate, up to 4 cm long

FB tightly overlapping, wide oval, longer than the sepals, keeled heavily scaled, usually red

Se lanceolate, pointed, 1·5–2 cm long, naked or slightly scaled, the posterior ones fused 7–8 mm high

Pe tubular, up to 4 cm long, blue

St protruding from the flower

Ha *T. juncea* inhabits an enormous area reaching from Florida, through Central America to Bolivia and Peru. Consequently, *T. juncea* has many

forms and is quite variable. In southern Ecuador and northern Peru it is among the characteristic plants of the deciduous dry forests at altitudes between 700 and 1,200 m.
CU Easy, bright and sunny, moderately damp. Robust, but at the same time very decorative.

Tillandsia karwinskyana Schult. fil., 1830 (Ill. 299)
(*karwinskyana* after W. F. von Karwinsky, a collector)

Pl stemless, 30–50 cm high with inflorescence, and before flowering forming numerous offsets at the base
Fo narrow, forming an almost upright rosette from 20–25 cm in diameter
LS triangular-oval, 2–3 cm wide at the base, 4–6 cm long, both sides grey scaled
LB narrow-triangular, long pointed, narrowing to an awl-shaped apex, 10–25 cm long, about 1·5 cm wide, densely silver grey scaled
Sc upright, 15–20 cm long, 4 mm thick, naked, red
SB upright, lying against the scape, longer than the internodes, but not completely covering the scape, about 3 cm long, green or red, light grey scaled, and with short upright blade
I upright, loose-compound with 5–8 spreading spikes, to 25 cm long, to 18 cm wide with naked, red axis
PB similar to the inflorescence bracts, shorter than the sterile basal portion of the spikes, the lower ones about 2·5 cm long, green or red, grey scaled
Sp about 9 cm long, with 3 cm longer sterile base, loosely 3–5 flowered, with knee-jointed red axis
FB loose, in two rows, shorter than the sepals, to 1·6 cm long, green or red, naked, not keeled
Fl lying tightly against the rachis, upright, to 3 cm long.
Se 1·8 cm long, the back ones about 5 mm high fused, naked, green, oval to elliptical, rounded tip
Pe 3 cm long, fused with a tube, pale green to yellowish green
Ha Mexico
CU Very fine small species; the red coloured inflorescence contrasting strongly with the silver-white rosette leaves. A light sunny position and little water are its requirements. Cactus cultivation suits it.

Tillandsia kirchhoffiana Wittm., 1889, non Mez, 1896 (Ill. 110)
(*kirchhoffiana(us)* after the botanist Kirchhoff)
Syn. *T. fournierii* Morren, 1889

Pl stemless, up to 1 m high, including inflorescence
Fo clustered into an almost upright rosette
LS long-oval, 10–12 cm long, 5·5–6·5 cm wide, dark lavender violet
LB narrowly lineal with curled edges, 1·5 cm wide above the sheath, average width 5–6 mm, 50–60 cm long, green
Sc slender, upright, 50–60 cm long, 6–7 mm thick
SB like the leaves, covering the stalk segments, sheaths long, green, surrounding the stalk, the blades long, narrow, recurved, green
I up to 40 cm long, loosely compound,

with 15 to 20 spread, long-stemmed spikes 10–20 cm long, about 1 cm wide, with 6 to 8 flowers

PB surrounding the lower part of the spikes with their red sheaths, the blades green, narrow, pendulous, in the lower part of the inflorescence they are longer than the spikes

FB loosely distichous, 2·5–3 cm long, 8–10 mm wide, lanceolate-pointed, red, as long as the sepals, sometimes a little longer

Se some 2 cm long, 4 mm wide, green with red tips, separate

Pe about 4 cm long, white at the base, dark blue in the upper part

St protruding from the flower

Ha Central America; epiphytic on trees, at heights of 1,500–2,000 m.

CU Moderately damp, semi-shady. Because of their almost black leaf sheaths it is a very decorative species, which, in its non-blooming stage can easily be confused with *T. punctulata* (see page 168).

Tillandsia lampropoda L. B. Smith, 1938 (Plate 39)
(*lampros* shining; *poda*(*us*) footed)

Pl stemless, 50–60 cm high with inflorescence

Fo forming an almost upright rosette

LS long-oval, 10–12 cm long, 6–7 cm wide, both sides dark lavender black

LB narrowly triangular, extending into a long point, 25–30 cm long, 3·5 cm wide above the sheath, average width 2 cm, close, short scales

Sc strong, upright, 20–25 cm long, 7 mm thick

SB closely surrounding the scape, longer than the stalk segments, similar to the leaves in the lower part, the upper ones without a sheath and with only a thread-like, brownish-red blade

I simple, sword-shaped, up to 20 cm long, 5–7 cm wide

FB distichous, tightly overlapping, broad-oval pointed, up to 6·5 cm long, distinct keel, stiff, yellow or greenish-red, scaled at the tip

Fl 5–6 cm long; their petals yellow, inclined into a tube, with slightly flared tips

Se separate, narrowly lanceolate, pointed, 3 cm long, naked

St slightly longer than the petals

Ha Guatemala, Honduras and in the southernmost part of Mexico (Lago de Monte Bello), on conifers and oaks

CU Moderately damp, light shade. Suitable for epiphyte trees. A very decorative plant in bloom, resembling a *Vriesea* with its sword-shaped inflorescence.

Tillandsia latifolia Meyen, 1843 (Ills. 17, 56, 57, 108)
(*latifolia*(*us*) wide-leafed)

Syn. *T. divaricata* Benth., 1845; *T. kunthiana* Gaud., 1846; *Platystachis kunthiana* Beer, 1857; *Platystachys latifolia* C. Koch, 1873; *T. grisea* Baker, 1887; *T. oxysepala* Baker, 1888; *T. murorum* Mez, 1913

T. latifolia is a very manifold and variable species, which propagates in the Peruvian coastal area primarily vegetatively by the formation of new shoots (Ill. 57).

Pl forming a stem or rosette, up to 60 cm high, very variable

Fo numerous, spiral, forming a fairly large, spreading rosette

LS indistinct, hardly delineated, unnoticeably merging into the blade, thickly grey-scaled

LB triangular, long-tapered, both sides heavily grey-scaled, very variable in length and width

Sc short, not longer than the leaves or up to 50 cm long (var. *divaricata*), upright or angling upward, round, slightly scaled to naked

SB in the lower part of the scape similar to the leaves with flared blade, in the upper part upright, surrounding the scape, longer than the stalk segments, green, heavily grey-scaled

I compound, either with close, almost upright spikes or open with spread spikes

PB always shorter than the spikes, the upper ones similar to scape bracts (var. *divaricata*), heavily grey-scaled

Sp various-shaped; 5–10 cm long, 0·5–1 cm wide, few flowers to many flowers, short-stemmed

FB either heavily grey-scaled or almost naked (var. *divaricata*), usually distichous, tightly overlapping

Se usually as long as the floral bracts, separate or the posterior ones fused up to 11 mm high, scaled or naked

Fl pink to lavender, tubular, 1–2 cm long

St enveloped in the flower

Ha Colombia, Ecuador, Peru. In Peru forming extensive stands in the coastal desert (Ill. 17), but also ranging up to 3,000 m, especially var. *divaricata*

CU Bright, sunny, but high humidity. A decorative species, that can be grown outside in the summer.

VARIETIES

T. latifolia var. *latifolia* grows frequently in thick mounds, turf or strands (Ill. 17).

T. latifolia var. *major* Mez, 1896 is much larger in all parts (Ill. 57). It too is among the plants characteristic of the sandy desert.

T. latifolia var. *divaricata* (Benth.) Mez, 1896 differs from the type by the flared, naked spikes (Ill. 52). It is limited to the higher locations in Peru.

Tillandsia leiboldiana Schlecht., 1844 (Plate 12)

Syn. *T. xiphophylla* Baker, 1888; *T. phyllostachya* Baker, 1888; *T. aschersoniana* Wittm., 1889; *T. rhodochlamys* Baker, 1889; *T. sparsiflora* Baker, 1890; *T. coccinea* Sesse et Moc., 1894; *T. lilacina* Mez, 1896. Closely related to *T. guatemalensis* (page 136) with which it is connected by transitional forms.

Pl stemless, 30–60 cm high including the inflorescence

Fo numerous, an upright funnel rosette

LS long-oval, 5·5–6·5 cm wide,

8–10 cm long

LB ribbon-like lanceolate, pointed, 3–4 cm wide above the sheath, 2·5 cm wide in the central part of the blade, soft, green

Sp arranged spirally on the axis, upright or flared, narrow, pointed, up to 8 flowers, about 6 cm long and 1 cm wide

FB distichous, tightly overlapping, oval-lanceolate, pointed, 2–2·5 cm long, 8 mm wide, naked, keeled, red or violet

Sc usually shorter than the rosette

SB similar to the leaves, tightly overlapping, surrounding the scape tightly

I loosely compound, 10–30 cm long

PB similar to the scape bracts, longer than the spikes, rolled into a dish shape at the base and surrounding the spikes, red or violet, with a recurved, green blade, the upper ones usually totally red or violet

Se lanceolate, 1·5–2 cm long, thin-skinned, separate

Pe blue, forming a tube about 3 cm long

St enveloped by the petals

Ha Southern Mexico to Costa Rica; epiphytic, from sea level to about 1,300 m. The plants from Costa Rica differ from the Mexican ones usually by their violet primary bracts and floral bracts

CU Damp, semi-shady; in a pot or epiphytic. A very attractive species at bloom time.

Tillandsia lepidosepala L. B. Smith, 1935 (Ill. 111)
(*lepidosepala*(*us*) scaled sepals; refers to the heavy scaling of the sepals)

Pl stemless or short-stemmed, usually growing in thick clumps, up to 15 cm high including inflorescence

Fo forming an upright to spread rosette

LS wide-oval, 1–1·5 cm long

LB lineally triangular, long-tapered, upright to flared, 7 mm wide above the sheath, 10–12 cm long with curled edges, flared, heavily silver grey-scaled

Sc very short, hidden in the rosette

SB similar to the leaves, as long as the inflorescence

I simple, laterally flattened, 3–5 cm long, tightly distichous, 2 to 5 flowers, hardly surpassing the rosette

FB lanceolate-pointed, 2–3·5 cm long, as long as the sepals or longer, thick, coarse scales

Fl green, 2 cm long; the petals 3 mm wide, upright, with slightly flared tips

Se lanceolate, long-tapered, up to 2 cm long, longitudinally ribbed, separate, green, heavily grey-scaled

St somewhat shorter than the flowers, about 12 mm long

Ha Central Mexico; epiphytic on oaks at 2,000 m altitude

CU Bright and sunny, moderate watering. A pretty, small species.

Tillandsia lindeni Regel, 1868 (Plate 30)
(*lindeni* after the Belgian nurseryman and writer J. Linden (1817–1898), also frequently written as *T. lindenii*)
Syn. *T. lindeniana* Regel, 1869; *T. lindeni* Morren var. *regeliana* Morren, 1870; *T. lindeni* var. *major* Dombrain, 1871; *Vriesea violacea* Hort., 1872; *T. lindeni* var. *rutilans* Linden, 1872; *T. lindeni* var. *intermedia* Morren, 1878; *Phytarrhiza lindeni* Morren var. *regeliana* Morren, 1879; *T. lindeni*

var. *violacea* Hort., 1896; *T. lindeni vera major* Duval, 1901.
We have already referred to the confusion among *T. lindeni, T. cyanea* and *T. anceps* on pages 89 and 114.
As L. B. Smith has already indicated, there is a relentless confusion concerning the name *T. lindeni*. In the sense used here and by Smith *T. lindeni* Regel is the species with the long scape, while the short scape *T. lindeni* Morren is identical with *T. cyanea* Linden ex Koch. There is also no uniformity in the spelling of the name; Regel published the long-scaped species twice (1868, 1869) under the name *T. lindeni*, which for unknown reasons he changed to *T. lindeniana* in 1869. But in 1869 E. Morren published the short-scaped species (known today as *T. cyanea*) also under the name *T. lindeni*. Then in order to avoid a confusion, Regel named Morren's *T. lindeni T. morreniana* Regel, which today stands as a synonym for *T. cyanea* (see page 113).

Pl stemless, up to 40 cm high including inflorescence
Fo numerous, forming a funnel rosette of 40–60 cm diameter
LS narrowly elliptical, 4–5 cm long, 2·5 cm wide, both sides brown with reddish longitudinal lines
LB 25–35 cm long, 1·5 cm wide at the base, lineal, long-tapered, green with red longitudinal lines, both sides slightly short-scaled
Sc upright, slender, partially covered by the leaves, much longer than in *T. cyanea* (see page 113)
SB tightly overlapping, upright, longer than the stalk segments, the lower ones similar to the leaves with a long blade, the upper ones with a short tip, green, red-striped
I simple, sword-shaped, up to 15 cm long, 5 cm wide
FB tightly overlapping, about 5 cm long, about 3 cm wide, elliptical, pointed, keel, somewhat longer than the sepals, pink or green, naked
Fl deep blue with a white throat; the petals about 7 cm long with the claw about 4 cm long, narrow-lineal, white, and the plate flared, about 3 cm long and 3·5 cm wide
Se 4 cm long, 8 mm wide, thin-skinned, lineal-lanceolate, pointed, separate, green with pink overtones, naked
St enveloped by the perianth tube
Ha Northern Peru and southern Ecuador; epiphytic
CU Moderately damp, semi-shady, usually a pot plant. Because of the large blue flowers it is a favourite, wide-spread and widely cultivated species, of which one seldom sees the wild species. The plants offered by nurserymen are usually select forms or hybrids.

VARIETIES
The following horticulture varieties of *T. lindeni* are recognized:
T. lindeni var. *luxurians* (Morren) L. B. Smith, 1951; is said to have axil inflorescences in addition to the terminal inflorescence; flowers simple.
T. lindeni var. *kautsinskyana* (Morren) L. B. Smith, 1951; Flowers more

or less double, up to 8 cm diameter; several inflorescences.

T. lindeni var. *duvali* (Duval ex André) L. B. Smith, 1951: is a hybrid between *T. cyanea* and *T. lindeni*, which possesses the wide sword from *T. cyanea* and the long scape from *T. lindeni*.

Tillandsia linearis Vell., 1825 (Fig. 38)

(*linearis* lineally thread-shaped; because of the thread-shaped leaves)
Syn. *Anoplophytum lineare* Beer, 1857; *T. selloa* K. Koch, 1873; *Phytarrhiza linearis* Morren, 1879; *T. setacea* sensu Baker, 1889, non SW., 1797.

Pl grass-like, stemless, 13–25 cm high including inflorescence

Fo upright to almost upright, up to 38 cm long, longer than the inflorescence

LS long-triangular, 2 cm long, 5 mm wide, light brown, heavily grey-scaled

LB lineal, very narrow, long-tapered, 1–2 mm wide above the sheath, thick, short, grey scales

Sc slender, upright, green, slightly scaled

SB upright, overlapping, longer than the stalk segments, with long, upright blade, grey-scaled

I upright, simple-spiked, narrowly lanceolate, 3–5 cm long, 6–10 mm wide, few flowers

FB upright, distichous, elliptical, pointed, 2cm long, longer than the sepals, longitudinally ribbed, reddish-brown, grey-scaled

Se separate, lanceolate, long-tapered, up to 1·9 cm long, keel, naked

Pe blue or violet, the claw narrowly lineal, the plate 1–1·5 cm wide and flared

St enveloped deeply in the flower tube, longer than the pistil

Ha Brazil; in thick turf, frequently growing on araucarias

CU No problems are likely.

Fig. 38 *Tillandsia linearis* Vell. (× $\frac{1}{3}$).

Tillandsia loliacea Mart., 1830 (Ill. 112)

(*loliacea*(*us*) Because of the similarity to darnel, *Lolium*)
Syn. *T. undulata* Baker, 1878; *T. quadriflora* Baker, 1889; *T. atrichoides* S. Moore, 1895

Pl small, up to 10 cm high including inflorescence, with a simple or branched short stem
Fo numerous, upright, spiral, sometimes turned to one side
LS indistinct, thin-skinned, surrounding the stem, merging into the blade, up to 6 mm long and 5 mm wide
LB narrow-triangular, awl-like, long-tapered, 2–5 cm long, up to 6 mm wide above the sheath, covered heavily with silver grey, separated scales
Sc slender, usually perpendicular, 1 mm thick, up to 7 cm long, slightly scaled
SB elliptical, pointed, longer than the stalk segments, upright, surrounding the scape, thick silver grey scales
I simple-spiked, up to 5 cm long, distichous
FB oval, pointed, 5–10 mm long, as long as or slightly shorter than the sepals, loosely overlapping, green, thick silver grey scales
Fl bright yellow, narrowly tubular, up to 6 mm long, the brownish-yellow tips of the petals somewhat recurved
Se lanceolate, pointed, naked, the posterior ones fused about 2 mm high
St enveloped by the perianth tube
Ha Dry areas of Brazil, Bolivia, Argentina, Paraguay and Peru; partially epiphytic, partially terrestrial
CU Easy; bright and sunny. A recommended, small, very variable species.

Tillandsia lorentziana Griseb., 1874 (Ill. 113)

(*lorentziana* after the German botanist P. Lorenz (1835–1881)
Syn. *T. lorentzii* André, 1891; *T. rusbyi* Baker, 1889

Pl up to 50 cm high including the inflorescence
Fo numerous, spiral, upright or recurved
LS not sharply delineated, merging into the blade, up to 3 cm wide and 2 cm long
LB triangular, very long-tapered, up to 2·5 cm wide above the sheath, up to 25 cm long, frequently shorter and narrower, both sides heavily grey-scaled
Sc slender, upright to angling upward, up to 20 cm long
SB upright, tightly surrounding the scape, as long as or longer than the stalk segments, green or reddish, heavily grey-scaled
I simple or loosely compound with 2 to 5 flared spikes, overall up to 25 cm long and 7 cm wide
Sp up to 6 cm long, up to 1·5 cm wide, 5 to 12 flowers
FB overlapping, distichous, up to 2·5 cm long, somewhat longer than the sepals, naked to slightly scaled, red or green
Fl white or blue, about 3·5 cm long, tubular, tips of the petals spread
Se 1·8 cm long, separate, naked, green or red
St enveloped in the upper part of the flower tube
Ha Brazil, Bolivia, Paraguay,

Argentina
CU Sunny, dry. Easily grown, very variable species. The various forms differ in regard to the length and breadth of the leaves, the length of the scape, and the colour (white or blue) of the flowers.

Tillandsia macdougallii L. B. Smith, 1949 (Plate 18)
(*macdougalii* after the Scottish nurseryman and collector MacDougall living in Mexico)

Can easily be confused with the species from the same area *T. benthamiana* (see page 95), *T. andrieuxii* (see page 90) and *T. oaxacana* (see page 159). With its red, pendulous inflorescences and white rosettes, it is a splendid species but not easy to grow.

Pl stemless, 20–25 cm high, forming rather large clumps
Fo numerous, forming a compact, silver grey rosette
LS broad-oval, 4–6 cm wide, 6–8 cm long, heavily covered with grey scales, partially bordered in lavender
LB narrow-triangular, very long-tapered, recurved above the sheath, 2–2·5 cm wide, 15–20 cm long, thick, coarse, grey scales, the edges slightly curled
Sc about 15 cm long, pendulous, sometimes covered by the leaves, heavily scaled
SB similar to the leaves, overlapping, the sheaths oval, with long blade, thin-skinned, green or pink scaled
I simple, pendulous, up to 20 cm long and up to 25 flowers
FB lanceolate-oval, long-tapered, 6–8 cm long, 3–4 cm wide, green or red, scaled, nerved
Fl loose, multifarious, blue, with tubes 5·5–6 cm long
Se 3–4 cm long, 1·5–2 cm wide, thin, heavily scaled, nerved, separate
St protruding from the tube
Ha Central Mexico (between Puebla and Tehuacan); epiphytic on conifers at altitudes between 2,500 and 3,000 m.
CU Bright and sunny, very sensitive to too much humidity. Not easy to grow.

Tillandsia magnusiana Wittm., 1901 (Plate 41, Ill. 5)
(*magnusiana* after the German botanist Magnus)

In its non-blooming stage it is easily confused with *T. plumosa*, but at bloom time the latter has a distinctly developed scape and green flowers. Nevertheless, L. B. Smith has 'united' *T. magnusiana* and *T. plumosa* into one species, a view which we do not share. The examples we collected in the province Chiapas, in the vicinity of Comitan near the Guatemalan border, differ so strongly from the type with their thin, short-scaled leaves, their sessile, simple spiked inflorescences and the violet flowers. *T. plumosa* from the area of Oaxaca has separated, feathery scales on the leaves, an elongated scape, a compound inflorescence, and green flowers. Therefore we consider *T. magnusiana* a separate species. Rohweder

(1965) considers the *T. magnusiana* found in El Salvador to be a variety of *T. plumosa*.

Pl stemless, growing in rather large turf areas, up to 15 cm high
Fo numerous, the sheaths forming a pseudo-bulb and the blades spreading out in all directions
LS oval-triangular, up to 1·5 cm wide, up to 1·5 cm long, naked at the base, in the upper part both sides long-scaled, pale
LB triangularly lineal, thread-like, long-tapered, up to 11 cm long, 5 mm wide at the base, average width 1 mm, heavily covered by coarse, separated, silver grey scales
Sc not visible, hidden in the rosette, about 1 cm long
SB shaped like the foliage
I simple, distichous, few flowers
FB 4·5 cm long, 2·5 cm wide, wide oval triangle, green, in the upper part grey-scaled, tightly overlapping
Fl violet, upright, about 4 cm long, tubular, surpassed by the stamens
Se 1·9 cm long, 5 mm wide, separate, the posterior ones fused 2 mm high, symmetrical
Ha Mexico, Guatemala, Honduras, El Salvador; epiphytic on oaks and conifers at altitudes between 1,100 and 2,000 m.
CU Little watering, very bright and sunny. With its white, fine to feathery-scaled leaves, one of the most beautiful of the small tillandsias.

Tillandsia makoyana Baker, 1889 (Ill. 9)

(*makoyana* after the collector Makoy)
Syn. *T. cucaensis* Wittm., 1891
Easily confused with *T. dasyliriifolia* (see page 114) and *T. utriculata* (see page 186)

Pl 50 cm–1 m high, including inflorescence; stemless
Fo very hard, forming an upright, almost cylindrical funnel rosette
LS broadly oval, 10–13 cm long, 7–9 cm wide, grey scaled on the outside, brown to dark lavender on the inside
LB triangular, awl shaped, long tapered, 3–5 cm wide above the sheath, up to 65 cm long, dull green to brownish green, both sides short scaled, longitudinally ribbed on the underside, the edges curled
Sc stout, upright, up to 1 cm thick
SB upright, the lower ones overlapping, with lineal blade, the upper ones lanceolate oval, pointed
I simple or panicled, upright
PB short, partially surrounding the basal, sterile section of the spikes
Sp up to 24 cm long, loosely arranged flowers, with slightly zigzag shaped, naked rhachis
FB broadly oval, blunt, shorter than the sepals; no keel
Fl blue or green, short-stemmed, upright, tight against the spike rhachis; tubular, up to 5 cm long
Se narrowly elliptical to egg-shaped, blunt
St protruding from the tube
Ha Mexico, Guatemala, Honduras, Costa Rica; epiphytic on conifers, oaks and cacti (Ill. 9) at altitudes from 1,500–1,900 m.
CU Bright, sunny, little watering. Can be grown along with cacti.

Tillandsia mallemontii Glaziou, 1894 (Fig. 39)
Syn. *T. linearis* sensu Baker, 1887, non Vell., 1825

Pl small, up to 25 cm high with inflorescence, growing in thick mounds; stem thin, elongated

Fo loose, distichous

LS surrounding the stem, up to 1·3 cm long and 3 mm wide

LB spread, thin, awl-shaped, thread-like, up to 11 cm long, 1 mm wide at the base; heavily covered with silver grey; separated scales

Sc slender, upright, up to 10 cm long, 1 mm thick, grey-scaled

SB lacking or 1 in the upper third of the scape; this one 3·5–4 cm long, lying close to the scape, its blade upright, heavily grey-scaled, nerved, green or brownish

I simple-spiked, 2 to 4 flowers, narrowly lanceolate, 2·5–3 cm long

FB close, hard, up to 1·3 cm long, 6 mm wide, oval-lanceolate, scaled, nerved, green or reddish-brown; shorter than the separate, naked sepals, which are up to 1·4 cm long and 3 mm wide

Pe large with 1·2 cm long, 1·5 mm wide claw and spread, almost disc-shaped plate 1 cm long, about 9 mm wide, dark blue

St enveloped in the flower

Ha Brazil; epiphytic in thick mounds on trees. In its non-blooming stage can be easily confused with *T. recurvata* (see page 171). The latter has small flowers

CU Easily grown, but hardly a decorative species.

Fig. 39 *Tillandsia mallemontii* Glaziou ($\times \frac{1}{2}$).

Tillandsia matudai L. B. Smith, 1949 (Ill. 6)
(*matudai* after the Mexican botanist Matuda)

Pl stemless, usually growing alone
Fo numerous, forming a thick, upright or flared, silver grey rosette
LS elliptical, slightly puffed, about 3 cm long, about 2 cm wide, heavily grey-scaled
LB narrowly triangular, long-tapered, with distinctly curled edges, both sides heavily covered with grey, flared scales, 14–16 cm long, 1 cm wide above the sheath
Sc very short, pendulous, partially hidden by the leaves
SB tightly overlapping, similar to the leaves, longer than the stalk segments, heavily covered with flared, grey scales
I pendulous, branched in a hand shape, with 5 to 7 spikes
PB similar to the scape bracts, shorter than the lanceolate, pointed, close, spread spikes, which have few flowers and are 4–5 cm long, 1–1·5 cm wide
FB distichous, elliptical, broadly pointed, tightly overlapping, up to 3·7 cm long, longer than the sepals; no keel; naked, green (in cultivation) or pink (in the native habitat)
Se separate, narrowly lanceolate, 3 cm long, a wing-like keel at the base, naked
Pe forming a white tube 4 cm long
St somewhat surpassing the petals
Ha Southern Mexico (Chiapas); epiphytic primarily on cypresses (2,000–2,200 m).
CU Bright, sunny, little watering. A pretty, silver white, small species.

Tillandsia mauryana L. B. Smith, 1937 (Plate 40)
(*mauryana* after the Mexican botanist P. Maury)

Pl stemless, growing in a turf, up to 10 cm high including the inflorescence
Fo numerous, forming a thick, spread, silver grey rosette
LS 1·5–2 cm wide, equally long, thick silver grey scales
LB narrowly lineal, awl-shaped, pointed, 1 cm wide above the sheath, 7–10 cm long, spread to recurved, thick silver grey scales
Sc very short, hidden by the leaves, 2–3 cm long, naked
SB tightly overlapping, similar to the leaves
I compound with 3 to 7 close spikes
PB oval-pointed, 2·5–3 cm long, 1–1·5 cm wide, cutaneous, heavily scaled, shorter than the 3 to 8 flowered spikes
FB up to 2·5 cm long, up to 1·5 cm wide, oval-lanceolate, cutaneous, scaled, longer than the sepals, which are 1·5 cm long, 5 mm wide, lanceolate, cutaneous, scaled and separate
Pe about 2·3 cm long, 4 mm wide, inclined into a tube, the tips somewhat flared, in the upper part dark green, in the lower part yellowish-green to white
St 1·5 cm long, enveloped by the petals
Ha Mexico. We found the plant on steep cliffs in the Valle de los Venados near Pachuca (north of Mexico City) at 1,600 m.
CU Like all white tillandsias, sunny and bright, little watering. A species worthy of growing, decorative.

Tillandsia meridionalis Baker, 1888 (rear bookjacket illustration)
(*meridionalis* noon bloomer)
Syn. *Anoplophytum refulgens* Morren, 1889; *T. hilaireana* Baker, 1889
A small tillandsia with silver grey rosette, which can be confused with *T. ixioides* (see page 143) and *T. didisticha* (see page 118) in its non-blooming stage. The species is somewhat variable in regard to the length and thickness of the inflorescence.

Pl stemless, growing in rather large clumps
Fo numerous, forming a dense, flared to upright, often asymmetrical rosette up to 10 cm in diameter
LS not distinct
LB 5–10 cm long, up to 2 cm wide at the base, narrowly triangular, long-tapered, hard, the edges curled, both sides heavily covered with silver grey scales
Sc slender, usually pendulous, up to 10 cm long, 2–3 mm thick
SB the basal ones similar to the leaves, the upper ones surrounding the scape with their sheath, 4–6 cm long, pink to red and heavily scaled, grey tip

I 4–5 cm long, 2–3 cm wide, simple spiked, with flowers loosely arranged in a spiral
FB about 2·5 cm long and 1·7 cm wide, broadly oval, moderately long, pointed, pink or red, heavily grey-scaled at the tip
Se 1·4 cm long, 6 mm wide, shorter than the floral bracts, separate, lanceolate-oval, abruptly pointed, thinly scaled
Pe up to 1·8 cm long, white, tips flared and broadly rounded
St enveloped in the flower
Ha Brazil, Argentina and Paraguay
CU Bright, sunny, little watering. Recommended.

Tillandsia micans L. B. Smith, 1956 (Ill. 307)
(*micans* sparkling, shining; because of the shiny flower bracts)

Pl forming a short stem, uniting into compact tufts, flowering to 50 cm high
Fo numerous, spirally arranged, forming a slender rosette
LS not clearly defined, 4–5 cm long, 3–4 cm wide, upperside densely brown scaled
LB to 25 cm long, 3 cm wide above the sheath, lanceolate, tapering to a long recurved tip, with upward curved margins, both sides densely silver grey scaled
Sc upright to arching: naked, round, green, 20–30 cm long, 1 mm thick, weakly grooved

SB densely overlapping, covering the scape, upright, the lower ones similar to the leaves, the upper ones broad-elliptical with short pointed tip, green, grey scaled, sometimes naked
I simple (seldom compound with 1 large and 2 small lateral spikes), lanceolate, laterally flattened, 15–20 cm long, 2–2·5 cm wide, upright arching or pendant
Fb densely overlapping in two rows, upright, 3·5–3·9 cm long, broadened to 1·9 cm, with blunt tip, firm; margins cutaneous, shiny green, naked, not keeled, longer than the sepals

Fl short, 3 mm long stalked, white
Se lanceolate, 2·6 cm long, thin with cutaneous margins, naked, free, the back ones keeled
Pe 4·5 cm long, flared to recurved plate for 6 mm
St and Pi enclosed within the flower; filament white, anthers 1 cm long

Ha South Peru, near Calca close to Cuzco at 3,100 m, and in the dry valley of the River Apurimac on steep sides of rock walls
CU a very decorative small species, requiring dry and light conditions in cultivation. Cactus cultivation suits it admirably.

Tillandsia monadelpha (Morren) Baker, 1887 (Ill. 86, Fig. 40)
(*monadelpha(us)* of one brother, i.e. the filaments of all the stamens are fused with each other)
Syn. *Phytarrhiza monadelpha* Morren, 1882; *T. graminifolia* Baker, 1887; *Catopsis alba* Morren, 1889; *T. monobotrya* Mez, 1919; *T. digitata* sensu Standl., 1927, non Mez, 1896

Pl stemless, 35 cm high with inflorescence
Fo up to 20 cm long, forming a dense rosette
LS oval
LB lanceolate, pointed, 10–15 mm wide
Sc upright or leaning to one side, slender, naked
SB much smaller than the foliage, lanceolate-elliptical, scaled at the tip
I simple-spiked, loose, up to 22 flowers, 13 cm long, 3–3·5 cm wide with zigzag narrowly winged rhachis
FB distichous, horizontally speed, 1·7 cm long, keel, slightly scaled to naked, longitudinally ribbed, green
Fl white, sessile, spread, 3 cm long; the

Fig. 40 *Tillandsia monadelpha*: stamen tube (*left*); cut open and spread out (*right*) (× 2).

petals separated
Se lanceolate-elliptical, keeled, naked, green, fused only slightly but to the same height
Ha from southern U.S.A. to Colombia; epiphytic, from the coast to about 300 m.
CU Shady and damp. *T. monadelpha* is closely related to *T. narthecioides* (see page 159).

Tillandsia multicaulis Steud., 1841 (Plate 8)
(*multicaulis* many-stemmed; refers to the numerous inflorescence paddles)
Syn. *T. caespitosa* Schlecht. et Cham., 1831, non Le Conte, 1828; *Vriesea schlechtendalii* Wittm., 1891.
T. multicaulis cannot be confused with other species. Along with *T. complanata* it is the only tillandsia with several inflorescences emerging from the leaf axes (see page 109). At bloom time they form a splendid sight. After bloom the plant perishes and pups continue the propagation.

Pl stemless, 30–35 cm high including inflorescence
Fo forming a large, green funnel rosette 30–40 cm in diameter
LS long-oval, 10–12 cm long, 5–6 cm wide, green to violet
LB strap-like, pointed, 20 cm long, 3–4 cm wide at the base, crisp green, hardly scaled
Sc several, originating from the leaf axes, slender, upright, 12–15 cm long, flattened
SB similar to the floral bracts but smaller, 3 cm long, tightly surrounding the scape, distichous, longer than the stalk segments, green, central nerve
I single paddle-shaped, pointed, 10–15 cm long, 4–5 cm wide, naked
FB tightly overlapping, distinctly keeled, shiny red, 5 cm long, the lower ones without flowers
Se shorter than the floral bracts, about 3·5 cm long and 1 cm wide, lineally elliptical, thin-skinned, greenish-white, separate
Pe 7 cm long, blue, slightly zygomorphic
St protruding from the bloom
Ha Central Mexico, Costa Rica, Panama; epiphytic, from sea level to the cloud forest region (2,000 m).
CU Semi-shady, damp, epiphytic or in peat.

Tillandsia myosura Griseb., 1878 (Fig. 41)

(*myosura*(*us*) mouse tail; perhaps because of the thin, mouse tail-like leaves)

The species is very closely related to *T. gilliesii* (see page 134), but is distinguishable by the roundish to awl-shaped leaves.

Pl stemmed, 10–30 cm high, forming thick mounds
Fo distichous, spreading horizontal to arched, 5–10 cm long, 4 mm in diameter, rounded awl-shape, pointed, upper side like a gutter, grey-scaled
Sc up to 10 cm long, perpendicular, 1 mm in diameter, grey-scaled, bearing a leaf-like scape bract at the base
I simple-spiked, 1 to 3 flowered
FB long-lanceolate, pointed, 1·5–1·7 cm long, 8 mm wide, green or reddish-brown, grey-scaled
Fl sessile or very short-stemmed, 1·8 cm long, very aromatic, yellow; petals tongue-shaped with flared tips
Se 1·2–1·5 cm long, 4 mm wide, blunt, reddish, naked
St enveloped in the flower
Ha Argentina, Bolivia
CU Bright, sunny, dry. A small, pretty, white tillandsia.

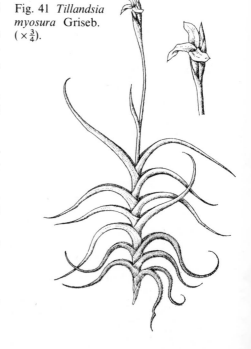

Fig. 41 *Tillandsia myosura* Griseb. (× ¾).

Tillandsia narthecioides Presl, 1827 (Ill. 87)
(*narthecioides* narthecium-like; reminiscent of the *Liliaceae narthecium*)

Pl stemless or short-stemmed, up to 45 cm high including the inflorescence
Fo upright, forming a dense rosette
LS barely distinguishable, about 3 cm long and 1·3 cm wide, merging into the blade, brown-scaled
LB 20–40 cm long, 5–8 mm wide at the basis, narrowly lineal, very long-tapered, densely scaled, green or reddish-green
Sc slender, perpendicular, shorter than the leaves
SB in the lower part of the scape similar to the leaves with recurved blades, in the upper part surrounding the scape, longer than the stalk segments, scaled at the tip, green or reddish
I simple-spiked, very loose, up to 30 flowers and up to 12 cm long, 2·5–3 cm wide, with a flexible rhachis
FB distichous, spreading about 6 mm from each other, lineally elliptical, up to 1·4 cm long and 5 mm wide, naked, indistinctly ribbed longitudinally, green to reddish-green
Fl white, open at night, very fragrant; the petals have plates 9 mm long and 7 mm wide, flared
Se about 8 mm long, 3–4 mm wide, green, naked, longitudinally ribbed
St enveloped in the flower; filaments not fused to each other
Ha Ecuador, especially where cacao is cultivated, from sea level to 400 m.
CU Similar to *T. monadelpha*, to which it is closely related. Interesting, small, night blooming species with strongly aromatic flowers.

Tillandsia oaxacana L. B. Smith, 1948 (Plate 19)
(*oaxacana(us)* named after the city Oaxaca (central Mexico), where the plant is found)
This species is closely related to *T. macdougallii* (see page 152), *T. benthamiana* (see page 95) and *T. andrieuxii* (see page 90), from which it can be distinguished by the chestnut brown sheaths.

Pl stemless
Fo numerous, forming a dense, pseudo-bulb rosette up to 20 cm in diameter
LS distinctly spread, broadly elliptical, 6 cm long, 5 cm wide, distinctly spoon-shaped, brown to dark brown
LB narrow triangular, 1·5–2 cm wide above the sheath, 15–20 cm long, heavily scaled
Sc short, arched pendulous, covered by the leaves
SB tightly overlapping, similar to the leaves, with pink sheaths and green blades
I simple-spiked, almost cylindrical, 12 cm long; flowers in a spiral.
FB overlapping, longer than the sepals, elliptical, pink, heavily grey-scaled, the lower ones with a long, blade-like tip
Se separate, elliptical, 2·5 cm long, thin, heavily scaled
Pe blue, upright, inclining into a 5 cm long tube
St surpassing the sepals
Ha Central Mexico; epiphytic on oaks and conifers at altitudes around 200 m.
CU Bright and sunny, moderately damp. A recommended pretty species.

Tillandsia oroyensis Mez, 1919 (Ill. 114)

(*oroyensis* named after the central Peruvian mining town Oroya)

Pl 50–70 cm high including the inflorescence, stemless

Fo rosette, up to 30 cm long, dense, short scales

LS long-oval, 10–12 cm long, 6–8 cm wide, both sides brown

LB narrowly triangular, long-tapered, 3–4 cm wide above the sheath

Sc sturdy, upright

SB loose, oval-elliptical with long pointed blade, upright, grey-scaled

I compound, panicle spiked (the basal branches frequently branched again)

PB the basal ones somewhat shorter than the upright to spread spikes, which have 6 loose flowers, are almost elliptical, pointed and 3·5 cm long

FB almost upright, 1–1·5 cm long, shorter than the sepals, no keel, with a strongly protruding nervation

Fl about 1·8 cm long, blue, enveloping the stamens

Se almost separate, oval, pointed, keeled, naked

Ha Central Peru, between Yarma and La Oroya, between 3,000 and 3,500 m on cliffs. The foliage can easily be confused with *Tillandsia latifolia* (see page 000)

CU Bright and sunny. Suitable only for fanciers with rather large greenhouses.

Tillandsia paleacea Presl, 1827 (Ill. 13, 14, 115)

(*paleacea*(*us*) chaf-like, covered with chaf; refers to the coarsely scaled leaves)

Syn. *T. fusca* Baker, 1878; *T. scalarifolia* Baker, 1887; *T. schenckiana* Wittm., 1889; *T. chilensis* Baker, 1889; *T. lanata* Mez, 1905; *T. favillosa* Mez, 1906

A greatly variable species. There are open forms with elongated axes and forms with a stubby shape and a short stem as well as short-leafed and long-leafed forms.

Pl stemmed, 10–70 cm long including the inflorescence, often growing in long strands (Ill. 13–14)

Fo spiral, with spreading blades

LS broadly oval or elliptical, partially naked, partially scaled

LB up to 12 cm long, 4–6 mm wide, spread or gutter-like, usually somewhat twisted, heavily covered with silver grey, coarse scales

Sc upright, slender, 15–20 cm long, naked or slightly scaled

SB upright, as long as or somewhat longer than the stalk segments, narrowly elliptical, scaled, with a thread-like tip

I simple, narrowly lanceolate, pointed, flattened, up to 5 cm long, 1 to 12 dense flowers

FB distichous, overlapping, oval or elliptical, 1·7 cm long, not keeled, scaled, sometimes naked

Se lanceolate, 1–1·7 cm long, separate, naked

Pe blue or violet, with a narrow claw and flat plate

St enveloped in the flower

Ha Colombia, Bolivia, Peru and

Chile, from the coastal desert up to 3,000 m. In Peru, in the area of the coastal desert, it forms extensive stands in the form of strands several metres long lying on the bare desert sand (see Ill. 13–14)

CU Bright, sunny, but in high humidity. An interesting species that can be grown on pure sand.

Tillandsia paraensis Mez, 1894 (Fig. 42)
(*paraensis* named after the Brazilian city Para (Belem))

Pl stemless, up to 50 cm high
LS large, oval, spoon-shaped, the exterior ones without a blade, overall forming an egg-shaped pseudo-bulb
LB narrowly triangular, pointed, with curled edges, and frequently with lighter horizontal banding
Sc almost upright, slightly scaled, 3 mm thick
SB elliptical, as long as or longer than the stalk segments, heavily scaled
I simple or composed of two spikes
PB similar to the upper scape bracts, much shorter than the spikes, which are up to 25 cm long, lineal, flattened and have 6 to 17 flowers
FB upright and overlapping, later loose, broadly elliptical or slightly pointed, up to 3·2 cm long, surpassing the sepals, no protruding keel, nerved, greenish-yellow or red, heavily scaled
Se elliptical, pointed, separate, up to 2·4 cm long, scaled
Pe up to 7 cm long, upright, red
St surpassing the flower
Ha Colombia, Peru, Bolivia and western Brazil
CU Bright, sunny, little watering; epiphytic.

Fig. 42 *Tillandsia paraensis* Mez (× ½).

Tillandsia pedicellata (Mez) Castel., 1945 (Fig. 43)
(*pedicellata(us)* equipped with a flower stem)
Syn. *T. coarctata* Gill, 1878 var. *pedicellata* Mez, 1896

Pl with short, branched stems uniting into thick mounds
Fo thick, overlapping, upright, multifarious, so that a leafed stem resembles a stem of club moss
LS cutaneous, surrounding the stem, 3 mm long, 4 mm wide
LB awl-like, triangular, needle-shaped, 6–7 mm long, 2 mm wide at the base, dull green, heavily grey-scaled
Sc missing at bloom time, only appearing with the ripening of the capsule, stretching 2–5 cm long
I a single-flowered, sessile spike
FB up to 9 mm long, green, cutaneous
Fl small, dark lavender, about 8 mm long
Se up to 8 mm long, cutaneous green, oval-lineal, rounded
St enveloped deeply in the tube, about 4 mm long
Ha Argentina and Bolivia in dry areas; epiphytic on trees, but also on cliffs. Can easily be confused with *T. bryoides* (see page 99); the latter, however, has yellow flowers

Fig. 43 *Tillandsia pedicellata* (Mez) Castell (life-size).

CU Bright, sunny, little watering. An interesting species reminiscent of moss or club moss. Because of its small size, it is recommended for terrariums.

Tillandsia peiranoi Castell., 1938 (Ill. 119)
(*peiranoi* after the discoverer of the plant, A. Peirano)

Pl small, short, stemmed, up to 15 cm high including the inflorescence, growing in clumps
Fo arranged in a tight spiral
LS cutaneous, surrounding the stem, about 5 mm long, 1·3 cm wide
LB triangular, pointed, rigid, succulent, 3–5 cm long, 1·4 cm wide at the base, dull green, grey-scaled, both sides prominently nerved, edges curled into a gutter

Sc upright, slender, naked, greenish-brown, about 1 mm thick, 3–5 cm long
SB lying close to the scape and surrounding it, as long as the stalk segments, greenish-brown, grey-scaled toward the tip, 7–8 mm long
I simple-spiked, loose, distichous, 3 to 8 flowers
FB 10–11 mm long, 4 mm wide, long-oval, pointed, cutaneous, greenish-brown, slightly scaled

Fl blue, short-stemmed, up to 1·5 cm long; the petals in the lower part forming a short tube, in the upper part flared
Se up to 11 mm long, as long as the floral bracts, or a little longer, 4 mm wide, separate, naked
St enveloped deeply in the tube, 7 mm long
Ha Argentina, growing in thick clumps on cliffs
CU Bright, sunny, little watering. A small, recommended species.

Tillandsia pendulispica Mez, 1896 (Ill. 117)

(*pendulus* hanging, *spica* spikes with hanging spikes; because of the thin, limp, pendulous flower spikes)
Syn. *T. scorpiura* Mez, 1905; *T. triangularis* Rusby, 1920

Pl stemless, growing in bundles, 50–80 cm high including the inflorescence
Fo numerous, forming a large funnel rosette with a slightly bulbous base
LS large, oval spoon-shaped, 8–9 cm long, 7 cm wide, dark brown
LB lineal to narrowly triangular, long-tapered, heavily grey, short-scaled, up to 30 cm long, up to 4 cm wide above the sheath
Sc thin, upright, naked, up to 25 cm long, 2 mm thick
SB lanceolate, pointed, upright, lying close to the scape, remote, shorter than the stalk segments
I loose, limp, panicled, up to 30 cm long and 10 cm wide
PB similar to the scape bracts, but blunt, usually shorter than the basal, stemmed segment of the panicle
Sp very narrow, up to 7 mm wide, multi-flowered, loose, distichous, almost upright, spread or pendulous, 5–6 cm long
FB upright, distichous, oval to rounded, 7 mm long, not keeled, cutaneous, nerved
Fl upright, sessile, very small, 7 mm long, yellowish-white
Se asymmetrical, about 5 mm long
St enveloped in the flower
Ha Northern Peru (Ayabaca and Huanuco), southern Peru (Puno) and Bolivia at altitudes between 1,500 and 2,000 m.
CU Moderately damp, semi-shady. Especially striking because of its fine, much branched inflorescence.

Tillandsia plumosa Baker, 1888 (Plate 23)

(*plumosa(us)* feathery; the leaves equipped with long, separated scales have the appearance of feathers)
T. plumosa in its non-blooming stage is easily confused with *T. magnusiana* (see page 152), but differs from it at bloom time by the stemmed inflorescences and the green leaves.

Pl stemless, growing in rather large clumps, up to 15 cm high including inflorescence
Fo before flowering spread in all directions, during and after flowering mostly recurved, densely covered with

white, feathery scales

LS long-oval, 1·5 cm long, 9 mm wide, pale, naked in the lower part, scaled in the upper half, succulent, overall forming a bulb

LB lineal awl-shaped, long-tapered, 8–10 cm long, 6 mm wide above the sheath, olive green, but densely covered with white, almost feathery scales

Sc variable in length, sometimes remaining in the rosette, sometimes surpassing the leaves greatly, upright to angling upward, up to 10 cm long, 2 mm thick, grey-scaled

SB similar to the leaves, surrounding the scape with the sheath and longer than the stalk segments; the blades narrow, long-tapered, upright, dense, grey, separated scales, usually distinctly growing to one side

I about 4 cm long, 2 cm wide, composed of 4 to 8 upright, head-shaped spikes, which have 2 to 4 flowers, are flattened and are up to 3 cm long and 1 cm wide

PB as long as or slightly shorter than the spikes, upright, narrowly lanceolate, pointed, 2 cm long, 5 mm wide, thin, green, grey-scaled on the outside

FB distichous, tightly overlapping, the lower ones sterile, up to 2 cm long, 7 mm wide, oval-lanceolate pointed, green at the base, pink at the top, scaled

Se narrowly lanceolate, pointed, 1·6 cm long, 5 mm wide, symmetrical, cutaneous, slightly scaled at the tip, separate, the posterior ones fused about 1 mm

Pe green, inclined into a tube, 2 cm long, 3 mm wide, with flared tips

St 1·7 cm long, reaching into the upper part of the tube, as long as the pistil

Ha Mexico, northern Oaxaca; epiphytic on *Arbutus* (*Ericaceae*) at 2,000 m.

CU Very bright and sunny, little watering. With its feathery white-scaled leaves, *T. plumosa* is one of the most beautiful Mexican species. Can be grown along with cacti.

Tillandsia pohliana Mez, 1894 (Ill. 118)
(*pohliana* after the Viennese botanist Pohl (1872 to 1934), who travelled extensively in Brazil)
Syn. *T. meridionalis* sensu Mez, 1896, non Baker, 1888; *T. windhausenii* Hassler, 1930; *T. latisepala* L. B. Smith, 1933

Pl stemless, 15–20 cm high

Fo numerous, forming a compact, broad, silver grey rosette

LS not distinct

LB lanceolate triangular to awl shaped, long tapered, 2·5 cm wide at the base (average width 1·5 cm), 20–25 cm long, densely silver white scaled with distinctly curled edges

Sc slender, pendulous, 8–10 cm long

SB similar to the leaves, tightly overlapping, surrounding the scape

I simple-spiked, loose, 5–7 cm long, 3 cm wide; flowers arranged in a spiral

FB broad, oval-triangular, pointed, about 3 cm long and 2 cm wide, dense silver grey scales, green to pale pink, the basal ones longer than the flowers, which are about 2·5 cm long and white

Se separate, broadly oval, with a small

point, 1·5 cm long, 1 cm wide, stiff, heavily scaled
St enveloped in the flower
Ha Brazil, Paraguay, Argentina and Peru. In Peru, the plant grows south of Cuzco in the dry valley of Quillabamba at 1,500 m.
CU Sunny, little watering. A species worthy of cultivating.

Tillandsia polita L. B. Smith, 1941 (Ill. 300)

(*polita*(*us*) shiny; because of the shiny flowerbracts)
The plant illustrated was found at 2,100 m in oak forest close to San Cristobal de las Casas (Chiapas), South Mexico.

Pl stemless, flowering between 45 and 65 cm high, growing into thick tufts
Fo numerous, outspread, forming a flat rosette from 40–50 cm in diameter
LS broad-oval to almost triangular, about 7 cm long, 5 cm wide, appressed brown scaled
LB narrow-triangular, long pointed, stiff, almost prickled, 1·5 cm wide at the base, 30–45 cm long, both sides appressed grey scaled
Sc upright, round, naked, 6 mm thick
SB constantly longer than the internodes, upright, fully covering the scape, the basal ones leaf-like, the upper ones with shorter blades, only slightly scaled, smooth, shiny red
I in several, nearly upright densely compound spikes, 11–15 cm long, about 5 cm wide

PB the upper ones similar to the scape bracts but much shorter than the spikes, almost upright, dense, spirally arranged, narrow-lanceolate, pointed, sword shaped with a few sterile flower bracts at the base, 5–7 cm long, 1·5–2 cm wide, 5 to 7 flowered
FB densely overlapping, in two rows, upright, 2–3 cm long, shiny red, smooth, naked, keeled, somewhat shorter than the sepals, at the tip only indistinctly scaled
Se narrow-lanceolate, pointed, about 3 cm long, the back ones fused for much of their height and keeled
Pe violet, fused into a tube, 5 cm long
St and Pi extending beyond the flower
Ha Mexico, Guatemala, Salvador
CU A very decorative species.

Tillandsia polystachya L., 1762 (Ill. 116)

(*polystachya*(*us*) many-spiked; because of the numerous spikes)
Syn. *Renealmia polystachya* L., 1735; *T. angustifolia* SW., 1788

Pl stemless, up to 70 cm high including inflorescence
Fo numerous, in a dense rosette
LS broad-oval or elliptical, brown-scaled
LB triangular, long-tapered, up to 35 cm long, 1–3 cm wide above the sheath, short scales, recurved in the upper section
Sc upright or angling upward, thin
SB leaf-like with upright, tightly overlapping sheaths and long, narrow, spread or recurved, green or reddish blades
I compound (seldom simple), almost cylindrical, up to 30 cm long

PB elliptical, pointed, shorter than the upright, sessile, lineal, pointed, flattened spikes of 3–10 cm length
FB tightly overlapping or loose, broad-oval with short tip, 1·5–2 cm long, approximately as long as the sepals, smooth or somewhat furrowed toward the tip, naked or slightly scaled, a slight keel at the tip
Se elliptical, smooth, naked, the posterior ones slightly fused
Pe united into a tube, 3 cm long, blue
St protruding from the flower
Ha From Mexico to Brazil and Bolivia; epiphytic on trees from 300–2,200 m. Because of the wide habitat, *T. polystachya* shows a great variability
CU Moderately damp, semi-shady.

Tillandsia prodigiosa (Lem.) Baker, 1888 (Ill. 1, 120)
(*prodigiosa(us)* marvelous, wonderful; probably refers to the pendulous inflorescences which are unusually long for tillandsias)
Syn. *Vriesea prodigiosa* Lem., 1869; *T. cossonii* Baker, 1887
MacDougall refers to the great variability of the plant, which we also noted; these are examples with extremely long inflorescences and those with shorter inflorescences, with blue or with green flowers.

Pl stemless
Fo numerous, forming a large, upright funnel rosette, the blades up to 50 cm long, 6 cm wide, close, short scales; the other rosette leaves pendulous
Sc sturdy, pendulous, 30 cm long
SB longer than the stalk segments, similar to the leaves, green or pink, short-scaled
I 50–70 cm long, pendulous, compound, with 20 to 30 spikes
PB similar to the scape bracts, longer in the lower part of the inflorescence, in the upper part shorter than the spikes, long-pointed, pink or green, short grey-scaled
Sp flared, arranged in a spiral on the axis, up to 8 cm long, up to 5 cm wide, 10 to 12 flowers
FB distichous, tightly overlapping, up to 5 cm long, green or pink, naked or scaled
Se about 3·6 cm long, shorter than the floral bracts, almost triangular, long-tapered, the posterior ones fused 4 mm high, naked
Pe 5·3 cm long, in a tube, blue or green
St protruding from the flower
Ha Central Mexico; epiphytic in cloud forests at 1,500–2,300 m.
CU Moderately damp, semi-shady. Very decorative with their long, pendulous inflorescences, but difficult in cultivation.

Tillandsia propagulifera Rauh, 1972 (Ill. 301)
(*propagulifera(us)* prolific; because of the vegetative propagation)
T. propagulifera grows into a large plant, but its self propagation makes it an interesting species. After flowering the rosette gradually dies but in the axils of the inflorescence bracts throughout the inflorescence young plants

originate and root on the mother plant's branches and stem, then fall to the ground and continue as new plants. This plant also appears capable of propagation by seed.

Pl stemless, flowering to 3 m high

Fo numerous, in a dense, to 80 cm high and 1·2 m in diameter rosette, erect; the leaf tips recurved

LS 13–15 cm long to 6·5 cm wide, both sides dark brown and densely appressed scaled, distinctly separated from the blades

LB to 60 cm long, 1·5 cm wide gradually extending to a narrow recurved tip, grey green, densely appressed white scaled

Sc to 80 cm long, 3 cm thick at the base, naked, green, when dry finely furrowed

SB loosely set, the basal ones similar to the leaves, with longer, threadlike blade, the upper ones shorter than the internodes, narrow-lanceolate, sharp pointed, cutaneous margined, quickly drying

I very loose, bipinnate to tripinnate, 1·7–2·2 m long; branches very thin, simple or divided, the basal ones to 80 cm long, with long, about 15 cm leaf, covered basal portion

PB much shorter than the sterile part of the branching arrangement of the

first order (see page 43), 2·5–3 cm long, to 0·8 cm wide, blunt, naked to scattered scaled, drying early and then nerved

Sp to 70 cm long, 0·5–1 cm wide, with thin, green, flat, fairly felexuose axis and to 10 cm long, sterile, leaf covered basal portion; internodes to 3 cm long

FB longer or shorter than the internodes, green, violet coloured tip, about as long as the sepals, keeled, pointed, about 2 cm long, 7 mm wide, upperside scattered scaled, hard, when dry clearly nerved

Fl sessile, very loosely set

Se thin, green, fused for 2 mm high at the base, 18–20 mm long, 4 mm wide, rounded tip with short spined tip

Pe about 2·5 cm long, narrow tongue shaped, violet

St and **Pi** extending beyond the flower

Fr to 4 cm long

Ha epiphytic on mesquite trees on the banks of the Rio Utcabamba near Milagro, North Peru at 450 m.

CU As a young plant the silver grey scales make it quite decorative.

Tillandsia pruinosa SW., 1797 (Plate 43)

(*pruinosa*(*us*) frosted; with its flared, silver grey scales the plant looks as if it is frosted)
Syn. *T. breviscapa* A. Rich., 1850; *Platystachys pruinosa* Beer, 1857; *Platystachys tortilis* Beer, 1857

Pl stemless, 8–20 cm high

LS oval to round, 2–4 cm long, spoonshaped, inside heavily covered with brown scales, outside scaled in coarse, silver grey to brown, overall forming a hollow pseudo-bulb

LB usually surpassing the inflorescence, lineal, long, pointed, twisted, heavily covered with coarse, separated, silver grey scales, up to 15 cm long with rolled edges, usually reddish-green

Sc short, hidden by the leaf sheaths

I simple or composed of 2 to 3 spikes, which are distichous, up to 7 cm long,

4 cm wide and have 5 to 15 flowers
FB upright, 2–2·6 cm long, longer than the sepals, keeled near the tip, close, spread, scaled, pink at bloom time
Se broad oval, blunt, 1·3–1·9 cm long, slightly scaled or naked, posterior sepals fused to 6 mm
Pe tubular, lineal, blunt or pointed, 3 cm long, blue

St protruding from the flower
Ha From the southern U.S.A. to Brazil and Ecuador; epiphytic at altitudes of 200–1,000 m.
CU Bright, sunny, little watering. Because of the dense scaling and the pseudo-bulb appearance it is a decorative species, which offers no difficulties in cultivation.

Tillandsia pueblensis L. B. Smith, 1934 (Ill. 122)
(*pueblensis* named after the central Mexican city Puebla)

Pl stemless, 16–24 cm high, growing in groups, since the older rosettes remain for several years
Fo forming an upright, often one-sided, hollow bulb rosette up to 16 cm long; the outer rosette leaves spread out into bladeless basal leaves
LS not clearly delineated
LB narrowly triangular, over 1 cm wide at the base, densely covered with coarse, grey scales, with distinctly curled edges and needle-sharp point
Sc upright, short
SB the lower ones similar to the leaves, the upper ones lanceolate, pointed, scaled
I simple-spiked, lineal-lanceolate, 5 to 7 flowers, up to 9 cm long and 1 cm

wide
FB upright, narrowly lanceolate, pointed, not keeled, distichous, over-lapping, 2·5–4 cm long, longitudinally ribbed, short grey scales, pink, usually green in cultivation
Se narrowly lanceolate, pointed, 2 cm long, keeled, naked, longitudinally ribbed, separate, the posterior ones fused up to 6 mm high
Pe about 4 cm long, violet, forming a tube
Ha Mexico; epiphytic on trees in dry areas of the highlands at 2,000 m.
CU Dry, bright, sunny. A very hard and robust species, which can be grown along with cacti.

Tillandsia punctulata Schlecht. et Cham., 1831 (Plate 42)
(*punctulata*(*us*) dotted; probably because of the fine spotting of the flower tube)
Syn. *T. tricolor* sensu Morren, 1879, non Schlecht. et Cham., 1831; *T. melanopus* Morren, 1896

Pl stemless, about 35–45 cm high including inflorescence, growing in groups; pups beginning with a short stolon segment
Fo numerous, spread or almost upright, the sheaths forming a large,

loose pseudo-bulb
LS spoon-like, up to 7 cm long, 3–4 cm wide, blackish-brown, long-oval
LB narrowly lineal, 1 cm wide above the sheath, 20–30 cm long, green, slightly scaled, with curled edges

Sc upright to arching upward, about 25 cm long, 8–10 mm thick

SB resembling the leaves, tightly overlapping, covering the stalk segments, the blades upright, reddish-brown; the upper scape bracts with only a short blade and bright red

I simple or compound, with 3 to 4 spikes, up to 13 cm long

PB similar to the upper scape bracts, red, as long as the short, almost upright spikes, which are 5–8 cm long, 2–3 cm wide

FB distichous, tightly overlapping, wide-oval, pointed, keel, 3–4 cm long, 1·5–2 cm wide, naked, green, longer than the 2·5–3 cm long, white, thin-skinned, winged sepals

Pe 4–5 cm long, 7 mm wide, white at the base, otherwise blackish-violet and finely dotted with white, white tipped, gathered into a tube

Ha From Mexico to Panama; epiphytic at altitudes of 300–2,300 m. In its non-blooming stage *T. punctulata* can be confused with *T. kirchhoffiana* (see page 145)

CU Moderately damp, semi-shady. At bloom time a colourful and decorative, easily grown species.

Tillandsia purpurea Ruiz et Pav., 1802 (Ill. 15, 16, 38)

(*purpurea(us)* purple; because of the purple scape bracts)

Syn. *T. azurea* Presl, 1827; *T. longebracteata* Meyen, 1843; *Anoplophytum longebracteatum* Beer, 1857; *Platystachys purpurea* Beer, 1857; *Phytarrhiza purpurea* Morren, 1889

In its non-blooming stage this species can easily be confused with *T. cacticola* (see page 101) or *T. straminea* (see page 176). According to our observations it seems to be linked with *T. cacticola* by transitional forms. In its non-blooming stage *T. punctulata* can be confused with *T. kirchhoffiana* (see page 145).

Pl stemless or stemmed and then up to 70 cm long

Fo numerous, spiral

LB narrowly triangular, long-tapered, recurved or spread, up to 30 cm long, 2·5 cm wide at the base (average width 1·5 cm), heavily grey-scaled with curled edges

Sc slender, upright, various lengths, round, 2–3 mm thick, scaled

SB as long as the stalk segments, narrowly elliptical, upright, surrounding the scape with the sheath and with spread and recurved, heavily scaled, green or reddish blades

I loosely compound, with 6 to 8 flared to almost upright spikes, overall 10–15 cm long, 7–8 cm wide, seldom simple

PB lanceolate, pointed, scaled, reddish, shorter than the distichous spikes which are open, 5–6 cm long, 2·5 cm wide and have 6 to 8 flowers

FB spreading, lanceolate, pointed, longer than the sepals, slightly protruding keel, naked, 1·5–2 cm long, longitudinally ribbed, lavender to purple

Se lanceolate, 1·5 cm long, green, naked, all fused equally 3 mm high

Pe fragrant, 2 cm long, white, the plates edged in blue violet

St enveloped in the flower tube
Ha Peru. From the coastal desert to 2,800 m, forming thick mounds or long strands

CU Bright, sunny, moderately damp. Can be grown on pure sand.

Tillandsia rauhii L. B. Smith, 1958 (Ill. 22)
(*rauhii* after Rauh, who first found this plant in the valley of the Rio Saña in northern Peru)

Pl stemless, 2–2·5 m high including inflorescence
Fo numerous, forming a large funnel rosette up to 1·5 m high and up to 1·3 m in diameter
LS broad-oval, up to 25 cm long, up to 20 cm wide, brown
LB up to 1·2 m long, strap-like, rounded, but short-tipped, grey green or wine red, with a waxy covering, the old leaves hang limply downward
Sc stout, up to 4 cm thick at the base, arched
SB tightly overlapping, the lower ones laf-like, upright
I compound, up to 1·5 m long, arched
PB broadly oval, usually shorter than the sterile bases of the lineal spikes, which are up to 70 cm long and 4 cm wide, frequently worm-like, flattened and multi-flowered
FB tightly overlapping, elliptical, blunt, 5 cm long, broadly rounded on the back, grey green, waxy-frosted
Fl up to 6 cm long, dark blue, usually appearing only on the underside of the spikes
Se elliptically blunt, separate, up to 3 cm long, the rear with a keel
St protruding from the flower
Ha As yet known only from northern Peru (in the Department of Piura) on steep cliffs at 700 m.
CU Moderately damp, semi-shady. *T. rauhii* is an easily grown species, which can be propagated by adventitious pups (see page 66). Even as a small plant *T. rauhii* is very decorative, since even in cultivation it retains the reddish colour and the waxy coating of the leaves.

Tillandsia rectangula Baker, 1878 (Ill. 124)
(*rectangula(us)* right-angled; because the leaves spread out in an almost right angle to the main axis)
Syn. *T. propinqua* Gay var. *rectangula* Griseb., 1879

Pl small, short, stemmed, up to 8 cm long, growing in thick mounds
Fo arranged in a spiral
LS cutaneous, surrounding the stem, 8–10 mm long, 5 mm wide
LB triangularly awl-shaped, very hard, bent back almost at a right angle, 1–2 cm long, 3–4 mm wide at the base, the edges curled, both sides heavily grey-scaled
Sc short, not surpassing the leaves, 3–5 mm long, thin
SB 1 to 3, similar to the leaves and surrounding the scape
I a single-flowered spike up to 1·5 cm long
Pe up to 1·5 cm long, brown to yellow, the tips flared
St enveloped in the flower
Ha Argentina. Epiphytic, growing in

thick mounds on cliffs
CU Bright and sunny. A small, very stiff species especially recommended to the fancier with limited room.

Tillandsia recurvata (L.) L., 1762 (Ill. 12, Fig. 44)
(*recurvata(us*) bent back; because of the recurved leaf blades)
Syn. *Renealmia recurvata* L., 1753; *T. uniflora* H. B. K., 1816; *Diaphoranthema uniflora* Beer, 1857; *Diaphoranthema recurvata* Beer, 1857

Pl short-stemmed, simple or slightly branched, 4–23 cm high (very variable in size), growing in thick turfs
Fo distichous
LS overlapping, surrounding the stem
LB upright to recurved, lineally awl-like, 3–17 cm long, 0·5–2 mm thick, heavily grey-scaled
Sc 3–8 cm long, 0·5 mm thick, upright silver grey-scaled, round
SB lineally lanceolate, often with a thread-like, short blade; usually only one scape bract directly under the inflorescence, green, heavily silver grey-scaled, 12–18 mm long
I simple, 1 to 5 close flowers
FB similar to the scape bract, densely silver grey-scaled, usually as long as the sepals, 8 mm long, green
Se almost separate, 4–9 mm long, usually naked, lanceolate, thin-skinned, longitudinally ribbed
Pe small, 1–1·3 cm long, narrow, with spread tips, pale blue (different from the yellow or brown flowers of *T. capillaris*)
St enveloped in the flower tube
Ha The species has a very wide range, from southern U.S.A. to Argentina and Chile. In its growth it is variable and can easily be confused with *T. mallemontii* (see page 154) and *T. capillaris* (see page 104). It forms thick mounds and grows epiphytically on trees, cacti and on cliffs
CU Bright, sunny, little watering. A non-glamorous but easily grown species

Fig. 44 *Tillandsia recurvata* (L.) L (×$\frac{3}{4}$).

Tillandsia remota Wittm., 1891 (Ill. 125)
(*remota(us*) remote; refers to the remote, i.e. loosely arranged spikes)

Pl up to 40 cm high, in grass-like mounds
Fo upright, grass-like
LS triangular, 1 cm long, 1 cm wide, outside dark brown
LB up to 40 cm long, up to 5 mm wide above the sheath, narrowly lineal, developing into almost bristle-like tips, green, sometimes reddish in the upper part, grey-scaled
Sc slender, upright, shorter than the leaves
SB upright, similar to the leaves, with long, thread-like blade, which is longer than the stalk segments
I composed of 3 to 5 spikes, which have few flowers and are almost upright and widely separated, overall 4–6 cm long, not surpassing the leaves
PB similar to the scape bracts, almost upright, green, slightly scaled to al-most naked, the sheaths not longer than the 2–3 cm long, about 5 mm wide, short-stemmed, dense spikes
FB distichous, overlapping, upright, 1 cm long, 4 mm wide, keel at the tip, loosely grey-scaled, green, sometimes reddish-brown
Fl upright, up to 1·4 cm long, almost sessile, white, opened wide; the tips of the petals flared or rolled back
Se up to 8 mm long, lanceolate, poin-ted, green, naked, the posterior ones fused for about 4 mm high
St protruding from the flower
Ha Guatemala, El Salvador; epi-phytic. In its non-blooming stage can easily be confused with *T. festucoides* (see page 126)
CU Moderately damp, semi-shady. Interesting species reminiscent of grass.

Tillandsia schiedeana Steud., 1841 (Ill. 123)
(*schiedeana* after the collector Schiede, active in Mexico)
Syn. *T. vestita* Schlecht. et Cham., 1831, non Wild., 1830

Pl forming loose clumps, with simple or slightly branched, very brittle stems; including the inflorescence these are up to 40 cm high
Fo arranged spirally, with spread blade
LS almost round, the covered part naked, otherwise grey-scaled
LB narrowly lanceolate, long-tapered, with curled edges, heavily grey or brown-scaled
Sc upright, shorter than the leaves
SB the basal ones leaf-like, the upper ones with a long, upright, thin blade, green, pink or red
I simple-spiked, few flowers, up to 7 cm long, 8 mm wide
FB tightly overlapping, up to 3 cm long, longer than the sepals, naked to scaled, green, pink or red
Se lanceolate-pointed, up to 2 cm long, naked, the posterior ones fused up to 5 mm high
Pe yellow or reddish-yellow, forming a tube 4–7 cm long, from which the stamens protrude
Ha From Mexico to Venezuela and Colombia. Very variable in form; there are forms with large, stiff leaves and forms with short, thin leaves. *T. schiedeana* usually grows in mounds epiphytically in conifer and oak forests, but also on cliffs at altitudes up to 2,000 m.
CU Sunny, light. Easy to grow; easy to bloom. Can be grown along with cacti.

Tillandsia schreiteri Lillo et Castell., 1929 (Ill. 126)
(*schreiteri* after the German natural scientist R. Schreiter (1877 to 1942)

Pl stemless, up to 30 cm high
Fo numerous, forming a dense, flat rosette
LS about 8 cm long and 5·5 cm wide, brown, gradually merging into the blade
LB about 3·5 cm wide above the sheath (average width 2 cm), 35–45 cm long, narrowly lineal to lanceolate, long-tapered, heavily covered with close scales, dull green, pliant
Sc 20–25 cm long, pendulous
SB tightly overlapping, surrounding the scape and surpassing it with their very long blades, heavily scaled
I dense, head-shaped, composed of 14 to 18 spikes, 5–6 cm long, 4 cm wide
PB at the edge of the flower head longer than the spikes, toward the middle as long as or shorter than the spikes, which are flattened, 4 cm long, 2·5 cm wide, upright, very close and have 2 to 3 flowers
FB about 2·5 cm long and 8 mm wide, cutaneous, as long as the sepals, slightly scaled, keel
Se about 2 cm long, 6 mm wide, narrowly lanceolate, cutaneous, slightly winged, separate
Pe white, united into a tube, with flared tips
St enveloped in the flower
Ha Argentina, in the vicinity of Tucuman; epiphytic
CU Moderately damp, sunny.

Tillandsia seleriana Mez, 1903 (Ill. 30)
(*seleriana* named after Seler the discoverer of the plant)

Pl stemless, growing in groups, since the old specimens remain for a long time, up to 30 cm high including the inflorescence
Fo few, dense, upright
LS spoon-shaped, 4–8 cm long, up to 7 cm wide, both sides heavily brown-scaled, overall forming a hollow pseudo-bulb, the exterior leaf sheaths with a very short blade
LB gutter-shaped, since their edges are distinctly curled, up to 20 cm long, forming a long tip, heavily grey-scaled
Sc short, up to 6 mm thick
SB similar to the leaves, surrounding the scape with the sheaths, surpassing the inflorescence with their heavily grey-scaled, green or pink blades
I 7–10 cm long, 5–6 cm wide, composed of 3 to 7 spikes
PB the basal ones with long blades, the upper ones shorter, pink to green, grey-scaled
Sp spread or upright, distichous, 3–5 cm long, 1·5–3 cm wide, stem 5–8 mm long, flattened
FB tightly overlapping, 2–2·8 cm long, 10–11 cm wide, keeled on the back, green or pink, thick silver grey scales
Se about 1·5 cm long and 6 mm wide, the posterior ones fused 5–6 mm high, cutaneous, green or red, slightly scaled at the base, otherwise naked
Pe blue, united into a tube 3·5 cm long and surpassed by the stamens
Ha Mexico, Guatemala; epiphytic on oaks and conifers up to 2,400 m. Quite variable in form; inhabited by ants in its natural habitat
CU Bright and sunny, little watering. A very pretty species eagerly sought by fanciers. Grown advantageously on epiphyte trees.

Tillandsia setacea SW., 1797 (Ill. 130)

(*setacea*(*us*) bristly; because of the narrow, bristly leaves)
Syn. *Renealmia recurvata* L., 1753; *T. tenuifolia* sensu auctt. plur., non L., 1753

Pl stemless, 20–30 cm high including the inflorescence, usually forming dense turf
Fo numerous, bristly, upright, longer than the inflorescence
LS triangular, up to 1·2 cm wide, 1·2–5 cm long, brown-scaled
LB grass-like, 20–40 cm long, 4 mm wide above the sheath, short scales
Sc upright, slender, 15–18 cm long, 2–3 mm thick
SB similar to the leaves, with long, upright blades, longer than the stalk segments, tightly overlapping, in the upper part of the scape usually red with grey scales
I dense, composed of 3 to 6 spikes, up to 5 cm long, up to 2 cm wide
PB almost upright, red, grey-scaled, with their blades surpassing the almost upright, short-stemmed spikes, which have 3 to 5 flowers and are up to 3·5 cm long and 7–8 mm wide
FB distichous, tightly overlapping, upright, reddish or green, keeled, 1·5–1·8 cm long, 8–9 mm wide, scaled, later naked
Se about 1·2 cm long, naked, the posterior ones fused to about 4 mm high, keeled, green
Pe blue, united into a tube 2 cm long and surpassed by the stamens
Ha Florida, Mexico, Guatemala, El Salvador, Venezuela and Brazil; epiphytic in forests or in dry bush areas
CU Moderately damp, sunny. A small tillandsia worthy of cultivation.

Tillandsia somnians L. B. Smith, 1961 (Ill. 60)

(*somnians* asleep; perhaps because of the partially resting (sleeping) buds of the inflorescence axis)

Pl stemless, up to 2 m high in flower
Fo forming a rosette of 25–30 cm diameter, 20–25 cm long, pliant, slightly scaled
LS broadly oval, 3–4 cm long
LB tongue-shaped, long-tapered, 2–3 cm wide, the dried tips recurved, green to purplish-red
Sc slender, up to more than 1 m long, 6 mm thick, round, naked, upright or arched
SB tightly overlapping, upright, the lower ones similar to the leaves, 5–6 cm long, the upper ones surrounding the scape with their sheaths, upright, with a short blade, longer than the stalk segments, axis frequently with growth buds that grow into new plants
I loosely compound, 5 to 10 spikes, up to 20 cm long
PB similar to the upper scape bracts, attractive but considerably shorter than the spikes, which are arranged multifariously, flared, lineally lanceolate, pointed, over 8 cm long, 1·5 cm wide, sterile at the base, laterally flattened and have 12 dense flowers
FB tightly overlapping, longitudinally lanceolate, pointed, 1·5–2 cm long, as long as or somewhat longer than the sepals, prominently keeled, stiff, green

to reddish, naked
Fl almost sessile, reddish-violet
Se lanceolate-long, 1·2 cm long, the posterior ones fused 4 mm high
Ha Northern Peru (Ayabaca); epiphytic up to 2,400 m.

CU Moderately damp, semi-shady. Because of its generous vegetative reproduction, a very remarkable species, but suitable only for the larger greenhouses.

Tillandsia sphaerocephala Baker, 1888 (Ill. 127)
(*sphaerocephala(us)* with a ball-shaped head; because of the spherical, head-shaped inflorescence)

Pl stemless, up to 40 cm high
Fo in a dense, upright, frequently one-sided rosette
LS large, elliptical, brown on the inside, grey on the outside
LB lineal awl-shaped, pointed, the edges curled into a gutter, dense, short scales, 20–40 cm long, 2–3 cm wide above the sheath
Sc angling upward to arched
SB tightly overlapping, similar to the foliage, the outer ones with a broadly oval sheath, and a long, triangular blade, the inner ones oval, blunt, with a short tip, not longer than the spikes,

heavily scaled
I spherical, with dense, flat spikes of few flowers
FB oval, 2–3 cm long, keeled, brown-scaled to naked, longitudinal rills
Se oval, pointed, keeled, 2–2·5 cm long
Pe blue or lavender, blunt at the tip
St not surpassing the flower tube, with twisted filaments
Ha Bolivia and southern Peru; epiphytic on rocks at altitudes between 3,000 and 3,500 m
CU Bright, sunny. Very decorative at bloom time with the head-shaped inflorescence

Tillandsia spiculosa Griseb., 1865 (Ill. 302 and 303)
(*spiculosa(us)* spiked; because of the spiked portion of the inflorescence

Pl stemless, flowering to 80 cm high
Fo few forming an upright and somewhat bulbous funnel shaped rosette
LS large, oval, dark brown, the upper part with red-brown irregular blotches, about 11–13 cm long, 8–10 cm wide
LB tongue-shaped, 20–30 cm long, 2–5 cm wide, the apex rounded and spine tipped, fresh green, sometimes with dark red spots or dark lilac coloured cross bands (var. *ustulata*)
Sc upright, exceeding the leaves, round, naked
SB elliptical, spined tipped, or with

short blade, as long as the internodes, the upper ones somewhat shorter
I loose, simple or double compound, in two rows, 30–40 cm long, 20 cm wide
PB narrow-oval, spine tipped, somewhat longer than the stalked portion of the branching, green, naked
Sp horizontally extended to spreading, lineal, often arched, dense, to 24 flowered, about 9 cm long, 1 cm wide, stalked with a sterile basal flower bract
FB broad-oval, 6–9 mm seldom 5 mm long, as long as the sepals or a little longer, not keeled, smooth, mostly

naked
Se asymmetric, broad-elliptical, naked, 7 mm long, greenish brown
Pe very small, insignificant, hardly exceeding the sepals, white to yellowish
St and **Pi** enclosed within the flower
Ha widely distributed from Panama to Venezuela, Colombia to Peru, epiphytic, predominating in mountain forests or cloud forests from 1,800–2,800 m.
CU a 'green' tillandsia of the Pseudocatopsis-group, easy growing and free flowering species; as with all 'green' Tillandsias requires high relative humidity. Var. *ustulata* is quite decorative.

VARIETIES

A very variable species; the following differences have been noted:
var. *micrantha* (Baker) L. B. Smith, 1970. syn. *Tillandsia micrantha* Baker, 1887; *Tillandsia chinchicuana* Harms, 1929. Flower bracts 4–5 mm long; mostly less or twice as long as the rachis segments; spikes at fruiting loose.
var. *spiculosa* syn. *Tillandsia bittoniana* Baker, 1889; *Tillandsia palmana* Mez, 1901; *Tillandsia micrantha* Baker, 1902 non Baker 1887; *Tillandsia spiculosa* var. *palmana* (Mez) L. B. Smith, 1930. Flower bracts 6–9 mm long, mostly more or twice as long as the rachis segments; spikes densely flowered, few at the apex. Blades singly coloured or with lightly blocked sheaths; inflorescence usually doubled branched (Ill. 302 and 303).
var. *ustulata* (Reitz) L. B. Smith, 1970. syn. *Tillandsia triticea* Burchell, 1888; *Tillandsia parkeri* Baker, 1888; *Tillandsia viridis* Baker, 1889; *Vriesea luschnathii* Mez, 1894; *Tillandsia triticea* var. *ustulata* Reitz, 1962. As above, but blades with irregular, dark red-brown cross banding; inflorescence usually simple branched.

Tillandsia straminea H. B. K., 1816 (Ill. 134)
(*straminea(us)* straw yellow; because of the straw yellow primary bracts)
Syn. *T. scoparia* Willd., 1830; *Platystachys scoparia* Beer, 1857

Pl stemless to short-stemmed, up to 50 cm high
Fo rosette-shaped, up to 25 cm long, heavily grey-scaled
LS not clearly delineated
LB narrowly lanceolate, about 1·4 cm wide, long-tapered
Sc slender, upright
SB overlapping, narrow, scaled, with a thin, straw yellow blade
I compound, seldom simple
PB lanceolate, naked, straw yellow, shorter than the loose spikes of 6 to 8 flowers
FB flared, oval-pointed, 1·8 cm long, as long as the sepals, keel

Fl up to 2·2 cm long, white, with a blue edge, fragrant, enveloping the stamens
Ha Southern Ecuador to central Peru, from the coastal desert up to 2,500 m. Can easily be confused with *T. pur-*

purea (see page 169). from which it differs by its straw yellow and naked primary bracts
CU As for *T. purpurea* (see page 169).

Tillandsia streptocarpa Baker, 1887 (Ill. 131)
(*streptocarpa*(*us*) with twisted fruit; because of the twisted capsule)
Syn. *T. tricholepis* Baker, 1887, non Baker, 1878; *T. bakeriana* Britten, 1888

Pl stemless to short-stemmed, up to 60 cm high including the inflorescence
Fo spiral, spread to recurved
LS broadly oval, 2–3 cm long, up to 2·5 cm wide, naked on the inside, brown-scaled on the outside
LB lineal, about 1·5 cm wide, up to 25 cm long, with distinctly curled edges at the base, awl-shaped and often rolled toward the tip, heavily silver grey-scaled, felt-like
Sc upright, slender, naked 20–25 cm long, 3 mm thick
SB the lower ones similar to the leaves, the upper ones surrounding the scape with their sheaths, the upper ones with a reduced blade, shorter than the stalk segments
I simple or panicle-spiked

PB similar to the scape bracts, scaled, nerved, soon drying, up to 3·5 cm long, much shorter than the spikes, which are upright at the base, then arched, and have 5 to 10 flowers
FB shorter than the sepals, lanceolate-pointed, naked, 1·5–1·7 cm long
Se separate, 1·5 cm long, naked, no keel
Pe with a narrow, white, about 8 mm long claw and a more or less 1 cm long, blue plate
St enveloped by the perianth tube
Ha Brazil, Paraguay and southern Peru; epiphytic up to 1,400 m.
CU Sunny, bright, little watering. A very pretty species that varies greatly in regard to size. There are very small and very sizeable specimens.

Tillandsia streptophylla Scheidw., 1836 (Ill. 128)
(*streptophylla*(*us*) with twisted leaves; because of the spirally twisted leaf blades)
Syn. *Vriesea streptophylla* Morren, 1873; *T. tortilis* Brongn., 1873
Quite variable in regard to the size of the pseudo-bulb, the length of the leaves and the rolling of the leaves.

Pl stemless, up to 45 cm high
Fo numerous
LS large, spoon-shaped, up to 10 cm long, both sides heavily brown-scaled, forming a large, hollow pseudo-bulb
LB lineally strap-shaped, long-

tapered, up to 5 cm wide above the sheath, up to 40 cm long, heavily covered with coarse, separated, grey scales, in the habitat spirally rolled (this trait is lost in cultivation, usually only the tips are rolled then)

Sc upright, 7–8 mm thick

SB similar to the leaves, longer than the stalk segments, surrounding the scape with the sheaths, the blades recurved and more or less rolled up, heavily scaled, with strong sunlight often red

I loose, compound with 8 to 14 upright or horizontal spikes, overall up to 20 cm long and 15 cm wide

PB similar to the scape bracts, the lower ones longer than the spikes, the upper ones shorter, green or red, heavily grey-scaled

Sp distichous, up to 10 cm long, 1·5 cm wide, up to 1·5 cm long stem

FB overlapping, about 2 cm long and 9 mm wide, green or red, heavily grey-scaled

Se about 2 cm long and 5 mm wide, lanceolate, elliptical, cutaneous, naked, separate, the posterior ones fused 5 mm high

Pe blue, united into a tube 4 cm long and surpassed by the stamens

Ha Mexico, Guatemala, Honduras and Jamaica; in Mexico from the coast up to 1,400 m.

CU Bright and sunny; moderate watering. With too high himidity the spiral rolling of the leaves disappears. A decorative species quite suited for planting on epiphyte trees and very much coveted by fanciers.

Tillandsia stricta Soland., 1813 (Plates 44 and 45)

(*stricta(us)* rigid, stiff; because of the stiff leaves)

Syn. *Anoplophytum strictum* var. *krameri* Andŕe 1888; *T. krameri* Baker, 1889; *T. meridionalis* Baker, 1889; *T. stricta* var. *krameri* Mez, 1894

Pl stemless to short-stemmed, up to 20 cm high, usually growing in thick clumps

Fo numerous, forming a rigid or pliant, flared to almost upright, frequently one-sided rosette

LS not distinct, gradually merging into the blade, 1–2 cm long, 1 cm wide, white, heavily grey-scaled

LB 6–18 cm long, 5–10 mm wide at the base, short to long-tapered, green, covered with silver grey scales on both sides

Sc usually pendulous, slender, 5–8 cm long, 2–3 mm thick

SB tightly overlapping, similar to the leaves, much longer than the stalk segments, upright, green, heavily grey-scaled

I pendulous, simple-spiked, loose, 4–6 cm long, 2–3 cm wide; flowers multifarious

FB the lower ones similar to the upper scape bracts, the upper ones broadly oval and pointed, about 2·5 cm long, 2 cm wide, scaled in the upper part, bright carmine red

Se shorter than the flower bracts, 1·2 cm long, 4 mm wide, not scaled, fused equally high about 1–2 mm at the base

Pe blue, up to 2 cm long, with somewhat flared tips

St enveloped in the flower

Ha Venezuela, Brazil, Trinidad, Paraguay, Guayana and Argentina; epiphytic on trees. In its non-blooming stage *T. stricta* could be confused with *T. pohliana* (see page 164) or *T. tenuifolia* (see page 181). Quite variable in regard to length, form and consistency of the leaves (pliant or hard, green or white-scaled)

CU Moderately damp, semi-shady.

Easy to grow, easy to bloom. At bloom time one of the most beautiful tillandsias, especially when the great masses are covered with glowing red inflorescences.

Tillandsia subconcolor L. B. Smith, 1970 (Ill. 304 and 305)
(*subconcolor* self coloured; because of the leaves being almost similar coloured on both sides)

Pl stemless, flowering to 1·20 high
Fo forming a loose, nearly upright rosette from about 1 m in diameter
LS longish-oval, 9–10 cm long, 6·5–7·5 cm wide, both sides densely brown scaled, distinctly formed
LB narrow-lanceolate, long pointed, 50–60 cm long, 5 cm wide at the base, 3·5 cm wide half way along; the margins slightly upward curved, dark green, both sides densely grey appressed scaled
Sc sturdy, upright, 70–80 cm long, 7 mm thick, round, green at the base, the upper part reddish green
SB the basal ones similar to the leaves, the upper ones enclosing the scape and with a more or less longer blade, upright, longer than the internodes, green to reddish green, densely grey scaled
I simple, but frequently with loose compound to 4–5 outspread to nearly upright spikes, 20–30 cm long and to 15 cm wide
PB similar to the upper inflorescence bracts, 5–6 cm long, much shorter than the sword-shaped, 15–20 cm long, 2–2·5 cm wide short stalked spikes
FB densely overlapping, 3 cm long, 1·7 cm wide, shiny green, hardly scaled, not keeled, weakly nerved
Se almost as long as the flower bracts, free, 2·7 cm long, 5 mm wide, thin cutaneous, pale green, the back ones fused for 1 cm of their height
Pe narrow-lineal, lanceolate, pointed, 3·5 cm long, 4 mm wide, white, violet upper part
St and **Pi** 2·2 cm long, enclosed within the flower
Ha South Peru, near Quillabamba, epiphytic in dry forests, 1,500 m.

Tillandsia subulifera Mez, 1919 (Ill. 121, Fig. 45)
(*subulifera(us)* bearing an awl; refers to the awl-like leaves)

Pl stemless, 15–20 cm high including the inflorescence
Fo not numerous, upright, forming a slender, cylindrical pseudo-bulb; the inner leaves up to 18 cm long, the outer ones scoop-shaped
LS 5–6 cm long, 1·5–2·5 cm wide, inside scaled-brownish
LB thick, almost succulent, gutter-shaped, narrowly awl-shaped to lineal, twisted near the tip, white-scaled, green or with white cross bands, 5 mm wide above the sheath, up to 7 cm long
Sc upright, slender, usually covered by the leaves, 8–9 cm long, 3–4 mm thick, scaled, round
SB similar to the leaves, upright, tightly overlapping, much longer than the stalk segments, surrounding the scape with the sheaths, scaled

I simple-spiked, upright, longitudinally lineal, 5–7 cm long, loose, 4 to 6 flowers
FB loose, distichous, oval, broadly pointed, up to 2·5 cm long, short-scaled, nerved, keeled near the tip
Fl upright, short-stemmed, reddish-brown with yellow tips, tubular, up to 3·2 cm long
Se oval, narrowly blunt, 2·3 cm long, separate, short-scaled, nerved
St as long as the flower tube
Ha Panama, Trinidad and Ecuador; epiphytic on trees
CU Moderately damp, bright. Recommended for planting on epiphyte trees. Small.

Fig. 45 *Tillandsia subulifera* Mez (life-size).

Tillandsia tectorum Morren, 1877 (Ill. 18, 129)

(*tectorum* growing on roofs; here probably rocks are meant, since *T. tectorum* grows only on rocks)
Syn. *T. argentea* C. Koch, 1867, non Griseb., 1866; *T. saxicola* Mez, 1906

Pl short or long-stemmed, up to 50 cm long
Fo numerous, spiral, separated, upright, to recurved
LS triangularly oval, up to 2 cm wide, 2·5 cm long

LB narrowly lineal, thread-like pointed, stiff, 1–1·3 cm wide above the sheath, up to 20 cm long, heavily covered in silvery white, feathery scales
Sc slender, upright, much longer than the leaves

SB similar to the leaves, overlapping, longer than the stalk segments, the tips almost upright, thread-like, feathery-scaled, the sheaths heavily silver-scaled, reddish

I composed of 5 to 10 close, upright to spreading spikes 5–6 cm long and with 5 to 10 flowers

FB distichous, overlapping, lanceolate, pointed, keeled, 1·2–1·3 cm long, scaled, soon naked, red, green or pink

Se 1 cm long, lanceolate, naked, separate, thin-skinned

Pe 1·8–2·1 cm long, united into a tube, the tips somewhat spread, pale blue, white in the upper third

St somewhat shorter than the flower tube

Ha Northern and central Peru, in the cactus area at 800 m, forming extensive stands among the boulders (Ill. 18). In northern Peru (Huanuco, Chachapoyas) striking, small forms were found by Rauh growing epiphytically on cliffs

CU Very bright and sunny, little watering. Can be grown among cacti hanging on wires with no planting medium. Because of its silvery white colour, a very pretty species worthy of cultivation.

Tillandsia tenuifolia L., 1753 (Plate 46)

(*tenuifolia*(*us*) fine-leafed; because of the slim leaves)
Syn. *T. pulchella* Hook., 1825

Pl variable, many forms, usually growing in dense turfs, stemmed, up to 25 cm long

Fo spiral, spreading or almost upright, when hanging it is turned to one side, very stiff

LS in the long-leafed forms narrow, not distinct, gradually merging into the blade, in the short-leafed forms broadly triangular

LB 4–13 cm long, narrowly lineal to awl-shaped, 2–7 mm wide above the sheath, long-tapered, short grey scales, with curled edges

Sc upright to angling upward, slender, naked, shorter or longer than the leaves

SB longer than the internodes, surrounding the scape with the oval sheaths, merging into a short, narrow, upright blade, grey-scaled, green or pink to red, the upper scape bracts bladeless, red and naked

I simple-spiked, with flowers arranged in a spiral

FB oval, short-tipped, longer than the sepals, slightly scaled to naked, pink to red

Se lanceolate, pointed, naked, 1 cm long, the posterior ones fused about 7 mm high

Pe about 2 cm long, blue or white, with flared tips

St not longer than the flower tube, folded

Ha Venezuela, Colombia to northern Argentina and in the Antilles; epiphytic or on rocks. In its non-blooming stage it can be confused with *T. bergeri* (see page 96). *T. aeranthos* (see page 86) and *T. araujei* (see page 90)

CU Sunny, bright, moderately damp. A robust species especially recommended to the beginner.

Tillandsia tetrantha Ruiz et Pav., 1802 (Ill. 132, 133)
(*tetrantha(us)* four-flowered; but the spikes have more than four flowers)

Pl stemless, 40–80 cm long, variable in size

Fo often forming a bulb-like funnel rosette

LS broadly elliptical to oval, about 8 cm long, up to 9 cm wide, dark brown

LB narrowly lanceolate, short or long-tipped, green, heavily grey-scaled, often spotted reddish

Sc usually pendulous, heavily grey-scaled

SB similar to the leaves

I usually pendulous, loosely compound, spikes arranged in a spiral

PB broadly oval, bowl-shaped, with more or less long tip, usually flared, heavily grey-scaled, longer or shorter than the spikes

FB much shorter than the sepals, broadly oval, heavily grey-scaled

Se asymmetrical, about 1 cm long, 7 mm wide, grey-scaled

Pe 1·4 cm long, yellow or orange with flared tips

St enveloped in the flower

Ha The varieties *tetrantha* and *aurantiaca* which we collected come from rather high altitudes. They grow mostly on *Polylepis* trees near the edge of the forest (2,800–3,500 m) and are somewhat difficult to grow in cultivation

CU Dry, sunny, but cool; suitable for epiphyte trees. Very decorative at bloom time.

VARIETIES

T. tetrantha var. *tetrantha* (Ill. 132)
Habitat Northern Peru
Floral bracts small or lacking; inflorescence loose, pendulous; spikes loose

T. tetrantha var. *scarlatina* (André) L. B. Smith, 1930
Syn. *T. aurantiaca* var. *scarlatina* André, 1888; *T. fulgens* var. *scarlatina* Mez, 1935
Habitat Ecuador, Colombia
Floral bracts about half as long as the sepals. Primary bracts bright red or rarely pale greenish yellow, longer than the spikes. Leaves leathery, stiff; inforescence elongated, loose.

T. tetrantha var. *miniata* (André) L. B. Smith, 1930
Syn. *T. aurantiaca* var. *densiflora* André, 1888; *T. fulgens* var. *densiflora* 1935
Habitat Ecuador, Colombia
Floral bracts about half as long as the sepals. Primary bracts bright red or rarely pale greenish yellow, longer than the spikes. Leaves pliant, soft; inflorescence usually small, dense.

T. tetrantha var. *aurantiaca* (Griseb.) L. B. Smith, 1930; Ill. 133

Syn. *Tussacia fulgens* Kl. 1857; *T. aurantiaca* Griseb., 1865; *Catopsis garckeana* Wittm., 1899; *T. fulgens* Mez, 1935
Habitat Venezuela, Ecuador, Colombia, northern Peru (Ayabaca)
Floral bracts about half as long as sepals. Primary bracts orange to dark brown, often shorter than the spikes. Inflorescence elongated, with loose spikes; scape bracts usually shorter than the stalk segments; sepals and petals orange.

T. tetrantha var. *densiflora* (André) L. B. Smith, 1930
Syn. *T. aurantiaca* var. *densiflora* André, 1888; *T. fulgens* var. *densiflora* Mez, 1935
Habitat Ecuador, Colombia
Floral bracts about half as long as sepals. Primary bracts orange to dark brown, often shorter than the spikes. Inflorescence short, dense; scape bracts overlapping.

Tillandsia tricholepis Baker, 1878 (Ill. 27b, c, 135)
(*tricholepis* with hairy scales)
Syn. *T. bryoides* Griseb., 1879 p. p.; *T. polytrichoides* Morren, 1880

Pl up to 20 cm long, stemmed, reminiscent of the large star moss *Polytrichum*, growing in clumps and strands
Fo spiral, scoop-like, upright or spread
LS cutaneous, surrounding the stem, 6 mm long and 4 mm wide
LB short, needle or awl-shaped, succulent, up to 1 cm long, green, heavily grey-scaled, 2 mm wide above the sheath, average width 1 mm
Sc thin, upright, up to 6 cm long, about 0·5 mm thick, naked
SB loose, upright, lying against the scape, as long as or shorter than the stalk segments, heavily green-scaled
I simple-spiked, 2 to 10 flowers, up to 3 cm long

FB loosely overlapping, about 5 mm long and 3 mm wide, oval, pointed, green, heavily grey-scaled, shorter than the sepals, which are up to 6 mm long, green, naked, lanceolate and long-tapered
Pe yellow green, up to 7 mm long, with flared tips
St enveloped in the flower, 2·5 mm long
Ha Brazil, Bolivia, Paraguay and Argentina; epiphytic on trees or on rocks
CU Moderately damp, bright. Easy to grow, small, especially suited for flower windows or terrariums. An interesting species reminiscent of large moss, but quite variable.

Tillandsia tricolor Schlecht. et Cham., 1831 (Plate 47)
(*tricolor* of three colours; because of the three colours in the inflorescence)

Syn. *Vriesea xiphostachys* Hook., 1861; *Platystachys complanata* Morren, 1872

Pl stemless, 30–40 cm high including the inforescence, sometimes forming short rhizomes, usually growing in clumps
Fo numerous, in a dense rosette, the exterior leaves recurved, green, in strong sunlight red
LS broadly oval, 3–4 cm long, 2·5–3·5 cm wide, both sides dark brown
LB lineal, long, thread-like tips, 1–1·5 cm wide above the sheath, 30–35 cm long, the edges curled into a gutter, green, few short scales
Sc perpendicular, slender, not much longer than the leaves
SB the lower ones similar to the leaves, the upper ones elliptical, long-tapered, in intensive sunlight bright red
I simple-spiked or compound
Sp lineally lanceolate, pointed, laterally flattened, 6–18 cm long, 1·8–2·5 cm wide
FB tightly overlapping, oval, pointed, 2–3 cm long, 1·4–1·8 cm wide, keel, slightly scaled or naked, longer than the sepals; in strong light the basal ones are bright red, the middle ones and upper ones straw yellow to green
Fl 3–3·5 cm long, blue
Se lanceolate, 2 cm long, 3–4 mm wide, separate, the posterior ones fused for about 1 cm high
St protruding from the tube
A variable species with many forms
Ha Mexico, Guatemala, Nicaragua, Costa Rica; epiphytic on trees in cloud forests at altitudes of 1,400–2,300 m.
CU Moderately damp. In the sun the scape bracts colour up bright red. A very pretty species suitable for epiphyte trees and flower windows.

Tillandsia triglochinoides Presl, 1827 (Ill. 88)
(*triglochinoides* the growth pattern is reminiscent of the arrow grass *Triglochin* (Fam. *Scheuchaeriaceae*))
Syn. *T. hartwegiana* Brongn., 1889

Pl stemless, up to 30 cm high including the inflorescence
Fo few, in a loose rosette
LS not distinct, about 1·5 cm long and 1·7 cm wide, heavily silver grey-scaled
LB lineal, long-tapered, both sides heavily silver grey-scaled, up to 25 cm long, 1·5 cm wide above the sheath (average width about 7 mm)
Sc slender, upright, shorter than the leaves, 2 mm thick
SB narrow, oval, the upper ones overlapping, lanceolate-pointed, surrounding the scape with the sheath, scaled at the tip
I simple-spiked, loose, narrowly lanceolate or lineal, pointed, laterally flattened, up to 13 cm long, up to 8 mm wide, with about 20 flowers or more; rhachis slightly zigzag
FB upright to almost upright, triangularly oval, not keeled, up to 1·9 cm long, up to 8 mm wide, naked, longitudinally ribbed
Fl upright to almost upright; the individual flowers spread only at the moment of ripeness, white, about 2 cm long; petals with flared plates, enveloping the stamens
Se lanceolate-oval, pointed, up to

12 mm long
Ha Southern Ecuador; epiphytic on trees or on rocks
CU Moderately damp, semi-shady. A pretty, small tillandsia, which can easily be confused with *T. narthecioides* (see page 159).

Tillandsia triticea Burchell, 1888 (Ill. 137)
(*triticea*(*us*) wheat-like, from *Triticum*, wheat; because of the segmented inflorescences, which look like wheat)

Pl stemless, 40–90 cm high including the inflorescence
Fo 20–30 cm long, close, dot-like scales, forming a small, narrow funnel rosette
LS broadly oval to elliptical, brown
LB tongue-shaped, the tip rounded and equipped with a thorn, 3–4 cm wide, with irregular, dark spots
Sc upright, naked
SB overlapping, narrowly oval, thorn-tipped, heavily scaled
I loosely compound, 20–30 cm long, naked
PB similar to the scape bracts, but much shorter and surrounding the base of the spikes

Sp flared, lineal, 1 cm wide, often with a long stem, with some sterile bracts at the base, 14 to 36 flowers
FB broadly oval, pointed, 7 mm long, as long as the sepals, not keeled, stiff, tightly overlapping
Se inverse egg-shaped, stiff
Pe yellowish, insignificant
St enveloped in the flower
Ha Colombia, Trinidad, Guinea, Brazil, Peru, Bolivia, in damp rain forests (in Peru up to 1,000 m)
CU Damp, warm, shady. Since the plant remains small it is suitable for terrariums. A very interesting species, which resembles a *Vriesea* more than a *Tillandsia* in its foliage.

Tillandsia unca Griseb., 1874 (Plate 48)
(*unca*(*us*) hooked, crooked; because of the somewhat crooked leaves)
Syn. *T. argentina* Wight, 1907

Pl short-stemmed, growing in thick mounds, up to 10 cm high including inflorescence
Fo spiral, close, almost upright, crooked or spread, dull green
LS cutaneous, surrounding the stem, up to 1·3 cm long and up to 1·4 cm wide
LB lanceolate awl-shaped, succulent, gutter-like, up to 7 cm long, 7 mm wide above the sheath, heavily grey-scaled
Sc short, slender, upright, naked, about 2 mm thick, usually covered by the leaves

SB longer than the stalk segments, cutaneous, surrounding the scape, green, slightly scaled to naked
I simple-spiked, 3–4 cm long, 1 to 6 flowers
FB distichous, overlapping, 1·8–2 cm long, 6–7 mm wide, narrowly lanceolate, long-tapered, cutaneous, naked
Se separate, thin-skinned, 1·4 cm long, narrowly triangular, lanceolate, long-tapered, naked
Pe bright carmine red, up to 2·9 cm

long with flared plates
St enveloped in the tube, 1·6 cm long
Ha Argentina (Cordoba, Tucuman);

terrestrial or growing on rocks
CU Dry, sunny. An interesting, small, hard tillandsia.

Tillandsia usneoides (L.) L., 1762 (Ill. 4, Fig. 4–6)

(*usneoides* Usnea-like; in its growth habit reminiscent of goat's beard *Usnea*)
Syn. *Renealmia usneoides* L., 1753; *Dendropogon usneoides* Raf., 1838; *Strepsia usneoides* Steud., 1841

Pl long, hanging, rootless strands up to 8 m long; elongated stems, about 1 mm thick
Fo about 5 cm long, thin, with cylindrical, silver grey blade
Sc lacking
SB lacking
FB oval with short tip, scaled, 6–7 mm long, thin-skinned, shorter than the 7 mm long, thin-skinned, naked, separate petals
Fl single, yellow green; petals 9–11 mm long, with flared tips
St enveloped in the flower, 5 mm long
Ha From the southern part of the U.S.A. over Central America to Argentina and Chile; epiphytic on trees, cacti, on cliffs, telegraph wires, hanging in long, rootless strands
CU Light, bright and sunny. Serves to vary the collection. A species of many forms: the var. *minor* is distinguished by thin, delicate growth, the var. *major* by stiff and thicker growth.

Tillandsia utriculata L., 1753 (Ill. 8, 136, Fig. 31b)

(*utriculata*(*us*) hose-shaped)

Pl stemless, 50 cm to 2 m high including inflorescence
Fo numerous, forming a large, spread rosette
LS large, broadly oval, inside violet brown, outside light brown
LB narrowly triangular, 2–7 cm wide over the sheath, 0·4–1 m long, short scales or naked
Sc upright, strong, usually not longer than the leaves, naked
SB the basal ones similar to the leaves, the upper ones surrounding the scape with their sheaths, longer than the stalk segments, ending in a narrow, flared blade, naked or short-scaled
I seldom simple, usually 2 or 3-fold
compound, many flowers, naked, upright; rhachis red or green, not zigzag-shaped but only contorted
PB much shorter than the sterile sections of the lateral branches, up to 4 cm long, often violet reddish
Sp up to 35 cm long, up to 14 flowers, loose, distichous
FB shorter than the sepals, up to 1·6 cm long, 11–12 mm wide, naked, green or red-edged, blunt, no keel
Se 1·4–1·8 cm long, 6–7 mm wide, blunt, separate, the posterior ones slightly fused, green, naked
Pe yellowish-white, united into a tube 3–4 cm long, surpassed by the stamens
Ha From the southern part of the

U.S.A. over Central America to Venezuela, from the coast up to 1,200 m (Mexico)
CU Bright, sunny, moderately damp.

Decorative at bloom time, but recommended only to owners of rather large greenhouses. Can easily be confused with *T. dasyliriifolia* (see page 114).

Tillandsia valenzuelana A. Rich., 1850 (Ill. 138)
Syn. *T. laxa* Griseb., 1864; *T. kunthiana* sensu Griseb., 1864, non Gaud., 1846; *T. brachypoda* Baker, 1887; *T. sublaxa* Baker, 1887; *T. polystachya* var. *alba* Wittm., 1889; *T. purpusii* Mez, 1916 p.p.; *T. domingensis* Mez, 1919

Pl 20–60 cm high including the inflorescence, stemless
Fo numerous, forming a rosette of 30–40 cm diameter
LS large, oval
LB lineal, long-tapered, usually flat, 1–2·5 cm wide, grey-scaled
Sc upright or angling upward, naked
SB similar to the leaves, the lower ones with a lineal blade, pink or red, oval, somewhat puffed, overlapping
I simple or composed of few spikes
PB similar to the upper scape bracts, their sheaths shorter than the spikes, but the blades, at least in the lower part of the inflorescence surpassing the flared, pointed, elongated, flattened spikes, which have 6 to 17 flowers and are 5–20 cm long, 1–2 cm wide; terminal spikes at the base with sterile bracts
FB upright or almost upright, elliptically long, blunt or thorn-tipped, 2 cm long, much longer than the sepals, almost naked, pink or red, sometimes with a keel toward the tip
Se longitudinally blunt, posterior ones somewhat fused
Pe lineal, 3 cm long, lavender or violet
St protruding from the flower
Ha Southern Florida, southern Mexico, Central America, Colombia, Venezuela, Bolivia
CU Bright, sunny. A recommended species.

Tillandsia vernicosa Baker, 1887 (Ill. 139)
(*vernicosa*(*us*) varnished, shiny; because of the shiny red floral bracts)
Syn. *T. drepanophylla* Baker, 1889; *T. polyphylla* Baker, 1889

Pl stemless to short-stemmed, up to 35 cm high including the inflorescence
Fo forming an almost upright to spread rosette
LS indistinct, 2–3 cm long, 2–2·5 cm wide
LB narrowly lineal-triangular, awl-like pointed, with edges curled into a gutter, hard, dull green, 1·5–2 cm wide above the sheath (average width 7 mm), up to 35 cm long, dense, short, grey scales
Sc upright, slender, 15–25 cm long, 4–5 mm thick
SB dense, overlapping, surrounding the scape with the sheaths, longer than the stalk segments, the blades briefly upright, green or reddish
I 10–15 cm long, with 5 to 10 dense, almost upright or arched, flared spikes
PB similar to the scape bracts, reddish, but much shorter than the lanceolate,

pointed spikes, which are up to 10 cm long, 1 cm wide and have up to 20 flowers
FB overlapping, distichous, 1·5–1·8 cm long, up to 1 cm wide, no keel slightly scaled, usually naked, shiny red, almost as long as the sepals, which are separate, up to 1·4 cm long, oval, rounded, hard and naked
Fl white; petals about 2·4 cm long, tips recurved
St enveloped in the upper third of the tube
Ha Bolivia, Argentina and Paraguay; epiphytic on trees in dry areas. In its non-blooming stage it can be confused with *T. didisticha* (see page 119)
CU Bright, sunny, little watering. Very hardy and therefore recommended to the beginner.

Tillandsia vicentina Standl., 1923 (Ill. 140)
(*vicentina* named after the volcano de San Vicente on San Salvador)

Pl up to 30 cm high
Fo in a dense rosette
LS 1·5–2 cm wide, oval, dark brown
LB narrowly triangular, 6–8 mm wide, with small, short scales at the tip, with coarse scales at the base, up to 35 cm long, longer than the inflorescence
Sc upright, 4 mm thick, scaled
SB tightly overlapping, longer than the stalk segments, with an upright, thread-like, scaled blade
I compound, with 3 to 14 flared spikes
PB similar to the scape bracts, their sheaths, however, only one third as long as the spikes, which are lanceolate-long, pointed, flattened, 4–7 cm long, 1–1·5 cm wide and have 6 to 8 flowers
FB distichous, tightly overlapping, oval, pointed, 2–2·5 cm long, 1–1·4 cm wide, keeled, scaled, longer than the lanceolate, pointed sepals, which are 2 cm long and whose posterior parts are fused 6 mm high
Fl blue; petals united into a 4 (5) cm long tube, which is surpassed by the stamens and the pistil
Ha Guatemala and El Salvador; epiphytic on conifers and oaks, as well as on boulders at altitudes of 1,500–2,700 m.
CU Bright, sunny. A pretty, small, recommended species.

Tillandsia violacea (Beer) Baker, 1887 (Ill. 143)
(*violacea*(*us*) violet; presumably refers to the violet flowers)
Syn. *Platystachys violacea* Beer, 1857; *Tillandsia foliosa* Hemsl., 1882
(The description has been prepared from the plant depicted in Ill. 143 and identified by L. B. Smith.)

Pl up to 30 cm high
Fo forming a compact funnel rosette, up to 40 cm long
LS broadly oval, 10–11 cm long, 8–9 cm wide at the base, dark grey brown to almost black violet
LB narrowly lanceolate, pointed, up to 4 cm wide above the sheath, short, grey scales
Sc pendulous, up to 20 cm long, waxy-frosted
SB similar to the leaves, green to red,

with lanceolate, long-tapered blade
I pendulous, loosely compound with 8 to 12 flared spikes
PB red, spread, similar to the scape bracts; the basal ones surpassing the spikes with their blades, the upper ones shorter than the spikes
Sp flared, short-stemmed, closely distichous, flat, 4 to 8 flowers, up to 8 cm long and 2 cm wide
FB tightly overlapping, upright, up to 3 cm long, 1·5 cm wide, rounded on the back, keeled only at the tip, stiff, naked, green, red-edged and waxy-frosted

Se 2·5 cm long, 1 cm wide, thin-skinned, the anterior ones separate, the posterior ones fused about 5 mm high
Pe inclined into a dark blue violet tube up to 4·5 cm long
St protruding far beyond the perianth tube
Ha Mexico, near Oaxaca in the valley of the Rio Y and near Perote; epiphytic on oaks and conifers at altitudes from 2,000–2,400 m.
CU Cool, moderately damp, sunny. A very decorative species at bloom time, whose pendulous inflorescence is reminiscent of *T. prodigiosa*.

Tillandsia viridiflora (Beer) Baker, 1888 (Ill. 213)
(*viridiflora(us)* green flowered; because of the green flowers)
Syn. *Platystachys viridiflora* Beer, 1857; *T. orizabensis* Baker, 1857.
This species was at one time placed under *T. grandis* Schlecht (see page 135) but recently it has been detached from *T. grandis* and is now a separate species.

Pl stemless, flowering to 1 m high
Fo numerous, forming a funnel shaped rosette from 50–60 cm in diameter
LS longish-oval, 13–15 cm long, 8–9 cm wide, dull green, grey scaled
LB tongue-shaped, long pointed, about 40 cm long, 3–4 cm wide, almost bare, upper side dark green, underside lilac-violet
Sc upright, sturdy, about 40 cm long, 1 cm thick, green, naked
SB the basal ones similar to the leaves with longer recurved blade; the upper ones with shorter blade, green to lilac, densely overlapping, much longer than

the internodes
I simple, loose, in two rows; to 14 flowered, to 25 cm long and 6 cm wide
FB loose upright-angled, firm, not keeled, broad-oval, rounded at the tip, to 3·5 cm long, green
Fl nearly upright, about 1 cm long stalked
Se elliptical, to 3·9 cm long, bare, firm, green, much fused at the base
Pe to 8 cm long, 1 cm wide, tongue shaped, flared or recurved, green
St and **Pi** about 10 cm long, much longer than the petals, at first upright, pendulous after flowering
Ha Mexico.

Tillandsia wagneriana L. B. Smith, 1963 (Plate 10)
(*wagneriana* named after R. Wagner an animal and fish collector in Peru)

Pl stemless, 40 cm high including the inflorescence

Fo up to 45 cm long, forming a rosette of 30–40 cm diameter, crisp green on the upper side, dull green on the lower side, often reddish

LS long-oval, 8–10 cm long, 6 cm wide, brown-scaled

LB strap-shaped, pointed, 3–4 cm wide at the base, flat; leaf edges partially wavy upper side naked, lower side with short, pale scales

Sc slender, upright, naked, 4–5 mm in diameter, 15–20 cm long

SB broadly oval, long-tapered, upright, longer than the stalk segments, surrounding the scape, but not completely covering, green or pinkish-lavender

I open compound, 4 to 13 flared spikes, 16 cm long, 10–15 cm wide

PB similar to the upper scape bracts, hardly longer than the lower, sterile part of the spikes, naked, delicate pink lavender

Sp lineal-lanceolate, pointed, up to 10 cm long, 2 cm wide, 8 to 15 flowers

FB tightly overlapping, distichous, oval, pointed, up to 2·5 cm long, somewhat longer than the sepals, keeled, thin, naked, pink lavender; their tips drying to brown even before the unfolding of the flowers

Se long-lanceolate, pointed, about 2 cm long, greenish with lavender tips, a slight keel

Fl deep dark blue; petals with a claw 2 cm long and 3 mm wide and a plate spread to recurved, 9 mm wide, 1 cm long

St deeply enveloped in the flower; filaments flattened into a ribbon, narrowed toward the tip; anthers 2 mm long, olive green

Pi very short, with large papillary knobs

Ha Peru, Amazon area (Nazareth, Iquitos). *T. wagneriana* is related to *T. hamaleana* (see page 137) and *T. dyeriana* (see page 121).

CU Damp, shady. Because of its delicate lavender-coloured floral bracts at bloom time, it is one of the most splendidly colourful species of Peru.

Tillandsia walteri Mez, 1906 (Ill. 141)

Pl stemless, 80–100 cm high including the inflorescence, similar to a *Catopsis* (see pages 211 and 212) in shape

Fo numerous, a narrow, upright funnel rosette with a diameter of 15–20 cm

LS indistinct, 15 cm long, 8–9 cm wide, bluish-violet on the upper side

LB 25–30 cm long, 7–8 cm wide at the base, lanceolate long-tapered, upright, pliant, green, short-scaled, frosted chalk white

Sc strong, upright, up to 70 cm long, shiny green

SB the lower ones similar to the leaves, green, the upper ones surrounding the scape with their sheaths, somewhat puffed, greenish-red, with a short, upright, green tip, which is almost completely reduced in the upper scape bracts, longer than the stalk segments.

I simple-spiked, 20–25 cm long, 4–5 cm wide

FB up to 5·5 cm long, about 4 cm wide, broadly oval, pointed, shiny reddish-green to raspberry red, short-scaled

Se as long as the floral bracts, about 4·5 cm long and 2 cm wide, cutaneous, greenish to transparent, green-nerved

Pe inclined into a tube, about 5·5 cm long and 1·2 cm wide, the lower $\frac{2}{3}$ white, the upper $\frac{1}{3}$ pink with slightly fluted edges

St about 5·5 cm long, white, anthers 9 mm long
Ha From central to southern Peru. Near Tarma *T. walteri* grows in rather large stands along with *T. cauligera*
CU Bright, sunny, little watering. A very decorative species at bloom time with the raspberry red bracts of the spiked inflorescences. With its mealy-frosted leaves it resembles a *Catopsis* more than a *Tillandsia*.

Tillandsia weberbaueri Mez, 1905 (Ill. 142)

(*weberbaueri* named after the German botanist A. Weberbauer, active in Peru)

Pl stemless, 30–50 cm high including the inflorescence
Fo 20–30 cm long, short, grey-scaled; scales have a brown centre
LS broadly elliptical, spoon-shaped, overall forming a hollow pseudo-bulb up to 7 cm in diameter, upper side lavender brown
LB lineal, 0·5 cm wide, long-tapered, with curled, wavy edges
Sc short, slender, perpendicular, naked to scaled
SB narrowly lanceolate, short or long-tapered, grey-scaled, remote, longer than the stalk segments
I open panicle, branched, 12–25 cm long, 8–15 cm wide
PB similar to the scape bracts, surrounding the base of the panicle branch
Sp stemmed, 2·5–4·5 cm long and up to 1 cm wide, open, 6 to 14 flowers
FB distichous, oval, pointed, slightly keeled, 2–3 mm long, smooth or ribbed, pale-scaled
Se surpassing the floral bracts, 3–4 mm long, asymmetrical, inverse egg-shaped, scaled, stiff, green
Pe slightly longer than the sepals, narrow, greenish-white
St enveloped in the flower
Ha From northern to southern Peru; epiphytic at altitudes between 900 and 1,700 m.
CU An interesting, small tillandsia, whose culture presents the same difficulties as with all South American species with pseudo-bulbs.

VARIETIES

T. weberbaueri is related to *T. pallidoflavens* Mez, *T. parviflora* Ruiz et Pav. and *T. commixta* Mez.
It differs from:
T. parviflora by its multi-branched inflorescences. In *T. parviflora* the inflorescences are simply branched and the flowers are yellow.
T. commixta by the longer scape bracts.
T. pallidoflavens by the narrower leaves. In the latter they are about 3 cm wide.

Tillandsia werdermannii Harms, 1928 (Ill. 308)

(*werdermannii* named after the botanist Prof. E. Werdermann)
T. werdermannii is affiliated to *T. latipolia*, *T. purpurea* and *T. paleacea*,

waste land and sand desert tillandsias; it is rather similar to *T. cauligera* (Ill. 80) differing from it by the larger, narrower, grey leaves.

Pl with long, decumbent, rooting, branched, densely leaved stems, forming large clumps, creeping strands or tufts, those at the rear dying but continuing to grow onwards
Fo numerous, spirally arranged, densely set
LS not clearly delineated, to 3 cm long
LB narrow-lanceolate, to 20 cm long terminating in a long narrowing tip, with channelled upward curved margins, densely grey scaled on both sides
Sc upright, 20–30 cm long, 4 mm thick, bare
SB densely set, upright, the basal ones similar to the leaves, the upper ones with shorter blades, densely appressed grey scaled

I simple sword shaped, 2 cm wide
FB to 4 cm long, extending well beyond the sepals, not keeled, naked
Fl very short stalked, blue?
Se lineal-lanceolate, wide pointed, 2·5 cm long, bare
Ha South Peru and North Chile (the environs of Tacna) between 700 m and 1,200 m high
In the sand where it grows *T. werdermannii* colonises areas each of about four kilometres square; the plant is propagated by the wind which tears off bits which grow-on where deposited.
CU In culture it requires a light sunny position and is suited to cactus conditions.

Tillandsia xerographica Rohw., 1953 (Ill. 7, Plate 49)
(According to Rohweder the species name refers to the delicate, yellowish, pastel tones of the floral bracts and scape bracts)

Pl stemless, 70 cm to 1 m high including the inflorescence
Fo forming a dense, spread, silver grey rosette of 80–90 cm diameter
LS broad, almost round, 10 cm long, 10 cm wide, short, light brown scales, somewhat spoon-shaped and overall forming a large, but not completely distinct pseudo-bulb
LB narrow, long and thread-like pointed, 6 cm wide above the sheath, up to 70 cm long, usually recurved; tips sometimes rolled into a spiral, the edges distinctly curled, heavily covered with silver grey, short scales
Sc sturdy, upright, 1·7 cm thick, 15 cm long, round, green, naked
SB the basal ones similar to the leaves,

the upper ones with a red sheath, grey-scaled and with a long, narrow, grey-scaled blade, tightly overlapping, up to 60 cm long
I open, compound, with 25 to 28 almost upright, spikes arranged in a spiral, overall 40 cm long, 10 cm wide
PB in the lower part of the inflorescence similar to the upper scape bracts; sheaths bright red, grey-scaled, in the upper part naked, bright red, with a very short or lacking blade
Sp stemmed, 10–12 cm long, 2·5 cm wide; stem flattened, 2–2·5 cm long with 2 sterile floral bracts
FB overlapping, distichous, 4–4·5 cm long, 1·6 cm wide, slight keel green to reddish-yellow, naked

Se about 4 cm long, 7 mm wide, the posterior ones fused 1·3 cm high, keeled, naked, green

Pe pale lavender, about 8 mm wide, somewhat recurved at the tip, surpassed by the stamens, which are 8·5 cm long and lavender in the upper part

Ha San Salvador and Mexico; epi-phytic on trees and boulders. The plants from San Salvador differ from those collected in Mexico by a rather short scape and a dense, shorter inflorescence

CU Very bright and sunny; spray only. A splendid, silver grey, but large species.

Tillandsia xiphioides Kerr., 1816 (Ill. 144)
(*xiphioides* sword-like; because of the sword-shaped inflorescence)
Syn. *Anoplophytum xiphioides* Beer, 1857; *Phytarrhiza xiphioides* Morren, 1879; *T. macrocnemis* Griseb., 1879; *T. odorata* Gill., 1887; *T. suaveolens* Lem., 1843; *T. sericea* Hort., 1843; *T. unca* Hicken, 1912, non Griseb., 1874.

Pl stemless to short-stemmed, up to 20 cm high including the inflorescence

Fo forming a rigid, open, upright or one-sided rosette

LS longitudinally triangular, gradually merging into the blade, 2–3 cm long, up to 2·5 cm wide, naked at the base of the inside, dense silver grey scales on the upper part and on the outside

I upright, simple-spiked, up to 10 cm long, up to 3 cm wide, up to 6 flowers

FB distichous, tightly overlapping, lanceolate, long-tapered, up to 7 cm long, 1–2 cm wide, upright, green to pale brown, naked to scaled

Se much shorter than the floral bracts, up to 4·5 cm long, 8 mm wide, lanceolate, pointed, cutaneous, naked, separate

Fl white, fragrant, 7–8 cm in size; petals with a narrowly lineal claw and a flared, twisted plate with wavy edges, dentate, 2·5 cm long, up to 1·5 cm wide

St enveloped in the flower

This species is quite variable in regard to size

Ha Argentina, Paraguay, Bolivia

CU Dry, bright and sunny. A species worthy of cultivation because of its large, white flowers.

Key for Identification of Tillandsias

It is hoped that these tables will facilitate the identification of the most common and most widespread tillandsias in cultivation. This represents about one third of all the known species.

Use of the tables

These have been designed so that even the beginner—having first studied Chapter 2 on morphology—can find the name of an 'unknown' species. The tables, based upon observation of the plant's characteristics, are used by checking the 'unknown' against the details in the table, until an identification is made, via a series of 'choices'. Sometimes, there may be a point when neither choice seems appropriate. In such instances, it is recommended to follow through both possibilities. If neither produces an identification, then the 'unknown' species is not described in this book, or a mistake has been made at one of the early stages in the use of the table. The appropriate table to use (Tables A to E) is first determined from the Main Table.

Main Table

Identification Stage		Plant Characteristic	Species
1	a	Plant leaves hardly scaled. Green (or red in strong sunlight). Chalky or blue frosted appearance. Usually uniting into a funnel rosette.	see **Table A**
	b	Plant leaves more or less heavily scaled. (Check with magnifying glass). Grey, white or brownish. Upper surface greenish, often less scaled than underside.	see **2**
2	a	Stemless plants. Leaves in rosette	see **4**
	b	Stemmed plants. Stem sometimes hidden by overlapping leaves	see **3**
3	a	Leaves distichous. (Sometimes twisting of axis gives illusion of spiral leaves)	see **Table B**
	b	Leaves polystichous (spiral)	see **Table C**
4	a	Leaf sheaths densely overlap. Roundish to spindle-shaped, hollow pseudo-bulb—or compact true bulb	see **Table D**
	b	Leaf sheaths in loose to dense rosette. No pseudo-bulb or true bulb	see **Table 3**

Table A *Plants with green or reddish, hardly scaled leaves*

Identification Stage	Plant Characteristic		Species
1	a	Inflorescence always single, simple or compound	see **3**
	b	Always multiple inflorescences; lateral, simple	see **2**
2	a	Inflorescences erect, with bright red floral bracts, sword-shaped; up to 5 cm wide, up to 15 cm long	*T. multicaulis*
	b	Inflorescences pendulous; spikes only 1–2 cm wide, 3–8 cm long; floral bracts green or red	*T. complanata*
3	a	Blades not grass-like	see **6**
	b	Leaves grass-like	see **4**
4	a	Inflorescence simple	*T. chaetophylla*
	b	Inflorescence compound	see **5**
5	a	Petals erect, forming a tube; blue violet	*T. festucoides*
	b	Petals with tips flared; flowers therefore open, white	*T. remota*
6	a	Inflorescence composed of several spikes	see **15**
	b	Inflorescence simple, consisting of only one spike	see **7**
7	a	Spike dense at bloom time, sword-shaped; no visible distance between the flowers (Fig. 14, 1)	see **11**
	b	Spike open at bloom time; visible distance between the flowers; flowers spreading to flared	see **8**

8	a	Petals greenish-yellow, narrowly strap-shaped	*T. grandis*
	b	Petals pure white, with broad, spread plates	see **9**
9	a	Inflorescence with slender, thin scape arching upward	*T. narthecioides*
	b	Inflorescence with strong, rigid, erect scape	see **10**
10	a	Flower bracts keeled; all stamens united into a tube (Fig. 40)	*T. monadelpha*
	b	Flower bracts not keeled; stamens not united into a tube	*T. cornuta*
11	a	Upper part of the petals spread	see **13**
	b	Upper part of the petals almost erect or hardly spread; flowers therefore more tubular	see **12**
12	a	Petals yellow	*T. lampropoda*
	b	Petals pink and white; flower bracts very large; in intense light bright raspberry-coloured, in cultivation frequently green	*T. walteri*
13	a	Petals narrowly lineal in the upper part	*T. anceps*
	b	Petals broadly plate-shaped in the upper part	see **14**
14	a	Scape very short; flowers uniformly dark blue	*T. cyanea*
	b	Scape elongated; flowers dark blue, with a white eye in the throat	*T. lindeni*
15	a	Inflorescence loosely compound; spikes visible	see **17**
	b	Inflorescence densely cylindrical to cone-shaped; flower bracts bright red; spikes mostly covered by the flower bracts	see **16**
16	a	Inflorescence cylindrical, thick; stamens protruding from the flowers; Mexico	*T. imperialis*
	b	Inflorescence a slender cylinder; stamens enclosed within the flower; Peru, Ecuador	*T. ionochroma*
17	a	Scape elongated, surpassing the leaves	see **19**
	b	Scape stubby; if elongated, does not surpass the leaves	see **18**
18	a	Inflorescence densely glomerate	*T. brachycaulos*

		(head-shaped); inner leaves of rosette bright red at bloom time	
	b	Inflorescence loose; spikes long and narrow, red	*T. flabellata*
19	a	Inflorescence multi-branched (panicle)	see **29**
	b	Inflorescence simple compound (double-spiked, Fig. 14, 2)	see **20**
20	a	Spikes with 2 (3) flowers	*T. biflora*
	b	Spikes with more than 3 flowers	see **21**
21	a	Lower primary bracts longer than spikes	*T. leiboldiana*
	b	All primary bracts shorter than spikes	see **22**
22	a	Stamens protruding from flower	see **27**
	b	Stamens enclosed within flower	see **23**
23	a	Flowers very small, insignificant, yellowish; sepals not symmetrical	*T. triticea*
	b	Flowers prominent; sepals symmetrical	see **24**
24	a	Scape several times longer than leaves, up to 1 m or more, often with new shoots on the scape; basal spikes recurved	*T. somnians*
	b	Scape not over 20 cm	see **25**
25	a	Flowers white	*T. dyeriana*
	b	Flowers dark bluish-lavender or dark blue	see **26**
26	a	Flowers dark bluish-lavender with a white eye; flower bracts greenish-brown, grey-scaled	*T. hamaleana*
	b	Flowers dark blue; flower bracts pale green (bright pink in strong light)	*T. wagneriana*
27	a	Leaves triangular, long-tapered; plants up to 70 cm high in bloom	*T. polystachya*
	b	Leaves strap-like, wide	see **28**
28	a	Petals greenish-white or green	*T. grandis*
	b	Petals dark blue	*T. rauhii*
29	a	Flowers small, white or greenish-white	see **31**
	b	Flowers prominent, of an other colour	see **30**
30	a	Flowers blue violet; plant medium sized (up to about 1 m); Central America	*T. guatemalensis*
	b	Flowers dark lavender to almost black; leaves chalky; plant very large,	*T. ferreyrae*

		over 2 m including inflorescence; northern Peru	
31	a	Pistil longer than the greenish-white petals	*T. insularis*
	b	Pistil shorter than white petals	see **32**
32	a	Flowers white, splayed; spikes 2–3 cm long	*T. ebracteata*
	b	Flowers yellowish-white, erect, lying close to the axis; spikes 5–10 cm long	see **33**
33	a	Flower bracts as long or longer than sepals; dense and prominently nerved	*T. pendulispica*
	b	Most flower bracts shorter than sepals; both heavily brown-scaled	*T. bakeri*

Table B *Stemmed plants with leaves that are not green, but are more or less heavily scaled, grey to white, and distichous (arranged in two rows)*

Identification Stage		Plant Characteristic	Species
1	a	Stalk segments distinctly elongated, up to 10 cm long, often twisted in corkscrew fashion; growth several metres long; thread-like, thin, pendulous, rootless; flowers single, yellowish-green	*T. usneoides*
	b	Stalk segments not noticeably elongated, not twisted in corkscrew fashion and not thread-like and thin	see **2**
2	a	Flowers large, attractive	see **9**
	b	Flowers small, insignificant	see **3**
3	a	Leaves thick, sickle-shaped	see **6**
	b	Leaves thin, awl-shaped	see **4**
4	a	Leaves shorter than the stem (length of last year's inflorescence); spread horizontally	see **7**
	b	Leaves longer than last year's stem segment; spreading to almost erect	see **5**
5	a	Flowers pale bluish-violet; scape and flower bracts scaled, latter sometimes lacking	*T. recurvata*
	b	Flowers yellowish, greenish or brownish; scape without bracts, naked, scaled only toward the tip	*T. capillaris*

6	a	Leaves above middle laterally compressed in cross-section	*T. gilliesii*
	b	Leaves above middle round in cross-section	*T. myosura*
7	a	Scape heavily scaled	*T. andicola*
	b	Scape slightly scaled (sometimes only at tip) to naked	see **8**
8	a	Flowers yellowish to yellowish-green or brownish	*T. capillaris*
	b	Flowers pale blue	*T. recurvata*
9	a	Flowers golden yellow	*T. crocata*
	b	Flowers bluish-lavender	see **10**
10	a	Leaves thin, thread-like, longer than inflorescence	*T. mallemontii*
	b	Leaves shorter, thicker, awl-shaped, surpassed by inflorescence	*T. bandensis*

Table C *Stemmed Plants with scaled, polystichous (in many rows) leaves*

Identification Stage		Plant Characteristic	Species
1	a	New shoot* elongated, more than 5 cm long	see **16**
	b	New shoot* short, no longer than 5 cm	see **2**
2	a	Plants with moss-like or club moss-like growth; short, scale-like leaves lying more or less closely to axis	see **14**
	b	Plants not of moss-like or club moss-like growth; leaves rather large, not scale-shaped	see **3**
3	a	Inflorescence composed of several spikes; flowers bluish-lavender	*T. streptocarpa*
	b	Inflorescence simple, not branched	see **4**
4	a	Inflorescence with a large, red flower	*T. funckiana*
	b	Inflorescence multiple flowered	see **5**
5	a	Flowers polystichous (arranged in spiral on axis), yellow	*T. ixioides*
	b	Flowers distichous (arranged in two rows on axis)	see **6**
6	a	Flowers large, attractive; petals 5 mm wide or more	see **10**

* The length of the axis from one year's to the next year's inflorescence.

	b	Flowers small, insignificant; petals no more than 5 mm wide	see 7
7	a	Inflorescence sessile, without a visible scape; flowers brown	T. rectangula
	b	Inflorescence with elongated, visible scape	see 8
8	a	Inflorescence with more than 3 flowers; flowers yellow or orange	T. loliacea
	b	Inflorescence with 1 to 2 flowers, seldom 3	see 9
9	a	Leaves rigid, with a bent, almost sharp tip, about 5 mm wide at base; flowers dark brown	T. funebris
	b	Leaves pliant, thin, almost thread-like; flowers greenish, yellowish or pale brownish	T. capillaris
10	a	Inflorescence on a short, hidden scape, appears to be sessile; flowers rose-coloured, leaves rigid	T. unca
	b	Flowers on a visible, elongated scape	see 11
11	a	Flowers arranged loosely on rhachis, bluish-lavender	T. peiranoi
	b	Flowers arranged densely on rhachis; inflorescence sword-shaped	see 12
12	a	Petals yellow, sometimes somewhat red at base, forming an erect tube	T. schiedeana
	b	Petals white or bluish-lavender	see 13
13	a	Petals white; round, broad plates wavy and dentate (toothed)	T. xiphioides
	b	Petals bluish lavender; edges of spread plates not dentate	T. streptocarpa
14	a	Leaves scale-like, somewhat spreading from axis; flowers greenish	T. tricholepis
	b	Leaves scale-like, lying close to axis	see 15
15	a	Flowers dark bluish-lavender; seed capsule stemmed	T. pedicellata
	b	Flowers yellow, sessile; seed capsule sessile	T. bryoides
16	a	Plants with a simple inflorescence	see 23
	b	Plants with a compound inflorescence	see 17
17	a	Leaf tips tendril-like, rolled into spiral	see 22
	b	Leaf tips not rolled	see 18
18	a	Inflorescence densely glomerate (head-shaped); scape very short	T. calocephala
	b	Inflorescence not densely glomerate,	see 19

but spikes sometimes close, scape
elongated

19	a	Flowers pink; inflorescences frequently with new shoots	*T. latifolia*
	b	Flowers purple or white, bluish-lavender tips	see **20**
20	a	Leaves densely covered with feathery, spread scales	*T. tectorum*
	b	Leaves with short appressed scales	see **21**
21	a	Primary bracts purple, heavily scaled	*T. purpurea*
	b	Primary bracts pale yellowish-green, not scaled	*T. straminea*
22	a	Flower bracts heavily scaled	*T. duratii*
	b	Flower bracts naked	*T. decomposita*
23	a	Inflorescence with only 1 large, red flower (seldom 2)	*T. funckiana*
	b	Inflorescence with 2 or more flowers	see **24**
24	a	Flowers distichous (in two rows)	see **28**
	b	Flowers polystichous (in many rows)	see **25**
25	a	Petals wavy and twisted, slate blue	*T. bergeri*
	b	Petals smooth, not wavy or twisted	see **26**
26	a	Flowers dark blue or grey blue	*T. aeranthos*
	b	Flowers white or light bluish-lavender	see **27**
27	a	Scape far surpassing short, secund (one-sided) leaves; flowers always white	*T. araujei*
	b	Scape rather short, usually not surpassing leaves; flowers bluish-lavender or white; leaves in some varieties secund (turned all to one side)	*T. tenuifolia* (syn *T. pulchella*)
28	a	Flowers arranged densely on rhachis, inflorescence therefore sword-shaped; rhachis not visible	see **34**
	b	Flowers arranged loosely on rhachis; latter therefore visible	see **29**
29	a	Flowers large, white or bluish-lavender	see **32**
	b	Flowers small, insignificant, greenish or brownish	see **30**
30	a	Plants with moss-like growth habit, with small, scale-like leaves, standing somewhat away from axis	*T. tricholepis*
	b	Plants not of moss-like growth habit, leaves not scale-shaped	see **31**

31	a	Leaves stiff, with arched, almost sharp tip, about 5 mm wide at base; plate of petals dark brown	*T. funebris*
	b	Leaves pliant, thin, almost thread-like; flowers greenish, yellowish or pale brownish	*T. capillaris*
32	a	Petals united into tube; white; stamens protruding	*T. albida*
	b	Petals spread, never pure white; stamens enveloped in flower	see **33**
33	a	Petals bluish-lavender with white base	*T. caerulea*
	b	Petals white with blue tips	*T. straminea*
34	a	Spike red at bloom time	see **38**
	b	Spike not red at bloom time	see **35**
35	a	Flowers bluish-lavender; petals spread	*T. paleacea*
	b	Petals yellow or white	see **36**
36	a	Flowers yellow, sometimes red at base; petals inclined into tube, erect	*T. schiedeana*
	b	Flowers white	see **37**
37	a	Petals smooth, not dentate (toothed) on edge	*T. diaguitensis*
	b	Petals wavy, dentate toothed) on edge	*T. xiphioides*
38	a	Leaves stubby-triangular to lanceolate	*T. cauligera*
	b	Leaves very narrow, lanceolate to awl-shaped, long-tapered	see **39**
39	a	Flowers white	*T. caulescens*
	b	Flowers bluish-lavender or yellow	see **40**
40	a	Flowers bluish-lavender	*T. dura*
	b	Flowers yellow, sometimes red at base	*T. schiedeana*

Table D *Plants with no stem; leaves scaled, their sheaths combined into a hollow pseudo-bulb or a true bulb*

Identification Stage	Plant Characteristic		Species
1	a	Plants with a true (not hollow), roundish bulb; leaf sheaths thick (with water tissue) and tightly overlapping (Fig. 9, 2)	see **29**

	b	Plants with a hollow, egg-shaped, cylindrical to spindle-shaped pseudo-bulb (Fig. 9, 1)	see **2**
2	a	Inflorescence simple	see **22**
	b	Inflorescence compound	see **3**
3	a	Tips of the leaves rolled into a spiral (a trait so noticeable on imported plants that can disappear in cultivation)	*T. streptophylla*
	b	Leaf tips not rolled into spiral (if tips are somewhat rolled, then blade is not thick)	see **4**
4	a	Inflorescence pendulous, since scape axis is arched downwards	see **11**
	b	Inflorescence erect (if directed downwards, then whole plant is directed downwards)	see **5**
5	a	Leaf blades not awl-shaped (round) and not gutter-shaped, at most, leaf edges may be somewhat curled	see **13**
	b	Leaf blades awl-shaped (round), forming tube open at top (cross section through blade)	see **6**
6	a	Leaves heavily covered with feathery, spreading scales	see **12**
	b	Leaves covered with short, coarse, but not feathery and spreading scales	see **7**
7	a	Pseudo-bulb spotted with greenish-brown spots partially flowing into each other	*T. butzii*
	b	Pseudo-bulb uniform in colour, not spotted	see **8**
8	a	Leaf sheaths and blades distinctly striped longitudinally (nerved) on outside	*T. baileyi*
	b	Sheaths and blades smooth, not nerved	see **9**
9	a	Flower bracts naked	*T. caput-medusae*
	b	Flower bracts scaled	see **10**
10	a	Pseudo-bulb longitudinally cylindrical; leaf blades usually bent backwards and tips sometimes spiral	*T. circinnata*
	b	Pseudo-bulb stubby egg-shaped; leaf blades usually erect (only old leaves bent backwards), but ends not spiral;	*T. bulbosa*

		leaf sheaths frequently edged in red	
11	a	Leaves greyish-white with chestnut brow sheaths; flowers violet	*T. oaxacana*
	b	Leaves green, underside heavily grey-scaled; flowers yellowish to orange	*T. tetrantha*
12	a	Leaves with a short, erect blade; plants up to 30 cm high; scape usually distinctly visible	*T. seleriana*
	b	Leaves with elongated, frequently twisted blade; plants rather small, up to 20 cm; scape short, usually not visible	*T. pruinosa*
13	a	Primary bracts and flowers densely distichous; spikes therefore sword-shaped	see **20**
	b	Primary bracts and flowers arranged loosely on rhachis; spikes therefore not sword-shaped	see **14**
14	a	Sepals symmetrical	see **19**
	b	Sepals asymmetrical	see **15**
15	a	Flower bracts longer than sepals	*T. crispa*
	b	Flower bracts shorter than sepals	see **16**
16	a	Flowers yellow	see **18**
	b	Flowers greenish-white	see **17**
17	a	Scape bracts as long as or almost as long as stalk segments (internodes)	*T. weberbaueri*
	b	Scape bracts small, upper ones hardly half as long as stalk segments (internodes)	*T. commixta*
18	a	Sepals heavily brown scaled, up to 7 mm long; petals 1 cm long	*T. adpressa*
	b	Sepals slightly scaled, 5–6 mm long; petals 6 mm long	*T. guanacastensis*
19	a	Flowers red; leaves frequently twisted and sometimes banded in greenish-white	*T. flexuosa*
	b	Flowers bluish-violet; leaves not twisted; scape very strong, up to 30 cm long	*T. kirchhoffiana*
20	a	Flowers small, yellowish; leaves narrowly lanceolate	*T. disticha*
	b	Flowers rather large, dark lavender or bluish-lavender	see **21**
21	a	Flowers dark lavender, almost black, with white tips; leaf sheaths blackish-brown	*T. punctulata*

	b	Flowers bluish-lavender; leaf sheaths rust brown; blades thread-like, long-tapered, usually recurved	*T. balbisiana*
22	a	Flowers white; leaf edges distinctly wavy	*T. crispa*
	b	Flowers of a different colour	see **23**
23	a	Flowers yellowish-brown with yellow tips; leaves usually with white cross bands	*T. subulifera*
	b	Flowers of a different colour	see **24**
24	a	Flowers bluish-violet or blackish-violet	see **26**
	b	Flowers red	see **25**
25	a	Flower bracts shorter than the sepals, slightly scaled to naked, very loose; leaves frequently twisted	*T. flexuosa*
	b	Flower bracts longer than the sepals, heavily scaled, at first overlapping, later loose	*T. paraensis*
26	a	Flowers bluish-violet or blue	see **28**
	b	Flowers violet or blackish-violet	see **27**
27	a	Flowers violet; scape short; uppermost foliage bright red at bloom time	*T. ionantha*
	b	Flowers blackish-violet with white edge; scape elongated; uppermost foliage not red	*T. punctulata*
28	a	Leaves covered with feathery scales; flowers bluish-violet; flower bracts green or pink at bloom time	*T. pruinosa*
	b	Leaves not covered with feathery scales; flowers blue	*T. baileyi*
29	a	Flowers yellow, very small	*T. disticha*
	b	Flowers of a different colour, rather large	see **30**
30	a	Scape short; inflorescence therefore sessile	see **33**
	b	Scape distinctly elongated	see **31**
31	a	Flowers green; leaves spread, covered with feathery scales	*T. plumosa*
	b	Flowers lavender or red	see **32**
32	a	Inflorescence simple, very open, with zigzag rhachis	*T. argentea*
	b	Inflorescence loosely compound; scape erect or pendulous, thin	*T. filifolia*
33	a	Flowers bright green	*T. atroviridipetala*

	b	Flowers violet	see **34**
34	a	The upper rosette leaves bright red at bloom time; plants usually growing in clumps or mounds	*T. ionantha*
	b	The upper rosette leaves not red at bloom time	*T. magnusiana*

Table E *Plants with more or less scaled leaves with sheaths standing away from axis; leaves overall forming a dense or loose (sometimes tubular) rosette*

Identification Stage		Plant Characteristic	Species
1	a	Inflorescence simple	see **52**
	b	Inforescence compound	see **2**
2	a	Flower bracts and blooms of the individual spikes dense at bloom time; spike is therefore sword-shaped or the inflorescence is glomerate (head-shaped)	see **11**
	b	Flower bracts and flowers of the individual spikes loose at bloom time (rhachis visible)	see **3**
3	a	Stamens shorter or as long as petals	see **6**
	b	Stamens surpassing flowers	see **4**
4	a	Flower stalk usually distinctly bent into a zigzag shape (Fig. 31a, 1)	*T. dasyliriifolia*
	b	Flower stalk not distinctly bent, usually only curved (Fig. 31a, 2–3)	see **5**
5	a	Rosette leaves spread out flat or recurved in upper part, 2–3 cm wide above the sheath, not distinctly ribbed longitudinally; flowers white	*T. utriculata*
	b	Rosette leaves more or less erect, forming a narrow funnel, leaf tips recurved; leaves 7–8 cm wide above sheath, very stiff, clearly ribbed longitudinally; flowers blue violet or green	*T. makoyana*
6	a	Sepals asymmetrical	see **9**
	b	Sepals symmetrical	see **7**
7	a	Flowers red	*T. geminiflora*
	b	Flowers of a different colour	see **8**
8	a	Flowers yellowish to light brown, with	*T. aureo-brunnea*

		dark brown spots or uniformly yellowish	
	b	Flowers white with blue tips	*T. purpurea*
9	a	Flowers orange to yellow	*T. tetrantha*
	b	Flowers yellowish-white to greenish-white	see **10**
10	a	Branching of the inflorescence simple; flowers yellow	*T. adpressa*
	b	Branching of the inflorescence branched again; flowers greenish-white	*T. bakeri*
11	a	Stamens protruding from flower	see **31**
	b	Stamens enveloped in the flower or as long as flower	see **12**
12	a	Inflorescence very dense, glomerate (like a head)	see **29**
	b	Inflorescence not glomerate (head-shaped); but spikes sometimes digitate (shaped like fingers of a hand)	see **13**
13	a	Petals white or cream-coloured with blue tip, or pale blue with white	see **21**
	b	Petals of a different colour	see **14**
14	a	Scape very short, hardly visible at bloom time; flowers green	*T. mauryana*
	b	Scape distinctly elongated; flowers not green	see **15**
15	a	Flowers bluish-lavender	see **19**
	b	Flowers red or pink	see **16**
16	a	Flower bracts shorter than sepals	*T. geminiflora*
	b	Flower bracts longer than or as long as sepals	see **17**
17	a	Scape arched to pendulous	*T. gardneri*
	b	Scape erect	see **18**
18	a	Flower bracts always naked, distinctly nerved	*T. gayi*
	b	Flower bracts scaled, becoming naked with age	*T. latifolia*
19	a	Leaves very narrow, up to 4 mm wide	*T. floribunda*
	b	Leaves 3–8 cm wide	see **20**
20	a	Flower bracts shorter than sepals	*T. oroyensis*
	b	Flower bracts as long as sepals	*T. deppeana*
21	a	Leaves densely covered with feathery, spread scales; flowers pale blue with white	*T. tectorum*
	b	Leaves with coarse or short scales (magnifying glass)	see **22**

22	a	Petals pure white	see **25**
	b	Petals of two colours	see **23**
23	a	Primary bracts naked; petals white with blue tips	*T. straminea*
	b	Primary bracts scaled	see **24**
24	a	Petals cream-coloured with blue tips	*T. cacticola*
	b	Petals white with blue tips	*T. purpurea*
25	a	Flower bracts heavily grey-scaled	*T. didisticha*
	b	Flower bracts naked or slightly scaled at first	see **26**
26	a	Petals 2·2–2·3 cm long, white, sometimes blue	see **28**
	b	Petals 0·6–1·3 cm long, white	see **27**
27	a	Petals about 1·3 cm long; tips spread to recurved; plant grass-like	*T. remota*
	b	Petals 0·6 cm long; tips only slightly spread; flower tubular; plant not grass-like	*T. ebracteata*
28	a	Flower bracts shiny, red; appear to be lacquered	*T. vernicosa*
	b	Flower bracts blunt, brownish-red or green; flowers white, sometimes bluish-violet	*T. lorentziana*
29	a	Flowers white	*T. schreiteri*
	b	Flowers bluish-lavender	see **30**
30	a	Inflorescence open, elongated, pendulous	*T. violacea*
	b	Inflorescence dense, glomerate (head-shaped), splayed, horizontal to pendulous	*T. sphaerocephala*
31	a	Flower bracts naked	see **39**
	b	Flower bracts scaled	see **32**
32	a	Scape very short, not visible; inflorescence apparently sessile	*T. carlsoniae*
	b	Scape elongated, visible	see **33**
33	a	Inflorescence dense; spikes erect and short, up to 4 cm long	*T. juncea*
	b	Inflorescence open; spikes loose to spread, longer than 4 cm, up to 15 cm long	see **34**
34	a	Flowers green	*T. bourgeaei*
	b	Flowers bluish-lavender	see **35**
35	a	Inflorescence pendulous, up to 1 m long	*T. prodigiosa*
	b	Inflorescence erect	see **36**

36	a	Leaves narrow, grass-like or bristly	*T. setacea*
	b	Leaves wider, long-tapered	see **37**
37	a	Leaf blades distinctly bent back; sepals fused posteriorly 1·5 cm high	*T. exserta*
	b	Leaf blades not noticeably bent back	see **38**
38	a	Flower bracts longer than the sepals	*T. vicentina*
	b	Flower bracts just as long as the sepals	*T. polystachya*
39	a	Flowers yellow, white or green	see **50**
	b	Flowers bluish-lavender, bluish-violet or lavender pink	see **40**
40	a	Leaves narrow, grass-like	see **49**
	b	Leaves not narrow and grass-like	see **41**
41	a	Inflorescence and scape very long, pendulous	*T. prodigiosa*
	b	Inflorescence erect	see **42**
42	a	Inflorescence composed of many spikes (10–25); leaves silvery white on both sides	*T. xerographica*
	b	Inflorescence composed of fewer than 10 spikes; leaves silvery white only on the underside if at all	see **43**
43	a	Spikes wider than 2 cm	see **47**
	b	Spikes no more than 2 cm wide	see **44**
44	a	Flower bracts as long as sepals	*T. polystachya*
	b	Flower bracts longer than sepals	see **45**
45	a	Flowers lavender pink	*T. concolor*
	b	Flowers bluish-violet	see **46**
46	a	Leaf sheaths and blades of uniform colour	*T. valenzuelana*
	b	Leaf sheaths brown, differing from that of blades	*T. fasciculata*
47	a	Petals 3–3·5 cm long	*T. tricolor*
	b	Petals 6 cm long	see **48**
48	a	Petals bluish-violet	*T. fasciculata*
	b	Petals lavender pink	*T. concolor*
49	a	Spikes arched away from axis	*T. festucoides*
	b	Spikes straight, erect	*T. setacea*
50	a	Flower bracts puffed up; scape bracts recurved, red or pale green	*T. intumescens*
	b	Flower bracts not puffed up; flowers green or white	see **51**
51	a	Flowers green; plant large; inflorescence erect	*T. bourgeaei*
	b	Flowers white; plants small; inflorescence pendulous with short scape	*T. matudae*

52	a	Flowers arranged densely on rhachis, flower bracts overlapping; distichous arrangement causing inflorescence to be sword-shaped	see **63**
	b	Flowers arranged loosely on rhachis; flower bracts remote from each other; rhachis therefore visible	see **53**
53	a	Flowers in a distichous arrangement on rhachis	see **60**
	b	Flowers polystichous (in many rows) on rhachis	see **54**
54	a	Flowers blue or violet	see **57**
	b	Flowers white or greenish-white	see **55**
55	a	Stamens protruding from the flower; petals greenish-white	*T. benthamiana*
	b	Stamens enveloped in the flower; petals white	see **56**
56	a	Flower bracts pinkish-red	*T. meridionalis*
	b	Flower bracts green to pale pink or brownish	*T. pohliana*
57	a	Stamens enveloped in flower	*T. stricta*
	b	Stamens protruding from flower	see **58**
58	a	Sepals 3–4 cm long	*T. macdougallii*
	b	Sepals no longer than 2·5 cm	see **59**
59	a	Leaf sheaths brown to dark brown, forming loose pseudo-bulb rosette	*T. oaxacana*
	b	Leaf sheaths silvery-grey; rosette often turned to one side	*T. andrieuxii*
60	a	Flowers white; petals spread	*T. triglochinoides*
	b	Flowers greenish, yellowish, orange or bluish-lavender	see **61**
61	a	Flowers small, yellowish or orange; plants forming small rosette	*T. loliacea*
	b	Flowers large, rosette large	see **62**
62	a	Rhachis distinctly bent (Fig. 31a, 1)	*T. dasyliriifolia*
	b	Rhachis not bent, only slightly arched (Fig. 31a, 2); rosette erect, funnel-shaped or tubular	*T. makoyana*
63	a	Leaves lineal to lanceolate or triangular, long-tapered	see **65**
	b	Leaves narrow, grass-like	see **64**
64	a	Leaf sheaths dark brown; flowers pale bluish-lavender, tubular	*T. chaetophylla*
	b	Leaf sheaths light brown; flowers bluish-violet; petals with wide, spread plates	*T. linearis*

65	a	Stamens protruding from flower	see **68**
	b	Stamens enveloped in flower	see **66**
66	a	Scape very stubby, not visible; inflorescence apparently sessile; flowers green, small	*T. lepidosepala*
	b	Scape visible, elongated	see **67**
67	a	Flowers white, large; plates of petals spread, their edges wavy	*T. xiphioides*
	b	Flowers bluish-violet; plates of petals small, not wavy; flowers polystichous, the basal ones surpassed by flower bracts	*T. stricta*
68	a	Flowers bluish-violet	see **71**
	b	Flowers lavender pink, yellow or green	see **69**
69	a	Flowers green; flower bracts red and distinctly nerved; inflorescence sword shaped	*T. achyrostachys*
	b	Flowers yellow or lavender pink	see **70**
70	a	Flowers yellow; broad sword shaped inflorescence	*T. lampropoda*
	b	Flowers lavender pink; narrow sword shaped inflorescence	*T. concolor*
71	a	Flower bracts naked to almost naked	see **73**
	b	Flower bracts always heavily scaled	see **72**
72	a	Leaves always erect, never spread or recurved	*T. pueblensis*
	b	Leaves distinctly bent back	*T. exserta*
73	a	Petals up to 3·5 cm long; blades up to 1·5 cm wide above sheath	*T. tricolor*
	b	Petals 4·5–6 cm long	see **74**
74	a	Scape distinctly elongated	*T. fasciculata*
	b	Scape very short or lacking	*T. ionantha*

Catopsis Griseb.

The genus *Catopsis* was established in 1864 by Grisebach as being separate from *Tillandsia*.

The number of identified species runs to about 20; there are a few new species that have yet to be described.

The range of the genus stretches from the southern part of the U.S.A., through Central America to Brazil and Peru.

Commercially, *Catopsis* has little significance since the flowers are small and insignificant. For botanical collections and for the plant fancier, the smaller species may be of interest when adding to a bromeliad collection.

The most important features of the genus are:

1 Leaves are never thorny or dentate, but always with entire edges, usually in a rosette or funnel shape, green, often frosted in chalk white.
2 Flowers are single-sexed or hermaphroditic.
3 Sepals are asymmetric in most species.
4 Petals are separate, often insignificant, yellow, white or greenish-white.
5 Ovaries are superior.
6 Fruit is a capsule
7 Seed is formed with a pappus (crown hairs) connected at the tip; the pappus is distinctly folded in the capsule.

The identification of the single-sexed *Catopsis* species is sometimes very difficult, since male and female plants are often different from each other in form.

The culture of *Catopsis* is quite simple. The plants are treated like the green tillandsias, i.e. moderately damp and semi-shady. Pour water into the funnel, because spraying or dipping is of no use in the mealy-scaled species. The following species, which represent only a limited selection, can be grown in a pot with a peaty medium or on epiphyte trees.

Catopsis berteroniana (Schult. f.) Mez, 1896
(*berteroniana* named after the Italian botanist C. G. Bertero (1789–1831))
Syn. *Tillandsia berteroniana* Schult. f., 1830

Pl stemless, 40–90 cm high
Fo up to 40 cm long, coated with white meal
LS very large, indistinct, gradually merging into the triangular blade 4–5 cm wide
Sc strong, upright
SB in the lower part of the scape overlapping and similar to the leaves, in the upper part oval and often further apart
I compound, panicled

PB short, broadly oval
Sp long-stemmed, loosely arranged flowers
FB oval-triangular, 2–3 mm long, shorter than the asymmetric sepals, which are 4–6 mm long
Fl hermaphroditic, white
Se not asymmetric, inverse egg-shaped, 1–1·2 cm long, stiff
Ha From southern Florida through Central America into eastern Brazil; epiphytic on trees or bushes.

Catopsis brevifolia Mez et Wercklé, 1904 (Ill. 145)
(*brevifolia*(*us*) short-leafed)

C. brevifolia, with its slender funnel rosette somewhat rounded at the base, is a very decorative species, whose leaves in the native habitat at rather high altitudes, as a result of the intensive sunlight, take on a distinctly leather brown colouration, which unfortunately disappears in cultivation.

Pl 20–30 cm high, stemless

Fo numerous, forming a dense, upright funnel rosette, dark green with or without meal-like powder

LS very large, oval or elliptical, 10–12 cm wide

LB narrowly triangular, long-tapered, about 2 cm wide, 15–20 cm long, narrow at the edge, white-edged

Sc slender, upright

SB in the lower part of the scape similar to the leaves, in the upper part shorter, almost upright, longer than the stem segments

I loose panicle

PB the basal ones longer or slightly shorter than the distinctly stemmed spikes, the upper ones much shorter

FB broadly oval, blunt, shorter than the sepals, longitudinally ribbed, thin, yellow orange, 6 mm long

Fl single-sexed, white; the petals slightly longer than the sepals, which are about 9 mm long, asymmetrical and yellow orange

Ha Mexico, Guatemala and Costa Rica; epiphytic on trees at altitudes of 1,000–2,500 m.

Catopsis floribunda (Brongn.) L. B. Smith, 1937

(*floribunda(us)* heavily flowered; because of the much-branched inflorescences)

Syn. *Pogospermum floribundum* Brongn., 1864

Pl stemless, 40–70 cm high including inflorescence

Fo numerous, upright, 20–40 cm long

LS longitudinally oval, gradually merging into the blade, 12–14 cm long, 6–7 cm wide

LB narrowly triangular, above the sheath about 3 cm wide

Sc upright to arched, slender, 30–40 cm long, 5 mm thick

SB similar to the leaves, narrowly triangular, upright to slightly spread, much longer than the stalk segments

I loosely compound, 20–40 cm long

PB narrowly triangular, shorter than the sterile sections of the branches

Sp open, upright to pendulous, 20–25 cm long, very thin, stemmed

FB oval-triangular, 2–3 mm long, shorter than the asymmetrical sepals, which are 4–6 mm long

Fl white, hermaphroditic, arranged in a spiral on the rhachis, almost upright

Se asymmetrical, 4–6 mm long

Ha From Florida to Venezuela; epiphytic in forests at altitudes around 1,500 m.

Catopsis hahnii Baker, 1887 (Ill. 146)

Syn. *C. oerstediana* Mez, 1896

Pl sometimes over 50 cm high

Fo 20–40 cm long, in a dense rosette

frequently coated with white meal

LS indistinct, gradually merging into

the blade, elliptical, approximately as long as the triangular, long-tapered blades, which are 3–4·5 cm wide
Sc upright or angling upward, strong
SB similar to the leaves, tightly overlapping, much longer than the stalk segments
I open, compound, 11–25 cm long
PB lanceolate-oval, long-tapered, almost upright, longer than the distinctly stemmed, dense spikes
FB oval-elliptical, blunt, shorter than the asymmetrical sepals, which are 6–9 mm long
Fl hermaphroditic, white; petals hardly longer than the sepals
Ha Southern Mexico, Guatemala, Honduras to Nicaragua; epiphytic in forests at altitudes from 1,300–2,500 m.
CU Of all the *Catopsis* species, *C. hahnii* is the one most commonly found in collections.

Catopsis morreniana Mez, 1896 (Ill. 52)
(*morreniana* named after the Belgian botanist and bromeliad researcher E. Morren, 1833–1886)
Syn. *C. bakeri* Mez, 1903

Pl 20–40 cm high
Fo 10–18 cm long, forming a pale to bright green rosette sometimes coated with white meal
LS longitudinally oval, 4 cm wide, gradually merging into the tongue-shaped, pointed blade, which is 1·5–2·5 cm wide with a narrow, white edge
Sc upright, slender
SB upright, overlapping, longer than the stalk segments
I open panicle, approximately 6–17 cm long, naked
PB usually shorter than the stemmed spikes, which are 4–7 cm long
FB oval, shorter than the asymmetrical sepals, which are 3–6 mm long, flared
Fl hermaphroditic or single-sexed, white
Ha From southern Mexico to Costa Rica; epiphytic, but also terrestrial from sea level up to 1,650 m.

Catopsis nutans (SW.) Griseb., 1864
(*nutans* nodding; because of the arched inflorescences)

Pl stemless, 14–40 cm high
Fo up to 24 cm long, forming a small, crisp green rosette partially coated with meal
LS not distinct, longitudinally oval, 7–8 cm long, 4 cm wide
LB almost triangular, 2·5 cm wide, short to long-tapered
Sc slender, usually arching
SB upright, remote, shorter than the stalk segments, 2 cm long
I simple or compound, with few open spikes up to 20 cm long
FB upright to flared, 5–7 mm long, shorter than the sepals, which are asymmetrical and about 1·5 cm long
Fl hermaphroditic, bright yellow, up to 2 cm long
Ha From southern Mexico to Venezuela and Ecuador; epiphytic. A

variable species, which is divided into the two varieties: var. *robustior* and var. *stenopetala*.

Catopsis sessiliflora (Ruiz et Pav.) Mez, 1896 (Ill. 147)

(*sessiliflora*(*us*) with a stemless flower)
Syn. *Tillandsia sessiliflora* Ruiz et Pav., 1802; *C. nutans* var. *erecta* Wittm., 1889

Pl stemless, 10–30 cm high
Fo forming a slender rosette
LS not distinct, gradually merging into the blade, longitudinally oval, 7–8 cm long, 3·5–4·5 cm wide
LB upright to spreading, in the upper part recurved, tongue-shaped, rounded at the tip, but with a short thorn, crisp green, partially coated with white meal, 1·2–2·5 cm wide
Sc upright, slender, 15–18 cm long
SB upright, remote, shorter than the stalk segments, 1·1–1·5 cm long

I up to 11 cm long, simple, open or composed of few, flared, loosely flowered, spiral spikes
PB much shorter than the sterile bases of the spikes
FB broadly oval, blunt, shorter than the asymmetrical, thin petals, which are 7–8 mm long
Fl hermaphroditic, white; the petals hardly longer than the sepals
Ha From southern Mexico to southern Brazil and Colombia.

Catopsis wangerinii Mez et Wercklé, 1904 (Ill. 148)

(*wangerinii* named for the botanist Wangerin from Königsberg (Kaliningrad))
Syn. *C. pusilla* Mez et Wercklé, 1916; *C. cucullata* L. B. Smith, 1934; *C. triticea*, L. B. Smith, 1934

Pl 15–20 cm high including inflorescence
Fo up to 18 cm long, green, partially coated with chalk white meal
LS not distinct, the broadened bases 6–7 cm long, 4 cm wide
LB 2·5 cm wide, 10–11 cm long, lineally lanceolate or triangularly pointed
Sc approximately as long as the leaves, 10–12 cm long, upright or arched, 2–3 cm thick
SB loose, somewhat longer than the stalk segments, upright, 1·5–2 cm long, green
I simple or composed of several spikes

PB similar to the upper scape bracts, shorter than the spikes, which are open or dense, multi-flowered and up to 8 cm long
FB spiral, almost upright, triangularly lanceolate, pointed, 1·3 cm long, 7 mm wide, thin, green, with a cutaneous edge, longer than the asymmetrical, green sepals, which are 8 mm long and 5 mm wide at the base
Fl 1·1 cm long, sessile, almost upright, spiral, smelling strongly of gilly flowers (sea stock); petals white, with recurved tips
Ha Southern Mexico and Central America.

Guzmania Ruiz et Pav

The genus *Guzmania* was established in the year 1802 by Ruiz and Pavon and was named after the Spanish apothecary A. Guzman. It comprises about 100 known species, of which only a small number is in cultivation. Unlike the tillandsias, this genus contains a number of species that have attracted the attention of nurserymen and play an important role as trade plants. They are grown *en masse* in many nurseries, especially the hybrids which have been created under nursery cultivation. For the plant fancier and flower fan, *Guzmania* is thereby of special interest, since many species possess beautifully coloured flowers and scape bracts or decoratively marked leaves. Because of their size, however, most members of the genus can only be grown in indoor glass window boxes (see page 77), small greenhouses or in botanical gardens. For terrariums, only relatively few small species can really be considered practicable.

The major growing habitat for guzmanias is in northwest South America, where they grow epiphytically and on occasions in tropical rain forests. By their nature they have softer leaves than the related tillandsias and require more attention to warmth, humidity and shade. As potted plants, or on epiphyte trees, they are kept relatively damp and shady.

When cultivation differs from the above information, this is given under the species description.

The general characteristics of the genus are:

1 Leaves are never dentate or thorned, but always entire, usually forming a water-holding funnel-shaped rosette.
2 Petals are fused into a tube, no basal scale.
3 Ovaries are superior.
4 Fruit is a capsule.
5 Seeds have a pappus (crown hair) attached to the base. The pappus lies straight in the capsule, not folded.

Guzmania angustifolia (Baker) Witt., 1889 (Plate 50)

(*angustifolia(us)* narrow-leaved; refers to the narrow leaves)
Syn. *Caraguata angustifolia* Baker, 1884; *G. bulliana* André, 1886; *G. caulescens* Mez et Sodiro, 1905

Pl forming clumps, with a stem that is branched, short or up to 20 cm long
Fo spiral, dense
LS distinct, oval, 2–2·5 cm wide, 2·5–3 cm long, sometimes scaled with brown dots, with thin, reddish-longitudinal lines
LB upright or separated, pliant, 8–12 cm long, 1–1·3 cm wide above the sheath, narrowly lanceolate, long-

tapered, green to reddish-green, especially on the underside covered with short, grey scales
Sc slender, usually hidden by the leaves
I simple-spiked, almost cylindrical, few flowers, up to 6·5 cm long, 4 cm thick, sterile toward the tip, arched
FB oval-elliptical, short to long pointed, upright, shiny red, sometimes with a dark tip, longer than the elliptical, blunt, thin-skinned, naked sepals, which are 1·5–2 cm long and fused

3 mm high
Pe bright yellow, up to 7 cm long, in the lower part fused into a tube, the tips separate, almost elliptical, blunt
St 4 mm shorter than the petals
G. angustifolia var. *nivea*: differs from the type by having white floral bracts
Ha From Costa Rica, Panama, Ecuador to Columbia; epiphytic on trees in rain forests at an altitude from 1,000–1,300 m.
CU One of the few small guzmanias suitable for terrariums.

Guzmania berteroniana (Roem. et Schult.) Mez, 1896 (Plate 51)

(*berteroniana* named after the Italian botanist C. G. Bertero, 1789–1831)
Syn. *caraguata berteroniana* Roem. et Schult., 1830; *Devillea speciosa* Balbis, 1830; *Tillandsia caraguata* Dietr., 1840; *Tillandsia devillea* Steud., 1841; *Caraguata grandiflora* Baker, 1889

Pl stemless, up to 40 cm high
Fo forming a large rosette up to 60 cm in diameter
LS and **LB** elliptical, 8–9 cm wide, 10–12 cm long, gradually merging into the 30 cm long blade, which is 5–6 cm wide above the sheath, lineal, short-tipped, green and partially reddish-brown on the underside
Sc short, upright, hardly surpassing the leaves
SB upright, tightly overlapping, much longer than the stalk segments, broadly triangular, 3·5 cm wide, 5 cm long
I a simple, cylindrical spike up to

25 cm long and 4 cm thick, but in cultivation usually much smaller; no flowers toward the tip
FB bright, cinnabar red, tightly overlapping, 3·5 cm wide, 4 cm long, broadly oval, pointed, thin
Se up to 2·2 cm long, 9 mm wide, broadly elliptical, rounded, fused 2 mm at the base
Pe yellow, 5·5–5·9 cm long, separate and flared in the upper part
St 4 mm shorter than the petals
Ha Puerto Rico. A splendid species at bloom time with its shiny red inflorescence.

Guzmania calothyrsus Mez, 1896 (Ill. 53)

(*calothyrsa(us)* with a beautiful staff; refers to the shape of the staff-like inflorescence. Strictly speaking, the correct form of the name should be *G. calothyrsa*.
Syn. *Anoplophytum calothyrsus* Beer, 1857, nom. nud; *Tillandsia calothyrsus* Poeppig, 1857, nom. nud.

Pl stemless, forming a funnel rosette up to 70 cm high

LS 12–13 cm long, 6–7 cm wide, pale brown

LB 40–50 cm long, 3 cm wide, lineal, long-tapered, green
Sc strong, upright, surpassing the leaves, brown-scaled, 5 mm thick
SB tightly overlapping, lanceolate, long-tapered, brown-scaled, upright, longer than the stalk segments
I simple, tightly cylindrical, bearing flowers up to the tip, 6–12 cm long, 3 cm thick
FB broadly elliptical, triangularly pointed, 2·5–3·5 cm long, 2 cm wide, tightly overlapping and arranged in a spiral, dark brown-scaled, thin
Se about 2·5 cm long, about 7 mm wide, fused 8 mm high at the base, narrowly elliptical, pointed, brown-scaled
Pe about 3·5 cm long; tips of the petals 8 mm wide, rounded, white, spread
St 2 mm shorter than the petals
Ha Northern Peru (Province Huanuco and Junin); epiphytic at an altitude of 1,700 m.

Guzmania conifera (André) André, 1896

(conifera(us) bearing a cone; because of the cone-shaped inflorescence)
Syn. Caraguata conifera André, 1888

Pl stemless, rosette, up to 1 m high including the inflorescence
Fo 60–80 cm long, short-scaled
LS indistinct, dark-scaled
LB tongue-shaped, pointed, 6–8 cm wide
Sc strong, upright
SB the basal ones similar to the leaves, the upper ones lanceolate, long-tapered, red
I simple, multifarious, densely cone-shaped, up to 13 cm long
FB tightly overlapping, stiff, triangularly pointed, 4–6 cm long, red with yellow tips
Fl sessile, 6·5–7 cm long; petals yellowish, fused quite high, blunt
Se almost triangular, 2·5–3 cm long, naked, stiff, dark brown with pale edges
Ha Peru and Ecuador; epiphytic and terrestrial at altitudes from 1,300–1,700 m.

Guzmania coriostachya (Griseb.) Mez, 1896

(coriostachya(us) bug-like spikes)
Syn. Caraguata coriostachya Griseb., 1865; Tillandsia nigrescens André, 1888; G. michelii Mez, 1903; G. strobilifera Mez et Werckle, 1905

Pl stemless, 40–100 cm high including the inflorescence
Fo forming an almost upright rosette 40–50 cm in diameter
LS narrowly oval, 8–9 cm long, 4–5 cm wide, scales as brown dots, brown at the base, violet-longitudinal lines on the outside
LB lineal, short or long-pointed, 20–50 cm long, 2–3 cm wide, green, underside indistinctly scaled, upperside naked
SB the basal ones similar to the leaves, upright, longer than the stalk segments, the upper ones surrounding the scape, green with violet-longitudinal lines
I simple-spiked, cone-shaped, cyl-

indrical, multi-flowered, 5–8 cm long, 2–3 cm thick
FB upright, tightly overlapping, almost round, with a broad, blunt tip, not keeled, naked
Se oval, blunt, 1·3–1·6 cm long, hard,

briefly fused
Pe white, 2 cm long, the tips elliptical, blunt, flared
St enveloped in the flower
Ha Costa Rica, Panama, Venezuela, Colombia and Ecuador; epiphytic.

Guzmania danielii L. B. Smith, 1953
(*danielii* named after the Colombian botanist H. Daniel)

Pl stemless, up to 1·5 m high
Fo numerous, splayed, dense, close scales, up to 80 cm long
LS elliptical, up to 20 cm long
LB 8 cm wide, strap-shaped, pointed
Sc perpendicular, 2 cm thick
SB similar to the leaves, tightly overlapping, almost perpendicular

I cylindrical, 20 cm, long
FB elliptical, densely scaled, longer than the sepals, which are blunt and fused 2 cm high
Pe about 7 cm long
Ha Colombia; epiphytic in cloud forests at 2,000 m.

Guzmania dissitiflora (André) L. B. Smith, 1934 (Ill. 149)
(*dissitiflora(us)* loosely flowered; because of the open arrangement of the flowers)
Syn. *Sodiroa dissitiflora* André, 1888

Pl stemless, 40–90 cm high when in flower
LS elliptical, distinct, dark brown at the base, pale green with reddish-longitudinal lines higher
LB 30–90 cm long, 7–12 mm wide, lineal, long-tapered, green
Sc slender, upright, much shorter than the leaves, up to 5 mm thick, naked
SB overlapping, upright, the lower ones similar to the leaves, the upper ones elliptical, pointed, bright red, naked

I simple-spiked, open
FB 3–3·5 cm long, shorter than the sepals, surrounding the sepals, 1·3 cm wide, red
Fl usually flared, about 8 cm long, stemmed, white, with flared petals
Se 4 cm long, fused into a tube up to 3 cm high, tips rounded, yellow, naked
St about 5 mm shorter than the petals
Ha Costa Rica, Panama and Colombia, from sea level to 1,000 m; epiphytic.

Guzmania donnell-smithii Mez, 1903 (Plate 52)
(*donnell-smithii* named after the bromeliad researcher Donnell-Smith)
Syn. *Schlumbergeria donnell-smithii* Harms, 1930

Pl stemless, rosette, 30–50 cm high
LS 10–12 cm long, 5–6 cm wide, longi-

tudinally elliptical, gradually merging into the blade, green with violet-longitudinal lines

LB up to 40 cm long, average width 3–3·5 cm, lineal-lanceolate, pointed, green grey-scaled on the underside, with violet-longitudinal lines at the base

Sc perpendicular, not surpassing the leaves, 6–7 cm thick

SB tightly overlapping, sometimes violet on the underside

I composed of 10–18 spikes with 3 flowers, close in the beginning, loose later, up to 11 cm long and 4 cm wide

PB up to 3·5 cm wide, 2 cm long, broadly triangular, the basal ones with a long tip (up to 2·5 cm), the upper ones only short-tipped, bright red with green tip

Sp open, 3 cm long, short-stemmed, spiraled around the axis

FB 1·2 cm long, 1·3 cm wide, almost round, stemmed, cutaneous

Se 2 cm long, 6 mm wide, lineal, rounded at the tip, briefly fused at the base, yellow

Pe yellowish, 2·5 cm long with separate tips about 5 mm long

St 2 mm shorter than the petals

Ha Costa Rica. At present *Guzmania donnell-smithii* is a favourite especially for commercial purposes.

Guzmania erythrolepis Brongn., 1856 (Ill. 150)

(*erythrolepis* red-scaled)

Pl stemless, forming a rosette 35–50 cm high

LS about 10 cm long, 7–8 cm wide, broadly elliptical, brownish-green

LB up to 40 cm long, 4 cm wide, lineally strap-shaped, pointed

Sc short, upright, 10 cm long, about 1 cm thick

SB similar to the leaves, tightly overlapping, longer than the stalk segments, triangular, long-tapered, blunt, grey green, 2·5 cm wide, 5–6 cm long

I simple, fat cone-shaped, flowers up to the tip, about 10 cm long, 4–5 cm diameter, surpassed by the leaves

FB red, broadly oval, rounded, the lower ones short-tipped, 3 cm long, 2–2·5 cm wide, tightly overlapping

Se 2 cm long, 1 cm wide, fused for 4 mm at the base, broadly rounded, white, pink at the tip

Pe white, 3–3·7 cm long, elliptical, rounded; the tips separate

St 5 mm shorter than the petals

Ha Costa Rica and the Antilles.

Guzmania lindenii (André) Mez, 1896 (Ill. 152)

(*lindenii* named after the Belgian nurseryman and writer J. Linden, 1817–1898)

Syn. *Massangea lindeni* André, 1878

Pl stemless, forming a large funnel rosette, up to 2 m high including inflorescence

LS broadly oval, 12–15 cm long, 10–12 cm wide, brown at the base, yellowish-green in the upper part, with dark green to brownish-green streaks arranged in bands

LB up to 70 cm long, 5–7 cm wide, lineal, pointed, yellowish-green, with

dark green to brownish-green streaks arranged in horizontal bands
Sc sturdy, upright to more than 1 m high
SB similar to the leaves, overlapping, oval-triangular, pointed, green with streaks
I branched into an open panicle
PB broadly triangular or elliptical, pointed, shorter than the egg-shaped or ellipsoid spikes, which are up to 6 cm long and cone-like

FB broad, egg-shaped, blunt, green, as long as the elliptical, pointed sepals, which are 1·3 cm long
Pe white, with flared tips
Ha Northern and central Peru; in cloud forests on the Pacific side, in the Chanchamayo valley up to 2,500 m.
A species that is quite beautiful even in its non-blooming stage. The foliage can easily be confused with *G. musaica* (see page 223).

Guzmania lingulata (L.) Mez, 1896 (Plates 6 and 7)
(*lingulata(us)* tongue-shaped; because of the tongue-shaped scape bracts)
Syn. *Tillandsia lingulata* L., 1753; *Caraguata lingulata* Lindl., 1827; *Caraguata splendens* Bouché, 1856; *Caraguata lingulata* var. *cardinalis* André, 1880; *Caraguata cardinalis* André, 1883; *G. lingulata* var. *cardinalis* André, 1896; *G. lingulata* var. *splendens* Mez, 1935; *G. cardinalis* Mez, 1935

Pl stemless, rosette-shaped, more than 30 cm high including the inflorescence
LS distinct, oval, densely brown-scaled, brown at the base, sometimes with thin, violet-longitudinal lines
LB 25–35 cm long, strap-shaped, pointed, green
Sc upright, sturdy, shorter than the leaves
SB tightly overlapping, the lower ones similar to the leaves, the upper ones lanceolate, usually red, flared and forming a kind of involucre for the flowers
I drawn together into a head, up to

7 cm wide, 10 to 50 flowers
FB lineal, shorter than the upright, short-stemmed, white to yellowish flowers, which are 4·5 cm long and whose lineal petals are drawn together into a cap at the tip
Se lineal, blunt, naked
Ha Central America, Colombia, West Indian islands to Guyana, Bolivia, and Brazil; epiphytic
CU A species of many forms and in many collections. There are many commercial cultivars and hybrids of this species.

Guzmania melinonis Regel, 1885 (Plate 55)
(*melinonis* honey yellow; because of the honey yellow flowers)
Syn. *Caraguata melinonis* Morren, 1889; *G. erythrocephala* Hort., 1889

Pl stemless, rosette-shaped, 20–25 cm high including the inflorescence

LS broadly elliptical, 6 cm long, 4 cm wide, green with narrow, reddish-

longitudinal stripes
LB 30–40 cm long, 2–4 cm wide, lineal, pointed or blunt, with a thorn tip, pale green, reddish-violet on the underside, grey-scaled
FB pink to orange, inverse egg-shaped, pointed, shorter than the blunt, yellow sepals, which are 2·5–4·5 cm long
Pe yellowish-white, up to 3·5 cm long, upright, the upper 9 mm not fused
Sc upright, naked, shorter than the leaves
SB overlapping, upright, broadly oval, pointed, red, longer than the stalk segments
I elliptical to cylindrical, 5–10 cm long, 2 cm wide, simple, bearing flowers up to the tip
FB red, broadly oval, almost as long as the flowers, thin
Se broadly elliptical, rounded, 1·5 cm long, 7 mm wide, naked, briefly fused
Pe yellow or white, 2·8 cm long, the tips separated about 7 mm, rounded
St 5 mm shorter than the petals
Ha French Guyana, Colombia, Ecuador, Peru and Bolivia; epiphytic on trees.

Guzmania minor Mez, 1896 (Plate 54)
(*minor* small)

Pl stemless and rosette-shaped, approximately 20–25 cm high including the inflorescence
LS longitudinally oval, 6 cm long, 3–4 cm wide, brown at the base, otherwise green or reddish
Sc short, upright, naked, shorter than the leaves, 5 mm thick
SB the basal ones similar to the leaves, upright, green or greenish-red, the upper ones lanceolate, long-tapered, flared, bright red or orange and forming a kind of involucre for the flowers
I few flowers, simple
FB similar to the upper scape bracts but narrower and thinner
Se lineal, pointed, about 2 cm long
Pe yellow with a white tip, lineal, about 3 cm long
St as long as the petals
Ha Costa Rica, Panama, Colombia, Ecuador and Brazil; epiphytic on trees. A very pretty, small species, that is quite variable in regard to shape, size and colour. It resembles *G. lingulata* and could be considered a variety thereof.

Guzmania monostachya (L.) Rusby, 1896 (Plate 9)
(*monostachi(y)a(us)* single-spiked)
Syn. *Renealmia monostachya* L., 1753; *Tillandsia monostachya* L., 1762; *G. tricolor* Ruiz et Pav., 1802

Pl stemless, 20–40 cm high including the inflorescence
LS longitudinally oval, 8–10 cm long, 5–6 cm wide, green or brownish
LB lineal, long-pointed, 25–30 cm long, 2·5 cm wide, pale green
Sc upright, 10–15 cm long, 1 cm thick, much shorter than the leaves, naked
SB oval-triangular, long-pointed, overlapping, longer than the stalk segments, pale green
I simple, cylindrical, 8–15 cm long,

2–3 cm wide, no flowers in the upper part, naked, in the beginning dense, becoming loose as it ages

FB oval, pointed, thin, 2–3 cm long, variously shaped and coloured depending on their position in the inflorescence: the basal ones 1·5–3 cm wide, long-tapered, the fertile floral bracts short-tipped, whitish, brown or reddish-nerved, the upper ones sterile and bright red

Fl white, up to 3 cm long; the tips of the petals separate

Se about 1·6 cm long, 9 mm wide, longitudinally oval, rounded at the tip, briefly fused at the base

Ha From Florida, the West Indies, Nicaragua to Venezuela, Colombia, Peru and Bolivia; epiphytic on trees or terrestrial, from sea level to 2,000 m. At bloom time a very decorative species, which is mass produced in the trade.

Guzmania mucronata (Griseb.) Mez, 1896 (Ill. 151)
(*mucronata(us)* equipped with a pointed tip; because of the sharply pointed floral bracts)
Syn. *Tillandsia mucronata* Griseb., 1864

Pl stemless, up to 40 cm high
LS broadly oval, 12–15 cm long, 9–10 cm wide, brownish to pale green
LB 30–40 cm long, 5–6 cm wide, lineal, rounded or pointed, with a short thorn tip, green, sometimes edged in brownish-red
Sc strong, upright, 9 mm thick, shorter than the leaves
SB tightly overlapping, similar to the leaves, upright, green
I simple, tight cone-shaped, up to 10 cm long, 4–5 cm thick, bearing flowers to the tip

FB stiff, tightly overlapping, broadly triangular, sharp-pointed, 5 cm long, 4 cm wide, yellowish-green with a brownish-red, narrow edge
Se about 3 cm long, about 1·1 cm wide, fused 6–7 mm high at the base
Fl yellowish-green, about 8 cm long, the upper 2·5 cm separate; tips of the petals flared at bloom time, 1·3 cm wide, rounded
St almost as long as the petals, which are surpassed by the pistil
Ha Venezuela; epiphytic on trees.

Guzmania musaica (Linden et André) Mez, 1896 (Ill. 41, Plate 56)
(*musaica(us)* mosaic-like; because of the beautiful leaf markings)
Syn. *Tillandsia musaica* Linden et André, 1873; *Caraguata musaica* André, 1877; *Massangea musaica* Morren, 1877

Pl stemless, rosette-shaped, 30–50 cm high
LS 10–12 cm long, 9–10 cm wide, oval, imperceptibly merging into the blade, greenish-brown
LB up to 60 cm long, 6–8 cm wide, broadly lineal, rounded at the tip, but with a short thorn, pale green with dark green to reddish-brown line markings running horizontal, underside violet to reddish
Sc upright, shorter than the leaves
SB tightly overlapping, broadly elliptical, short to long-tapered, pink

I simple-spiked, in a dense head, 12 to 25 flowers, naked
St shorter than the petals
Ha Panama and Colombia; epiphytic in rain forests and mangrove swamps from sea level to an altitude of 500 m.

Because of the leaf markings, *G. musaica*, in its non-blooming stage, can easily be confused with *G. lindenii* (see page 220). Even when not in bloom it is an extremely decorative plant.

Guzmania nicaraguensis Mez et C. F. Baker, 1903 (Plate 53)
(*nicaraguensis* from Nicaragua)
Syn. *G. bracteosa* sensu Donn. Smith, 1909, non André, 1896

Pl stemless, rosette-shaped, up to 30 cm high
LS 9–10 cm long, 5–6 cm wide, brownish-green
LB up to 25 cm long, 2 cm wide, lineal, rounded, pointed, green underside with thin, reddish-longitudinal stripes
Sc about 8 cm long, 5 mm thick, upright and slightly arched, shorter than the leaves
SB tightly overlapping, upright, longer than the stalk segments, 3–4 cm long, oval-triangular, pointed
I simple-spiked, cylindrical, 7–10 cm long, up to 2 cm thick, bearing flowers to the tip
FB elliptical, blunt or broadly pointed, about 5 cm long, 1·5–2·5 cm wide, thin, red
Se about 2·3 cm long and 8 mm wide, blunt, briefly fused at the base
Pe yellow, about 6·5 cm long, the upper 1·2 cm separate, the tips 5 mm wide, blunt, flared
St 5 mm shorter than the petals
Ha Costa Rica, Guatemala, Honduras, Nicaragua and Panama; epiphytic on trees.

Guzmania sanguinea (André) André, 1896 (Plate 3)
(*sanguinea(us)* blood red; because of the colour of the inner rosette leaves at bloom time)
Syn. *Tillandsia sanguinea* André, 1879; *Caraguata sanguinea* André, 1883; *G. crateriflora* Mez et Wercklé, 1905

Pl stemless, up to 20 cm high, in a broad, flat rosette
LS 9–10 cm long, 6–7 cm wide, light brown with violet-longitudinal stripes
LB 15–25 cm long, 3–4 (to 8) cm wide, broadly pointed to rounded; the inner rosette leaves shorter, at bloom time bright red, or orange and yellow, spotted with green
I sunken into the rosette, simple, 7 to 12 flowers
FB shorter than the sepals, 1·5–1·8 cm long, 1·4 cm wide, thin-skinned, similar to a calyx
FL with a flattened stem about 8 mm long, yellow, about 6 cm long; petals separate in the upper part, flared and rounded
Se about 1·8 cm long, 6 mm wide, briefly fused at the base
St 5 mm shorter than the petals
Ha Costa Rica, Trinidad, Ecuador; epiphytic from sea level to 1,000 m.
CU With its colourfully spotted,

partly bright red centre leaves, *G. sanguinea* at bloom time is certainly one of the most beautiful guzmanias.

After bloom the beautiful colour of the inner leaves fades.

Guzmania strobilantha (Ruiz et Pav.) Mez, 1896
(*strobilantha(us)*) cone-flowered; because of the cone-shaped inflorescence)
Syn. *Bonapartea strobilantha* Ruiz et Pav., 1802; *Tillandsia strobilantha* Poir., 1817; *Acanthospora conantha* Spreng., 1825; *Misandra strobilantha* Dietr., 1830; *Anoplophytum strobilanthum* Beer, 1857; *Tillandsia conantha* Baker, 1889; *G. parviflora* Ule, 1907

Pl stemless, over 60 cm high
LS oval
LB 35–45 cm long, 2–2·5 cm wide, lineal, long-tapered
Sc slender, upright
SB overlapping, similar to the leaves
I simple, cone-like, spherical or ellipsoid, up to 5 cm long
FB broadly elliptical, rounded, pointed, 1·5 cm long, stiff, naked, striped
Se elliptical, blunt, 1·2–1·5 cm long, fused 2 mm high at the base
Pe yellowish-white
St as long as the petals
Ha Northern Peru (Provinces of Lareto, Huanuco and Junin), at altitudes from 600–1,000 m.

Guzmania variegata L. B. Smith, 1960 (Ill. 153)
(*variegata(us)*) colourful)

Pl stemless, 50–80 cm high when in flower
LS broadly oval, 15–18 cm long, 11 cm wide, tongue-shaped, pointed, dark green, upperside grey-scaled
Sc strong, upright, shorter than the leaves, 20–30 cm long, 1·5 cm thick, naked
SB upright, longer than the stalk segments, similar to the leaves, with spread blades
I loosely compound, with 8 to 12 spread to horizontal, short spikes, 30–60 cm long, 15 cm wide
PB bright red, similar to the scape bracts, horizontal, surrounding the spikes in the shape of a bowl; the basal ones up to 10 cm long and 4 cm wide
Sp rounded-ellipsoid, 4–5 cm long, about 10 flowers, seated at the base on a thickened, disc-shaped stem
FB about 3 cm long and 2 cm wide, slightly keeled, almost as long as the elliptical, blunt, green sepals, which are 2–8 cm long, 9 mm wide, briefly fused at the base, thin, longitudinally ribbed and with a slight keel
Pe white, about 4 cm long
Ha Ecuador and northern Peru; epiphytic on trees and terrestrial
CU Very decorative at bloom time.

Guzmania vittata (Mart.) Mez, 1896 (Ill. 154)

(*vittata*)*us*) banded; because of the markings on the leaves)

Syn. *Bonapartea vittata* Mart., 1830; *Caraguata vittata* Baker, 1889

Pl stemless, up to 55 cm high
Fo 35–55 cm long, 1·5–2·5 cm wide, lineal, pointed, green, with broad, dark latitudinal bands and broadly oval sheaths, forming a large funnel rosette
Sc perpendicular, slender, naked
SB tightly overlapping, lanceolate, long-tapered
I broadly egg-shaped, composed of few spikes or simple and spherical, 3–4 cm long
PB triangularly oval, pointed, usually shorter than the sessile spikes, which have 10 to 15 dense flowers
FB oval, 1·3 cm long, hard, shorter than the broadly elliptical, blunt sepals, which are 1·5 cm long
Pe white, over 2 cm long
Ha Colombia and Brazil; epiphytic at altitudes from 200–500 m.

Guzmania zahnii (Hook. f.) Mez, 1896 (Plate 57)

(*zahnii* named after the collector Zahn, who collected mainly in Central America.

Syn. *Caraguata zahnii* Hook f., 1873; *Tillandsia zahnii* Hort., 1873; *Encholirion zahnii* Hort, 1910; *Schlumbergeria zahnii* (Hook f.) Harms, 1930.

Pl stemless, up to 60 cm high including the inflorescence
LS longitudinally oval, imperceptibly merging into the blade, 16–17 cm long, 7–8 cm wide, pale green with reddish-longitudinal stripes
LB up to 60 cm long, 2·5–3·5 cm wide, lineal, pointed, upper side dark green, lower side brownish-green
Sc upright, 6 mm thick, only slightly longer than the leaves
SB similar to the leaves, upright, much longer than the stalk segments, tightly overlapping, reddish
I composed of 10 to 15 clustered spikes, 10–15 cm long, 4 cm wide, loose at the base, dense toward the tip
PB yellow, with a red tip; the basal ones longer than the upper spikes, which are short-stemmed, arranged in a spiral and have 10 to 15 flowers
FB about 1·3 cm long, broadly elliptical, rounded, keeled, shorter than the sepals, which are up to 1·8 cm long, 7 mm wide, narrowly elliptical, rounded, yellow and green-tipped and have a slight keel
Fl yellow, about 3 cm long
Ha Costa Rica. *G. zahnii* is a very beautiful species of commercial value. There is a series of cultivars of this plant in the nursery and florist trade.

Vriesea Lindl.

The genus *Vriesea* was established in 1843 by J. Lindley and named after the Dutch professor of botany H. de Vries (1807–1862). The number of

Plate 1   *Tillandsia ionantha* Planch.

Plate 2   *Tillandsia brachycaulos* Schlechtend.

Plate 3   *Guzmania sanguinea* (André) André.

Plate 4   *Aechmea mariae-reginae* Wenol.

Plate 5   *Tillandsia intumescens* L. B. Smith.

Plate 6   *Guzmania lingulata* (L.) Mez.

Plate 7   *Guzmania lingulata* (L.) Mez. inflorescence shown from above.

Plate 8   *Tillandsia multicaulis* (Linden et André) Mez.

Plate 9   *Guzmania monostachya* (L.) Rusby.

Plate 10   *Tillandsia wagneriana* L. B. Smith.

Plate 11   *Vriesea carinata* Wawra.

Plate 12   *Tillandsia leiboldiana* Schlechtend.

Plate 13   *Tillandsia bourgeaei* Baker.

Plate 14   *Tillandsia imperialis* Morren.

Plate 15   *Tillandsia cacticola* L. B. Smith.

Plate 16 *Tillandsia andrieuxii* (Mez) L. B. Smith.

Plate 17 *Tillandsia benthamiana* (Beer) Klotsch.

Plate 18 *Tillandsia macdougallii* L. B. Smith.

Plate 19 *Tillandsia oaxacana* L. B. Smith.

Plate 20   *Tillandsia achyrostachys* Morren.

Plate 21   *Tillandsia biflora* Ruiz et Pav.

Plate 22   *Tillandsia atroviridipetala* Matuda.

Plate 23   *Tillandsia plumosa* Baker.

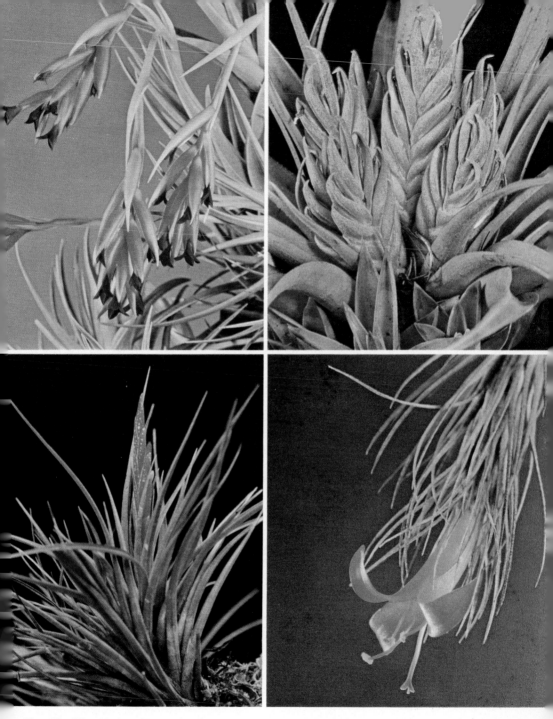

Plate 24   *Tillandsia aeranthos* (Loisel.) L. B. Smith.

Plate 25   *Tillandsia carlsoniae* L. B. Smith.

Plate 26   *Tillandsia dura* Baker.

Plate 27   *Tillandsia funckiana* Baker.

Plate 28 *Tillandsia anceps* Lodd.

Plate 29 *Tillandsia cyanea* Linden.

Plate 30 *Tillandsia lindeni* Regel.

Plate 31 *Tillandsia hamaleana* Morren.

Plate 32　*Tillandsia concolor* L. B. Smith.

Plate 33　*Tillandsia ixioides* Griseb.

Plate 34　*Tillandsia fasciculata* SW.

Plate 35　*Tillandsia flabellata* Baker.

Plate 36   *Tillandsia gardneri* Lindl.

Plate 37   *Tillandsia geminiflora* Brongn.

Plate 38   *Tillandsia ionochroma.*

Plate 39   *Tillandsia lampropoda* L. B. Smith.

Plate 40   *Tillandsia mauryana* L. B. Smith.

Plate 41   *Tillandsia magnusiana* Wittm.

Plate 42   *Tillandsia punctulata* Schlechtend. et Cham.

Plate 43   *Tillandsia pruinosa* SW.

Plate 44 *Tillandsia stricta,* white-leafed form.

Plate 45 *Tillandsia stricta,* green-leafed form.

Plate 46 *Tillandsia tenuifolia.*

Plate 47 *Tillandsia tricolor* Schlechtend. et Cham.

Plate 48   *Tillandsia unca* Griseb.

Plate 49   *Tillandsia xerographica* Rohw.

Plate 50   *Guzmania angustifolia* (Baker) Wittm.

Plate 51   *Guzmania berteroniana* (Roem. et Schult.) Mez.

Plate 52   *Guzmania donnell-smithii* Mez.

Plate 54   *Guzmania minor* Mez.

Plate 53   *Guzmania nicaraguensis* Mez et C. F. Baker.

Plate 55   *Guzmania melinonis* Regel.

Plate 56 *Guzmania musaica* (Linden et André) Mez.

Plate 57 *Guzmania zahnii* (Hook. f.) Mez.

Plate 58 *Vriesea cylindrica* L. B. Smith.

Plate 59 *Vriesea olmosana* L. B. Smith.

Plate 60　*Vriesea maxoniana* (L. B. Smith) L. B. Smith.

Plate 61　*Vriesea platynema* Gaud. var. *platynema.*

Plate 62　*Vriesea psittacina* (Hook) Lindl. var. *psittacina.*

Plate 63　*Vriesea splendens* (Brongn.) Lem.

Plate 64   *Aechmea bracteata* (SW.) Griseb.

Plate 65   *Aechmea caudata* Lindm. *var. variegata* M. B. Foster.

Plate 66   *Aechmea chantinii* (Carr.) Baker.

Plate 67   *Aechmea comata* (Gaud.) Baker.

Plate 68 *Aechmea fasciata* (Lindl.) Baker.

Plate 69 *Aechmea dealbata* Morren.

Plate 70 *Aechmea kertesziae* Reitz.

Plate 71 *Aechmea distichantha* Lem. var *distichantha*.

Plate 72 *Aechmea fulgens* Brongn. var. *discolor* (C. Morren) Brongh.

Plate 73 *Aechmea miniata* (Beer) hort. var. *discolor* (Beer) Beer.

Plate 74 *Aechmea nudicaulis* (L) Griseb. var. *nudicaulis*.

Plate 75 *Aechmea orlandiana* L. B. Smith.

Plate 76   *Aechmea ornata* (Gaud.) Baker.

Plate 77   *Aechmea ferruginea* L. B. Smith.

Plate 78   *Aechmea   pineliana*   (Brongn). Baker.

Plate 79   *Aechmea veitchii* Baker.

Plate 80   *Aechmea recurvata* (Klotsch) L. B.
Smith var. *recurvata* (after exposure to intense
sunlight).

Plate 81   *Aechmea recurvata* var. *benrathii*
(Mez) Reitz.

Plate 82   *Aechmea racinae* L. B. Smith.

Plate 83   *Aechmea weilbachii* Didr.

Plate 84   *Ananas comosus* (Lindl.) Schult.
var. *variegatus* hort.

Plate 86   *Billbergia leptopoda* L. B. Smith.

Plate 85   *Pseudananas   macrodontes*
(Morren) Harms.

Plate 87   *Billbergia macrocalyx* Hook.

Plate 88 *Billbergia nutans* H. Wendl. var. *nutans*.

Plate 90 *Billbergia vittata* Brongn.

Plate 89 *Billbergia pyramidalis* (Sims) Lindl. var. *pyramidalis*.

Plate 91 *Billbergia zebrina* (Herb.) Lindl.

Plate 92   *Fascicularia bicolor* (Ruiz and Pav.) Mez.

Plate 94   *Canistrum lindenii* (Regel) Mez var. *roseum* (Morren) L. B. Smith, *f. procerum* Reitz.

Plate 93   *Bromelia pinguin* L. (from 'Les Cèdres').

Plate 95   *Bromelia balansae* Mez.

Plate 96   *Aechmea cylindrata* Lindm.

Plate 97   *Ochagavia lindleyana* (Lem.) Mez.

Plate 98   *Cryptanthus bromelioides* Otto and Diedr. var. *tricolor* M. B. Foster.

Plate 99   *Cryptanthus fosterianus* L. B. Smith.

Plate 100   *Neoregelia coriacea* (Antoine) L. B. Smith.

Plate 101   *Neoregelia carolinae* (Mez) L. B. Smith var. *tricolor* L. B. Smith.

Plate 102   *Neoregelia concentrica* (Vell.) L. B. Smith.

Plate 103   *Neoregelia farinosa* (Mez) L. B. Smith.

Plate 104   *Neoregelia olens* (Hook f.) L. B. Smith.

Plate 105   *Neoregelia marmorata* (Baker) L. B. Smith.

Plate 106   *Neoregelia sarmentosa* (Regel) L. B. Smith var. *chlorosticta* (Baker) L. B. Smith.

Plate 107   *Neoregelia spectabilis* (Moore) L. B. Smith.

Plate 108   *Gravisia aquilega* (Salisb.) Mez.

Plate 109   *Hohenbergia stellatae* Schult. f.

Plate 110   *Neoregelia princeps* (Baker) L. B. Smith.

Plate 111   *Nidularium billbergioides* (Schult. f.) L. B. Smith.

Plate 112   *Nidalarium fulgens* Lam.

Plate 113   *Nidularium innocentii* Lem. var. *lineatum* (Mez) L. B. Smith.

Plate 114   *Nidularium rutilans* Morren.

Plate 115   *Nidularium scheremetiewii* Regel.

Plate 116   *Nidularium seidelii.*

Plate 117   *Nidularium terminale* (Vell.) Ule.

Plate 118   *Orthophytum navioides* L. B. Smith.

Plate 119   *Orthophytum rubrum* L. B. Smith.

Plate 120   *Portea kermesina* C. Koch.

Plate 121   *Portea leptantha* Harms.

Plate 122 *Quesnelia humilis* Mez.

Plate 123 *Quesnelia marmorata* (Lem.) R. W. Read.

Plate 124 *Quesnelia quesneliana* (Brongn.) L. B. Smith; part of the inflorescence.

Plate 125 *Streptocalyx poeppigii* Beer.

Plate 126 *Hechtia epigyna* Harms.

Plate 127 *Hechtia rosea* Morren.

Plate 128 *Pitcairnia atrorubens* (Beer) Baker.

Plate 129 *Pitcairnia heterophylla* (Lindl.) Baker.

Plate 130 *Pitcairnia nigra (Carr.)* André

Plate 131 *Puya sp.* (aff. *chilensis*) Chile, south of Illapal.

Plate 133 *Puya medica* L. B. Smith.

Plate 132 *Puya lanata* (H.B.K.) Schult. f.

Plate 134 *Puya nana Witten.*

known species cannot be stated exactly since new ones are constantly being discovered during collection trips in South and Central America—and research on living plants also causes constant reassessment of the taxonomy. Specifically, many tillandsias have to be assigned to the genus *Vriesea* because of the presence of a ligule at the base of the petals (see also page 48). These little scales can only be detected with any certainty on fresh flowers and only with great difficulty on dried specimens. They are practically the only point of distinction between the vrieseas and the tillandsias. In growth habit, many vrieseas resemble tillandsias to such a degree that they cannot be classified by examination of the foliage alone. There is thus a strong case for combining the two genera.

Of the 150 or so species that have been described, only a few are in cultivation. The number of species suited to commerce is relatively small, but these are grown in great numbers in specialty nurseries. Many species that are not suitable for the trade serve as hybridizing partners for new varieties.

The main habitat of vrieseas is Brazil; but they are also spread in smaller numbers over all of Central and South America.

The most important features of the genus are:
1 Leaves are never with thorns or dentate, but always entire, usually in a rosette, funnel-shaped, green or grey-scaled.
2 Petals are separate and bearing scales on the inner side of the base (Fig. 16, 2,L).
3 Ovaries are superior.
4 Fruit is a capsule.
5 Seeds are equipped with a pappus (crown hair); the hairs are straight in the capsule, not folded.

Most vrieseas are cultivated like the green tillandsias and guzmanias, with semi-shady and moisture. The rosette funnels should be kept full with rain water. They can be grown in a pot or on an epiphyte tree. The 'grey', tillandsia-like vrieseas need more light; they are grown primarily as epiphytes and need little watering.

Vriesea appenii Rauh, nov. spec., 1969 (Ill. 163)
(*appenii* named after Hans von Appen of Lima, Peru, who generously supported Rauh's expeditions)

Pl stemless, 80–100 cm high including the inflorescence
Fo numerous, forming a dense, almost upright rosette 80–90 cm in diameter
LS distinct, both sides brownish-black, 11–12 cm long and 8–10 cm wide
LB narrowly lanceolate, narrowing gradually into a long point, 4–5 cm wide above the sheath, up to 50 cm

long, both sides heavily grey-scaled
Sc upright, 30–35 cm long, surpassing the rosette
SB similar to the foliage, longer than the stalk segments, surrounding the scape with their sheaths, heavily grey-scaled
I up to 35 cm long and up to 18 cm wide, composed of 20 to 25 open spikes
PB the basal ones similar to the upper scape bracts, with a long, almost thread-like blade, the upper ones with only a short tip and shorter than the sterile section of the spikes
Sp sterile section 2–3 cm long, upright, fertile section almost horizontal from the axis of the inflorescence, 12–14 cm long, about 1 cm wide, 10 to 20 flowers
FB tightly overlapping, rounded on the back, about 1·7 cm long, 1 cm wide, oval triangular, red, heavily grey-scaled, hardly nerved

Fl upright
Se 1·2 cm long, 4 mm wide, thin, separate, red-tipped and heavily grey-scaled
Pe dark blue with recurved tips, smooth-edged, 1·8 cm long, 4 mm wide, a 9 mm long scale at the base
St and **Pi** enveloped in the perianth tube
Ha Northern Peru, Valley of Olmos near the pass Abra Porculla, 2,000 m; epiphytic on trees (Holotype Rauh 15420)
CU Bright and sunny, little watering. *V. appenii* at bloom time is a splendid plant, whose large inflorescence lasts several months. It is closely related to *V. heterandra* (André) L. B. Smith (see page 240 and Ill. 164), which ranges from Colombia over Ecuador to Bolivia.

Vriesea barclayana (Baker) L. B. Smith, 1951 (Ill. 311)

(*barclayana(us)* named after H. G. Barclay, collector)
Syn. *Tillandsia barclayana* Baker, 1887; *T. lateritia* André, 1888

Pl stemless, flowering 50–60 cm high
Fo numerous, forming a narrow funnel shaped rosette
LS 9–10 cm long, 7–8 cm wide, both sides chestnut brown scaled, elliptical, overall forming a slightly onion shaped swelling
LB narrow-triangular, long pointed, 30–40 cm long; 4 cm wide at the base, recurved, both sides densely grey scaled
Sc upright, 20–30 cm long, 6 mm thick, round, green, bare
SB upright, densely set, lying against the scape, the basal ones with shorter blade, the upper ones pointed or blunt tipped, scaled
I simple, loose sword shaped:

10–30 cm long; 3·5 cm wide; rachis straight, winged and enclosing or surrounding the flower bases, densely scaled
FB in two rows, overlapping, reversed-egg shaped (obovate), triangular-pointed, 3–4 cm long, as long as or somewhat longer than the sepals, scaled towards the tip, not keeled, carmine red
Se free, 2·5 cm long, elliptical, pointed or spine tipped, smooth, naked
Pe violet, 3·5 cm long, with two large scales at the base
St as long as or longer than the petals
Ha South Ecuador and North Peru, epiphytic in deciduous tree forests
CU *V. barclayana* is a 'grey' vriesea

sometimes mistaken for a tillandsia; it should be grown in similar conditions.

VARIETIES

var. minor is distinguished from the type by having all its parts smaller, and also from North Peru another form is known whose flower bracts which dry to a straw colour before flowering commences.

Vriesea bituminosa Wawra, 1862

(*bituminosa(us)* asphalt-like; probably because of the smell of the flowers)
Syn. *Tillandsia platynema* sensu Baker, 1888, non Griseb., 1864; *V. platynema* sensu Wittm., 1891

Pl 50–75 cm high including the inflorescence
Fo numerous, forming an attractive funnel rosette 40–50 cm in diameter
LS not distinct, gradually merging into the blade, 9–10 cm wide, 13–14 cm long, brown on both sides as the base
LB broadly lineal, broadly rounded, with a thorn tip, up to 70 cm long, 7–8 cm wide, slight markings, lower side pale lavender, leaf tips lavender brown
Sc strong, upright, 45–60 cm long

SB longer than the stalk segments, upright
I simple, sword-shaped, up to 50 cm long, 8 cm wide, broadly lineal, sometimes more than 40 flowers
FB oval, 3–4 cm wide, up to 4·5 cm long, stiff, no keel
Fl spread, up to 5·5 cm long, dirty yellow green
Se up to 3·5 cm long, almost oval, no keel
Ha Brazil.

Vriesea carinata Wawra, 1862 (Plate 11)

(*carinata(us)* boat-shaped, having a keel; because of the floral bracts, which have a distinct keel)
Syn. *V. brachystachys* Regel, 1866; *V. psittacina* var. *brachystachys* Morren, 1870; *V. psittacina* var. *carinata* Morren, 1882; *Tillandsia carinata* Baker, 1888; *Tillandsia psittacina* sensu Morton, 1893

Pl up to 30 cm high including the inflorescence
Fo 15 to 20 leaves, forming a small rosette 15–20 cm in diameter
LS 4–5 cm wide, 5–6 cm long, elliptical
LB strap-shaped, lanceolate, pointed with a thorn tip, 18–20 cm long, 2–3 cm wide above the sheath, crisp green
Sc upright, slender, 10–15 cm long, 3 mm thick

SB lying close to the scape, longer than the stalk segments, green or red
I simple sword-shaped, dense, 4–5 cm long, 4–5 cm wide, almost square
FB 3·5–4 cm long, 1·5 cm wide, with a keel, red at the base, yellowish-green toward the tip
Fl yellow, spread, 5 cm long
Se almost lanceolate, pointed, 3·1 cm long, 7 mm wide

St surpassing the petals
Ha Brazil, epiphytic in the rain forest
CU *V. carinata* has been in cultivation for a long time and more commonly known as *V. psittacina*. The inflores- cences are made attractive by the large, distinctly keeled, red and greenish-yellow floral bracts. A recommended species for cultivation.

Vriesea cereicola (Mez) L. B. Smith, 1958 (Ill. 158)

(*cereicola*(*us*) perched on cereus (cactus))
It is very closely related to *V. rauhii* L. B. Smith, 1958. The latter differs by its roundish, very narrow spikes and the short floral bracts, which reach only 2·7 cm.
Syn. *Tillandsia cereicola* Mez, 1906; *V. lopezii* L. B. Smith, 1953

Pl stemless, up to 45 cm high including the inflorescence
Fo numerous, forming an upright or spread rosette 40–70 cm in diameter
LS 6–8 cm long, 5·5 cm wide, longitudinally oval, dark brown-scaled, not distinct
LB 35–45 cm long, narrowly lanceolate, 4 cm wide above the sheath, long-tapered with curled edges, grey green, lower side heavily scaled
Sc thin, upright to angling upward, 25 cm long
SB longer than the stalk segments, about 4 cm long, surrounding the scape with their sheaths, with a short, recurved blade
I composed of about 4 finger-like spikes, which are 20–30 cm long, flattened and have 12 to 20 flowers
PB similar to the scape bracts, 3–4 cm long, naked, surrounding the sterile bases of the spikes
FB overlapping, about 3·5 cm long, 7 mm wide, narrowly lanceolate, thin, pink
Se shorter than the floral bracts, 6 mm wide, 2 cm long, greenish-white, thin, naked, separated
Pe greenish, sprinkled with fine, pale violet spots, reddish-violet when dried, about 3·5 cm long, 5 mm wide, 2 stiff scales at the base
St and Pi surpassing the bloom
Ha Northern Peru (Valley of the Rio Saña) at 800 m; epiphytic on cacti, trees and on cliffs
CU Sunny, little watering. An interesting species reminiscent of tillandsias.

Vriesea chrysostachys Morren, 1881, (Ill. 157)

(*chrysostachys* gold or yellow-spiked; because of the yellow floral bracts)
Syn. *Tillandsia chrysostachys* Baker, 1886; *Vriesea cryptantha* Hort., 1889; *Tillandsia trinitensis* Baker, 1889

Pl stemless, 55–65 cm high including the inflorescence
Fo numerous, forming a rosette 50–60 cm in diameter
LS not distinct, heavily brown-scaled at the base on both sides, 7–8 cm long
LB 30–40 cm long, 5–6 mm wide, lineal, pointed, dull bluish-green, laven-

der on the lower side at the base
Sc 35–40 cm long, upright, 7 mm thick, naked
SB tightly overlapping, longer than the stalk segments, upright, lying closely to the scape, the lower ones broadly oval, long-tapered, the upper ones blunt, brownish-red at the base, green toward the tip
I simple-spiked or composed of 1 to 3 spikes
Sp distichous, multi-flowered,

10–15 cm long, 2·5 cm wide
FB tightly overlapping, upright, rounded on the base, triangular, pointed, 2·5–3·5 cm long, stiff, naked, dull yellow, surpassing the flowers, which are about 2·5 cm long; the flowers are therefore not visible
Se lanceolate-oval, 1·5 cm long, yellowish, thin-skinned
St enveloped in the flower
Ha Colombia, Trinidad and Peru.

Vriesea corcovadensis (Britt.) Mez, 1894
(*corcovadensis* named after the mountain Corcovado, south of Rio de Janeiro)
Syn. *Tillandsia ventricosa* Wawra, 1880, non Griseb., 1865; *Tillandsia corcovadensis* Britt., 1888; *Tillandsia oligantha* Baker, 1889; *Vriesea rubida* Morren, 1894; *Vriesea ventricosa* Mez, 1896

Pl stemless, 20–30 cm high
Fo numerous, forming a pseudo-bulb-like, thickened, loose rosette 10–15 cm in diameter
LS distinct, 4–5 cm long, 3–4 cm wide, somewhat scooped out, lavender brown
LB narrowly lineal, pointed, up to 20 cm long, 1 cm wide above the sheath, dull green, short grey scales
Sc slender, upright, 10–15 cm long
SB similar to the leaves, upright, longer than the stalk segments

I simple or compound, loose, 7–10 cm long
FB red, about 3 cm long, approximately as long or somewhat longer than the narrowly elliptical sepals, which are up to 2·5 cm long
Fl yellowish-white, arranged in 3 to 4 lines around the axis
Ha Brazil
CU In its non-blooming stage it can easily be confused with *V. flammea* (see page 235).

Vriesea crenulipetala (Mez.) L. B. Smith, 1955 (Ill. 239, Fig. 46a)
(*crenulipetala*(*us*) toothed petals; because of the toothed petals)
Syn. *Tyllandsia crenulipetala* Mez. 1919; *T. attenuata* Rusby, 1920.

Pl stemless, flowering to 30 cm high; offsets with short stolons emerging from between the lower leaves
Fo numerous, in an upright rosette to 25 cm in diameter
LS distinctly set, roundish-oval, 2·5–3 cm long, to 3 cm wide, both sides

dark-leather brown scaled
LB narrow-lanceolate, to 25 cm long; 7·5 cm wide above the sheath, narrowing to a long thread like tip, both sides densely grey scaled. Scales brown centred
Sc thin, to 13 cm long, shorter than the

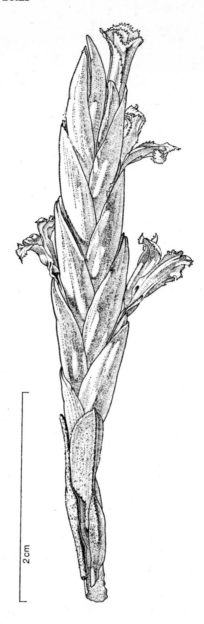

Fig. 46a *Vriesea crenulipetala* (Mez) L. B.
Smith; part of the inflorescence (see Ill.
312).

leaves but much longer than the spikes
SB densely set, the basal ones similar to the leaves, with longer thread like blade, the upper shorter, longer than the internodes, densely scaled
I upright, surpassing the leaves, with 3–5 arched, angled or upright spikes each about 5 cm long
PB 1·2–1·5 cm long, about as long as the leaf covered sterile lower part of the spikes; these 3–5 cm long, 1·2 cm wide, sword shaped
FB densely set, carmine red, bare, only somewhat scaled towards the weakly keeled apex, to 1·5 cm long, with indistinctly emerging nerves
Fl upright, to 1·8 cm long, sessile, pale-greenish, white when dry
Se about as long as the flower bracts, 13–15 mm long, pale green, rounded tip, free nearly to the base, the back ones indistinctly keeled
Pe exceeding the sepals by 0·8 cm, with flared toothed plates; scales about 1 cm high at the base
St and **Pt** enclosed deeply within the flower
Ha Colombia, epiphytic in cloud forest of the Cordillera Santa Marta in high places between 1,200 and 2,100 m.
CU A very attractive small species, reminding one of a tillandsia; in its habitat it grows in groups formed by the short stoloned offsets.

Vriesea cylindrica L. B. Smith, 1951 (Plate 58)
(*cylindrica(us)* cylindrical; because of the cylindrical inflorescence)

Pl stemless, up to 80 cm high including the inflorescence
Fo numerous, forming an upright to spread rosette 60–70 cm in diameter
LS oval, with a cutaneous edge, somewhat scooped out, 9–10 cm long, 5–7 cm wide, both sides brown-scaled, not clearly distinct from the blade
LB lineal, long-tapered, 40–50 cm long, 4 cm wide above the sheath, dull green, both sides grey-scaled
Sc upright, 15–20 cm long, about 8 mm thick, green, naked
SB similar to the upper rosette leaves, much longer than the stalk segments, surrounding the scape with the brownish-red, scaled sheaths, the recurved blade narrowly lanceolate, pointed, dull green, grey-scaled
I cylindrical, up to 30 cm long and 2 cm thick, composed of 13 to 18 upright spikes
PB upright, similar to the upper scape bracts, the sheaths approximately as long as the spikes; the basal ones with a short, recurved blade, the upper ones bladeless, reddish-brown to brick red or tan, scaled only on the edge
Sp arranged in a spiral, upright, lying against the axis of the inflorescence, short-stemmed, 6–8 cm long, 2–3 cm wide, flattened, 8 to 12 flowers
FB tightly overlapping, shorter than the sepals, about 2·4 cm long, about 1·2 cm wide, oval-triangular, rounded on the back, keeled toward the scaled tip, thin, yellowish-orange
Se separate, about 2·2 cm long and 8 mm wide, thin, naked, with a scaled, orange tip
Pe about 4 cm long and 6 mm wide, green, dark blue on the edge and on the recurved tips
St and **Pi** protruding from the flower
Ha Dry forests of northern Peru, up to 2,000 m.
CU Sunny, little watering. A very pretty species reminiscent of tillandsias.

VARIETIES

Vriesea cylindrica is very closely related to *V. harmsiana* and *V. tillandsioides.*

Vriesea harmsiana (L. B. Smith) L. B. Smith, 1951 has distinctly keeled, thinly scaled flower bracts 3–4 cm long and multi-flowered spikes.

Vriesea tillandsioides L. B. Smith, 1963, has keeled floral bracts, but they are only 2·3 cm long and densely scaled. The spikes have few flowers (see page 251). *V. tillandsioides* and *V. harmsiana* are also found in northern Peru.

Vriesea erythrodactylon (Morren) Morren, 1896 (Ill. 159)
(*erythros* red, *dactylus* fingered; because of the red, finger-like floral bracts)
Syn. *V. psittacina* var. *erythrodactylon* Morren, 1882; *Tillandsia carinata* sensu Baker, 1889; *V. duvaliana* sensu Wittm., 1891, non Morren, 1884; *V. carinata* sensu Mez, 1894

Pl 40–45 cm high including the inflorescence
Fo crisp green, forming a rosette 35–45 cm in diameter
LS broadly oval, 7–8 cm long, 6 cm wide, brownish-lavender on the outside
LB lineally tongue-shaped, up to 30 cm long, 2·5–3·5 cm wide; against the light it shows fine, horizontal markings
Sc upright, round, 4 mm thick
SB upright, green, longer than the stalk segments and overlapping
I simple, forming a distichous, flat spike up to 15 cm long and 7 cm wide
FB up to 6 cm long, red, the bases tightly overlapping; the tips separate and bent upwards into a sickle shape, red, green at the base
Se about 3·2 cm long, elliptical, tip broadly rounded
Fl yellow
St protruding from the flower
Ha Brazil.

Vriesea espinosae (L. B. Smith) L. B. Smith, 1968 (Ills. 21, 155–156)
(*espinosa(us)* thornless; probably because the lower leaves of the offshoots do not end in a spine)
Syn. *Tillandsia espinosae* L. B. Smith, 1951

Pl stemless, inflorescence 15–18 cm high; offshoots beginning with 5–15 cm long, horizontal or oblique shoots, forming dense clumps
Fo numerous, forming a compact, flat rosette
LS broad, short, hardly distinct from the blade
LB lineally triangular, long-tapered, rigid, 6–7 cm wide at the base, heavily grey-scaled
Sc upright, slender, slightly scaled, not

much longer than the leaves
SB elliptical, blunt, upright, dense, overlapping, red, short scales
I a simple, lineally lanceolate, flattened, dense spike 8 mm wide, 7 cm long and having 6 flowers
FB upright, longitudinally elliptical, blunt, 3 cm long, longer than the sepals, no keel, bright red, slightly scaled
Se separate, elliptical, blunt, 1·2 cm long, rounded on the back, no keel
Pe upright, 3 cm long, blue

St enclosed, 2 cm long
Ha Southern Ecudador and northern Peru; epiphytic in dry (savannah) forests between 500 and 800 m; in dense clumps completely surrounding the branches of their host (Ill. 21)
CU Bright, sunny and dry; acclimatizing is difficult. A small, recommended species with a tillandsia-like appearance. Can be grown along with cacti. The plants from Ecuador seem to be larger than those from Peru.

Vriesea fenestralis L. B. Smith, 1941, non Hort. Duval, 1902
(*fenestralis* window-like, perforated as in a grid; because of the grid-like markings on the leaves)
Syn. *Tillandsia fenestralis* Hook. f., 1886; *V. fenestrata* Hort., 1894; *V. hamata* L. B. Smith, 1941

Pl up to 1 m high including the inflorescence
Fo forming a large funnel rosette
LS broadly oval, 9–10 cm long, 9–10 cm wide, brown at the base
LB broadly lineal, rounded at the tip, but with a thorn, usually recurved, up to 40 cm long, 8 cm wide above the sheath, yellow green, with a net-like or grid-like marking, lower side partially red-spotted
Sc stout, upright, 50 cm long
I simple, loosely distichous, 20 to 30 flowers, up to 30 cm long, up to 9 cm

wide
FB oval, green, 3 cm long, no keel, shorter than the sepals, which are up to 3·6 cm long, oval, rounded, green and have no keel
Fl yellowish-green, spreading, up to 4·5 cm long
St shorter than the petals
Ha Brazil.
CU Because of the grid-like markings on the leaves *V. fenestralis* is a very decorative and promising species even in its non-blooming stage.

Vriesea flammea L. B. Smith, 1941, non Hort., 1902
(*flammea(us)* fire red; because of the bright red floral bracts)

Pl stemless, up to 40 cm high including the inflorescence
Fo numerous, upright to recurved, forming a rosette 30 cm in diameter
LS distinct, about 7 cm long and 4 cm wide, dark brown
LB narrowly triangular, long-tapered,

up to 25 cm long, 1 cm wide above the sheath, dark green, dense, grey scales on the underside, thin, grey scales on the upper side
Sc 20–22 cm long, 3 mm thick, upright, slender
SB similar to the leaves, overlapping,

upright, surrounding the scape with the sheaths, blades longer than the stalk segments
I 10 cm long, 4–5 cm wide, simple, open at bloom time and thereafter, otherwise dense
Fl arranged in 4 to 5 lines on the axis of the inflorescence, almost upright, 3·5–4 cm long, white
FB 2·5 cm long, 2 cm wide, no keel, bright red
Se 2·5 cm long, 9 mm wide, oval
St and Pi protruding from the flower
Ha Brazil. Closely related to *V. corcovadensis* (see page 231) and linked to it by transitional forms

Vriesea fosteriana L. B. Smith, 1943 (Ills. 313, 314 and 315)
(*fosteriana* named after M. B. Foster, American bromeliad breeder and collector)
There are several forms or varieties under the names: 'Red chestnut', 'forma nova' and 'seideliana', with stronger colouration and more intensive cross-lining. *V. fosteriana* blooms at night. The flowers begin to open at evening time, open during the night and close during the early morning. A repugnant smell is given off, like the odour of putrid fruit. In its habitat pollination is effected by bats.

Pl stemless, with inflorescence to 1·50 m high
Fo numerous, forming a dense funnel shaped rosette more or less 1 m in diameter
LS broad-oval, distinctly set, 10–15 cm long, 13–15 cm wide, both sides dark brown scaled at the base
LB broad-tongue shaped, 50–60 cm long, 9–10 cm wide, wide rounded with small spine tipped apex; on the underside irregularly crosslined red brown, these forming separated bands of lines; the zones between the bands green or yellowish; lines on uppersides of young leaves dark green
Sc upright, sturdy, round, fine longitudinally ribbed, 1 m long, to 1·5 cm thick, naked, green, spotted brown red.
SB spirally arranged enclosing the scape, densely set, upright with short blade, 6–10 cm long, green, densely spotted red brown
I simple, loose, sword shaped in two rows, 20–25 cm long, 10 cm wide, 18 to 25 flowered; rachis angled, green, spotted red brown, bare
Fl short stalked, 5 cm long; buds upright, at the time of flowering horizontal, recurved after flowering
Se 3 cm long, 1·5 cm wide, braked, green, spotted red brown; rounded tip
Pe broad-oval, 4 cm long, 2·3 cm wide, pale yellow, reddish brown tip
St and Pi not exceeding the petals
Ha Brazil
CU Decorative, vegetatively like *Vriesea hieroglyphica* and *Guzmania musaica*. Requires shade, warmth and humid conditions.

Vriesea friburgensis Mez, 1894 (Ill. 160)
(*friburgensis* named after the city of Freiburg, Germany)
Syn. *V. tweedieana* sensu F. Müller, 1893

Pl up to 1 m high including the inflorescence

Fo forming a dense rosette about 40 cm in diameter

LS broadly oval, 11–12 cm long, 9–10 cm wide, brown outside

LB uniformly green, up to 20 cm long, 5–6 cm wide, broadly lineal, rounded, but short-tipped with recurved ends

Sc upright, stout, 45 cm long, 5 mm thick

SB similar to the leaves, upright, longer than the internodes; the blades spread and recurved, green

I 40–50 cm long, loosely compound with 5 to 10 spikes

PB similar to the scape bracts, green, shorter than the spikes, but longer than the sterile section of the spikes, which are arranged in a spiral, almost upright to flared

FB spread, up to 3·5 cm long, keeled to not keeled, yellow

Fl distichous, up to 5·5 cm long, spread to flared, yellow

Se elliptically lanceolate, tip rounded, 3 cm long, yellow

St protruding from the flower

Ha Argentina and Brazil.

VARIETIES

var. *friburgensis* branching upright or almost upright; inflorescence distinctly longer than wide and all floral bracts not keeled, up to 3·5 cm long; flowers all flared.

var. *paludosa* (L. B. Smith) L. B. Smith as above with upper floral bracts keeled, about 2·5 cm long; upper flowers often overlapping.

var. *tucumanensis* (Mez) L. B. Smith branching spread to recurved; flowers flared.

Vriesea gigantea Gaud., 1843 (Ill. 42 and 162)

(*gigantea*(*us*) gigantic, very large)

Syn. *Tillandsia gigantea* Griseb., 1864; *Tillandsia tesselata* Linden, 1873; *Vriesea tesselata* Morren, 1882; *Tillandsia reticulata* Baker, 1887; *Vriesea reticulata* Mez, 1894; *Vriesea mosenii* Mez, 1894; *Vriesea alexandrae* Hort., 1903

Pl up to 2 m high including the inflorescence

Fo numerous, forming a large, funnel-shaped rosette up to 1 m in diameter

LS broadly oval, 12–13 cm long, 12–13 cm wide, dark brown

LB broadly lineal, short or long-tapered, up to 50 cm long, up to 8 cm wide, marked on both sides with a yellow green checkerboard pattern, explaining the name *tesselatus* (checkered; Ill. 42)

Sc sturdy, upright, not surpassing the leaves

I up to 1·5 m long, composed of about 20 spikes arching upward, open, multi-flowered and up to 30 cm long

FB about 2·4 cm long, upright, distichous, shorter than the sepals, one-sided at bloom time, green

Fl almost upright, distichous, one-sided at bloom time, yellowish-green, about 4 cm long; the tips of the petals flared or recurved

Se up to 3·5 cm long, oval-lanceolate, almost pointed, green
St only slightly surpassing the petals
Ha Brazil.
CU like *V. fenestralis* (see page 235), a very attractive species even when not in bloom; the inflorescence and the flowers themselves are hardly attractive.

Vriesea gladioliflora (Wendl.) Ant., 1880

(*gladioliflora(us)* gladiolus flowered; because of the large flowers)
Syn. *Tillandsia gladioliflora* Wendl., 1863; *V. princeps* Hort., 1877; *V. gladioliflora purpurascens* Ant., 1880; *V. gladioliflora* var. *purpurascens* Ant., 1884

Pl up to 1 m high including the inflorescence
LS elliptical, not distinct, brown-scaled
LB tongue-shaped, broadly pointed or blunt, thorn-tipped, 6–8 cm wide, up to 60 cm long
Sc stout, upright
SB overlapping, longer than the stalk segments, elliptical, pointed
I upright, simple-spiked, tightly distichous, multi-flowered, 20–40 cm long, up to 5 cm wide
FB upright, distichous, overlapping, broadly oval, blunt or broad, almost pointed, 4·5–5·5 cm long, no keel, naked, green, reddish toward the tip, becoming wrinkled as they dry
Fl one-sided during bloom time, large, up to 5 cm long and greenish-white
Se broadly elliptical, blunt, up to 2·5 cm long
St not protruding from the flower
Ha Costa Rica and Panama.

Vriesea guttata Linden et André, 1875 (Fig. 46b)

(*gutta(us)* spotted; because of the spotted leaves)
Syn. *Tillandsia guttata* Baker, 1888; *Tillandsia duvaliana* Baker, 1889

Pl stemless, up to 30 cm high including the inflorescence
Fo 10 to 15 leaves in a dense, upright rosette about 15 cm in diameter
LS not distinct, longitudinally oval, gradually merging into the blade, 6–7 cm wide, 10 cm long, brown to lavender
LB 15–20 cm long, 4–5 cm wide above the sheath, tongue-shaped, broadly rounded, pointed, upright, green with irregular lavender to dark red spots
Sc upright to arched, slender, dark green, naked
SB overlapping, upright lying close to the scape, as long as the stalk segments or somewhat longer, green to lavender, with white, chalky scales
I pendulous, loose, simple, up to 20 cm long and 4 cm wide, about 20 flowers
FB open, distichous, almost upright to spread, up to 3·5 cm long, 2·5 cm wide, shorter than the sepals, green, with chalky, white scales
Se 3 cm long, 1·2 cm wide, lanceolate, rounded, thin-skinned, greenish, with

chalky, white scales
Pe yellow, 4 cm long, lineal, with rounded tips

St protruding from the flower
Ha Brazil.

Fig. 46b *Vriesea guttata* Linden et André (× ⅓).

Vriesea hainesiorum L. B. Smith, 1963 (Fig. 47) (*hainesiorum* named after its discoverers, A. Lee and Bruce L. Haines)
This species is closely related to *V. viridis* Smith et Pitt. (*Thecophyllum viride* Mez et Wercklé, 1904), from which it differs by the very loose, few-flowered inflorescences, the thinner inflorescence axis and the larger floral bracts.

Pl stemless, 20–35 cm high
Fo forming a rosette 15–25 cm in diameter
LS longitudinally elliptical, 6–8 cm

long, about 3·5 cm wide, short, grey scales, partially edged in red

LB narrowly lanceolate, long-tapered, up to 20 cm long, 2–3 cm wide above the sheath, green or reddish, grey-scaled, edges usually curled

Sc pendulous, thin, about 20 cm long

SB red, longer than the internodes, surrounding the scape with the sheaths, with upright, narrowly lineal, long-tapered blades

I very loosely compound, about 8 cm long with few, short-stemmed spikes with 2 flowers

FB yellow, 1·5–2 cm long, up to 1·5 cm wide, shorter than the sepals, which are 2 cm long, 1 cm wide, broad-edged, and yellow

Fl 2·5–3 cm long, 7 mm wide, yellow with enclosed stamens

Ha Costa Rica; epiphytic at altitudes from 1,500–1,700 m.

CU A pretty, small species. Grow in semi-shade.

Fig. 47 *Vriesea hainesiorum* L. B. Smith (× ⅓).

Vriesea heterandra (André) L. B. Smith, 1951 (Ill. 164)

(*heterandra*(*us*) with varying stamens)

Closely related to *V. heterandra* is Rauh's recent discovery *V. appenii* (see page 227). It differs from *V. heterandra* by its much larger and very loose inflorescence; the spikes, which are up to 16 cm long and protrude horizontally from the axis (Ill. 163); the floral bracts, which are rounded on the back, slightly nerved and heavily scaled; and the deep dark blue, smooth edged petals.

Syn. *Tillandsia heterandra* André, 1888

Pl stemless, 35–40 cm high including the inflorescence

Fo numerous, a dense, almost upright rosette 20–30 cm in diameter

LS distinct, very broad, oval or elliptical, 6–8 cm long, 5–7 cm wide, dark brown

LB narrowly triangular, long-tapered, 20–30 cm long, 2–3 cm wide above the sheath, dark green, thick, short scales

Sc slender, angling upward, 10–15 cm long, about 6 mm thick

SB similar to the leaves, upright, longer than the stalk segments, with a long, pointed blade, dark green, short grey scales

I 10–20 cm long, 3–4 cm wide, loosely compound with 10 to 15 simple spikes

PB the basal ones similar to the scape bracts with long-tapered blade, the upper ones with a short tip; sheaths of the primary bracts are much shorter than the spikes, dark green, heavily grey-scaled

Sp arranged in a spiral, stemmed, 5–7 cm long, 1 cm wide, the stemmed part of the spike lying close to the axis of the inflorescence, the flower-bearing segment arched away from the axis, flattened

FB tightly overlapping, distichous, oval, pointed, keeled in the upper section, distinctly nerved, 1·4 cm long, 7 mm wide, brownish-red, grey-scaled, somewhat longer than the sepals, which are longitudinally elliptical, pointed, naked, 1 cm long, 4 mm wide

Pe about 1·7 cm long with plates that are 4 mm wide, flared, finely dentate on the edge, delicate blue and fade to pink or red

St enveloped in the flower

Ha Colombia, Ecuador and Bolivia; epiphytic. *Vriesea heterandra* is one of those 'grey' vrieseas, that can easily be confused with tillandsias

CU Bright, sunny, moderate amounts of water.

Vriesea hieroglyphica (Carr.) Morren, 1884 (Ill. 165)
(*hieroglyphica*(*us*) like handwriting; because the leaf markings resemble hieroglyphs)

Syn. *Massangea hieroglyphica* Carr., 1878; *Tillandsia hieroglyphica* Baker, 1888; *Massangea santoviensis* Hort., 1889; *Massangea tigrina* Hort., 1889

Pl over 1 m high including the inflorescence

Fo forming a funnel rosette 80–100 cm in diameter

LS not distinct, imperceptibly merging into the blade, 8–10 cm long, 10–12 cm wide, dark brown

LB 50–80 cm long, 8–10 cm wide above the sheath, broadly lineal, roun-

ded, with recurved tip, green, with irregular, darker cross banding

Sc upright, sturdy, as long as the leaves

SB the basal ones similar to the leaves, the upper ones with a short blade; sheaths somewhat scooped out and flared, longer than the stalk segments, green with dark green spots

I up to 60 cm long, loosely compound with approximately 20 spikes
PB similar to the upper scape bracts, shorter than the stemmed part of the loose to dense, distichous, upright to spread spikes
FB upright, loosely distichous, broadly elliptical-oval, up to 3 cm long, yellowish-green

Fl distichous, at bloom time all turned to one side, up to 6 cm long, yellow, the petals with flared tips
Se up to 3 cm long, 1·3 cm wide, yellowish-green
Ha Brazil
CU Because of the leaf markings this is a very decorative plant of great commercial value.

Vriesea hitchcockiana (L. B. Smith) L. B. Smith, 1951 (Ill. 309)
(*hitchcockiana* named after the collector A. S. Hitchcock)
V. hitchcockiana is very near to *V. rauhii* and *V. cereicola* distinguished from both by the sepals being longer than the flower bracts. *V. rauhii* has thin nearly round spikes, *V. cereicola* strong flattened wide spikes. All 3 species belong to the 'grey' vrieseas vegetatively like tillandsias.
Syn. *Tillandsia hitchcockiana* L. B. Smith, 1930

Pl stemless, flowering to 50–60 cm high
Fo numerous, forming an upright funnel shaped rosette from 30–60 cm high and 50 cm in diameter; the outer leaves small, scale shaped
LS not clearly distinguished, 13–15 cm long, 5–6 cm wide, upper side dark chestnut brown scaled
LB narrow-triangular, gradually terminating in a long tip, with upward curved margins, to 50 cm long, 4·5 cm wide above the sheath, both sides densely appressed scaled
Sc upright or pendulous, 30–50 cm long, 6–8 mm thick, wine reddish covered by the scape bracts
SB the basal ones similar to the leaves, the upper ones shortly tipped, densely scaled, longer than the internodes, lying against the scape
I loose, 30–40 cm long, 10–20 cm wide, with 4 to 6 pendulous to outspread,

nearly digitate spikes
PB upright, to 5·5 cm long, elliptical, wine red, naked
Sp 10–35 cm long, narrow-swordshaped, 2–2·7 cm wide, 18–25 flowered with 3–8 cm long, sterile leafy bases
FB at first dense, together with the flower loose, upright-angled, shorter than the sepals, pointed, not keeled on the back, naked, dark purple, somewhat grey-waxen frosted
Fl about 3·2 cm long, violet
Se 16–25 mm long, free, not keeled, pointed
Pe forming a narrow tube with two scales at the base
St and Pi extending beyond the flower
Ha South Ecuador (Loja) and North Peru (Olmos Valley) in deciduous dry forests, between 800 m and 1,200 m high
CU Full light and a little water.

Vriesea incurvata Gaud., 1843 (Ill. 55)
(*incurvata*(*us*) turn inward; because of the incurved floral bracts)

Syn. *Tillandsia incurvata* Baker, 1888; *V. rostrum-aquilae* Mez, 1894; *V. duvaliana* sensu Alexander, 1936, non Morren, 1884

Pl up to 40 cm high including the inflorescence
Fo numerous, forming a rosette of 40–60 cm diameter
LS longitudinally oval, 5–6 cm wide, 10–13 cm long, pale green
LB up to 30 cm long and up to 3 cm wide, pointed, crisp green and shiny
Sc upright, short, not surpassing the leaves, round, 4 mm thick
SB tightly overlapping, lying close to the scape, upright, longer than the stalk segments, green to reddish
I simple, sword-shaped, 20–25 cm long, 4 cm wide
FB distichous, tightly overlapping, 4·5–5·5 cm long, distinctly keeled beginning at the middle of the back, with a flat, saber-shaped, bent, blunt tip, yellow, orange and red, much longer than the sepals
Se oval-lanceolate, pointed, not keeled, 2–3 cm long
Fl 5 cm long, yellow with green tips
St protruding from the flower
Ha Brazil
CU At bloom time just as decorative as *Vriesea splendens*.

Vriesea malzinei Morren, 1874 (Ill. 166)
(*malzinei* named after Omer de Malzine)
Syn. *Tillandsia malzinei* Baker, 1879

Pl rosette-shaped, up to 60 cm high including the inflorescence
Fo up to 30 cm long, 4–5 cm wide, lineal, short tipped, upper side greenish red, lower side dark maroon
Sc upright, approximately as long as the leaves
SB longer than the stalk segments, the basal ones similar to the leaves, the upper ones scoop-shaped with a very short blade, maroon
I an upright, cylindrical spike up to 17 cm long, with 10 to 24 flowers
FB red or yellow green, broadly oval, rounded at the tip, no keel, up to 2·5 cm long, shorter than the sepals, arranged in a spiral
Fl white, upright, 7–8 cm long
Se up to 3·2 cm long, elliptical, rounded at the tip
St enveloped in the flower
Ha Mexico (near Cordova) on boulders.

Vriesea maxoniana (L. B. Smith) L. B. Smith, 1959 (Plate 60)
Syn. *Tillandsia maxoniana* L. B. Smith, 1939; *V. icterica* Cast., 1945

Pl stemless, 20–50 cm high including the inflorescence
Fo numerous, forming a rosette of 40–50 cm diameter
LS longitudinally oval, 7–8 cm long, 5 cm wide, green, imperceptibly merging into the blade, which is 25–35 cm long, 2–3 cm wide, strap-shaped, lanceolate, pointed, green, naked and equipped with a thorn tip
Sc upright, shorter than the leaves, 15–20 cm long, 5 mm thick, green, naked
SB tightly overlapping, upright, lying

close to the scape, 3–4 cm long, 1·5 cm wide, with a short tip 1 mm long, green, naked
I simple, distichous, up to 12 cm long and 7 cm wide; rhachis visible
FB 5–5·5 cm long, 2 cm wide, distinctly keeled, naked, yellowish-green at the base, green toward the tip
Se 4 cm long, 1 cm wide, keeled, yellow
Pe 4·5 cm long, 1 cm wide, yellow
Ha Argentina and Bolivia; epiphytic.

Vriesea olmosana L. B. Smith, 1966 (Plate 59 and Ill. 167)
(*olmosana* named after the northern Peruvian town of Olmos)

Pl stemless, 60 cm high including the inflorescence
Fo numerous, forming a more or less spread, dense rosette of about 50 cm diameter
LS longitudinally elliptical, 10–11 cm long, up to 7 cm wide, dark brown-scaled
LB narrowly lineal, long-tapered, 4·5 cm wide above the sheath, average width 2·5 cm, up to 50 cm long, dense, short, grey scales
Sc slender, upright or arched, depending on the growth pattern
SB tightly overlapping, surrounding the scape, longer than the stalk segments, with a long, narrow blade
I composed of about 10 spikes, 30–40 cm long; rhachis arched
PB broadly oval, hardly surpassing the lower, sterile part of the spikes, green to red, heavily scaled, especially at the tip
Sp upright to almost upright, the lower ones up to 14 cm long, 10 to 12 flowers
FB tightly overlapping, distichous, upright-lanceolate, pointed, keeled, 2·5 cm long, yellow green at the base, red toward the tip, as long as the separate, lanceolate, broadly pointed sepals, which are 2·3 cm long and have no keel
Pe lineal, 3·5 cm long, dark blue with narrow, white edges and flared tips
St protruding from the flower
Ha Northern Peru (in the vicinity of Olmos); epiphytic in the dry forest up to 1,000 m. Because of the shape and the dense grey scaling of the leaves, it can easily be confused with a *Tillandsia*
CU Sunny, moderately damp. At bloom time very decorative.

Vriesea patula (Mez) L. B. Smith, 1955 (Ill. 168)
(*patula*(*us*) spread, open; because of the form of the rosette)
Syn. *Tillandsia patula* Mez, 1906

Pl stemless, up to 40 cm high including the inflorescence
Fo forming a dense, more or less spread rosette thickened into a pseudo-bulb at the base
LS broadly oval, 5–6 cm long, 5 cm wide, dark brown
LB 10–15 cm long, 2·5 cm wide above the sheath, triangular, long-tapered, heavily grey-scaled
Sc very short, shorter than the leaves
SB overlapping, elliptical, scaled, the lower ones with a long, pointed blade
I simple or compound, with 2 to 5

upright or spread spikes up to 16 cm long and 1·5 cm wide with many flowers
PB much shorter than the basal, flowerless section of the spikes
FB 2·5 cm long, somewhat longer than the sepals, upright, distichous, not keeled, raspberry red and grey-scaled
Fl upright, about 3·5 cm long, greenish-yellow
Se about 2·1 cm long, separate, elliptical, blunt, naked
St surpassing the petals
Ha Peru (Province of Tarma); epiphytic at altitudes from 1,900–2,400 m.
CU Very decorative, reminiscent of a tillandsia; small. Grow it epiphytically, sunny and bright.

Vriesea platzmannii Morren, 1875 (Ill. 169)
Syn. *Tillandsia platzmannii* Baker, 1888

Pl up to 1 m high including the inflorescence
Fo forming an upright rosette, somewhat constricted in the middle, 50 cm high and 30 cm in diameter
LS distinct, broadly oval, about 14 cm long, 9 cm wide, dark brownish-lavender, with distinctly protruding nerves, somewhat grey-frosted
LB 3–4 cm wide, 23–25 cm long, lineal, rounded, thorn-tipped, dull green, grey-frosted, with distinctly protruding nerves
Sc upright, 60–70 cm long, 5 mm thick, longitudinally ribbed, green to brownish-lavender
SB the lower ones similar to the leaves and longer than the stalk segments, the upper ones shorter than the stalk segments and blending from dull green into brownish-lavender, grey-frosted
I simple, turned to one side, 15–20 cm long, 5 to 10 flowers
FB loosely distichous, upright, 3–4 cm long, 2·5 cm wide, pointed, keeled, brownish-lavender, grey, waxy scales
Fl yellow, loosely distichous, at bloom time and later turned to one side, stem about 7 mm long
Se 3 cm long, 1·4 cm wide, asymmetrical, stiff, dark brown in the lower part, greenish-yellow toward the tip
St enveloped in the tube, 3·5 cm long
Ha Brazil. A very pretty species recommended to the fancier.

Vriesea platynema Gaud., 1843 (Plate 61)
(*platy* flat, broad; *nema* thread; broad-threaded, presumably because of the broadly lineal leaves)
Syn. *Tillandsia platynema* Griseb., 1864; *V. corallina* Regel, 1870; *Encholirion corallinum* Linden, 1871; *Tillandsia bicolor* Niederl., 1890, non Brongn., 1829

Pl up to 1 m high including the inflorescence
Fo numerous, forming a rosette up to 60 cm in diameter
LS 12–13 cm long, 10–11 cm wide, broadly oval, dark brown
LB broadly lineal, rounded, thorn-tipped or long-tapered, up to 45 cm long, up to 6 cm wide, uniformly coloured, dull green, violet or multicoloured with a stripe pattern, dark violet toward the tip

Sc upright, strong, up to 60 cm long
SB tightly overlapping, longer than the stalk segments, surrounding the scape with the sheaths, the tips spread, green or red
I simple, loosely distichous, multi-flowered, up to 40 cm long and 8 cm wide
FB oval-triangular, up to 3 cm long, up to 2 cm wide, red or yellow, much shorter than the 3·5 cm, yellow, narrowly oval sepals
Fl green, at bloom time spread to recurved, 3·5–4 cm long
Ha West Indies and in the eastern part of South America. A species of many forms.

VARIETIES

var. *platynema* Flower bracts red; leaftips rounded with a thorn tip; leaves of uniform colour (Plate 61)
var. *flava* Reitz, 1952 Flower bracts yellow; leaf tips rounded with a thorn tip; leaves uniform in colour
var. *rosea* Mez, 1894 Leaf tips long-tapered; leaves uniform in colour, green to violet
var. *striata* Wittm., 1894 Leaves with pale stripes
var. *variegata* Reitz, 1952 Leaves red violet on the underside, green on the upper side, toward the tip pale-striped and darker violet

Vriesea psittacina (Hook.) Lindl., 1843 (Plate 62)
(*psittacina*(*us*) parrot-coloured; because of the colour combination of floral bracts and flowers, which is reminiscent of the colourful plumage of a parrot)

Pl up to 50 cm high including the inflorescence
Fo forming a loose rosette of 50–60 cm diameter
LS broadly oval, 7 cm wide, 12 cm long, pale green
LB lineal, pointed, up to 40 cm long, 2·5 cm wide at the base, crisp green, shiny
Sc upright, about 35 cm long and 5 mm thick, round
SB overlapping, upright, lying close to the scape, longer than the stalk segments, red with a short green tip
I simple, open, up to 30 cm long and 8 cm wide; rhachis visible, shiny red
FB loose, distichous, spreading at bloom time, twice as long as the axis segments, no keel, red with yellow, red or green
Fl almost horizontal from the axis, sessile, up to 6 cm long, yellow with a green tip
Se lanceolate-elliptical, up to 4 cm long, yellow
St protruding from the flower
Ha Brazil, Paraguay
CU *V. psittacina* is a very decorative plant at bloom time, but it is seldom found in cultivation as a pure species plant. Much more common are hybrids of it with other plants. In most bromeliad collections *V. carinata* Wawra (syn. *V. psittacina* var. *carinata* Morren (see page 229) is labelled *V. psittacina*.

VARIETIES
var. *psittacina* (Plate 62)
Syn. *Tillandsia psittacina* Hook., 1828; floral bracts red with yellowish tips, no keel
var. *rubro-bracteata* Hook., 1859
Syn. *V. krameri* Morren, 1884; floral bracts uniformly red. Known only in culture
var. *decolor* Wawra, 1880; floral bracts uniformly green, some with a keel

Vriesea racinae L. B. Smith, 1941 (Ill. 171)
(*racinae* named after Mrs. Racine Foster, wife of the bromeliad collector and grower M. B. Foster)

Pl up to 15 cm including the inflorescence
Fo 10 to 15 leaves, forming a small rosette 10–13 cm in diameter
LS distinct, broadly oval, 5–6 cm long, 4–5 cm wide, green, irregularly spotted with reddish-brown
LB tongue-shaped, thorn-tipped, recurved in the upper part, 5–6 cm long, 2·5 cm wide, green, irregularly spotted with reddish-brown
Sc slender, upright to 12 cm long and up to 4 mm thick, green
SB 1·5–2 cm long, 1·3 cm wide, upright with a short tip, lying close to the scape, green, spotted with reddish-brown
I simple, open, few flowers
Fl white, usually oriented toward one side, short-stemmed, 4 cm long
FB 2 cm long, 2 cm wide, green, reddish-brown spots, almost round
Se 2 cm long, 1·2 cm wide, broadly rounded, green, reddish-brown spots
St and **Pi** enveloped in the flower
Ha Brazil. A pretty, small species suitable for terrariums.

Vriesea rauhii L. B. Smith, 1958
(*rauhii* named after W. Rauh)
A very interesting species closely related to *V. patula* and *V. cereicola* and resembling a *Tillandsia* in growth shape. Only the scales at the base of the petals indicate its position in the genus *Vriesea*

Pl up to 60 cm high including the inflorescence
Fo forming a rosette up to 40 cm in diameter
LS triangularly oval, 6–8 cm long, 3–4 cm wide, dark brown scales
LB lineally triangular, up to 30 cm long, 2 cm wide, green, short grey scales
Sc upright to arched, slender, naked
SB upright, overlapping, surrounding the scape, elliptical, scaled, reddish-green
I up to 38 cm long, composed of few lineal spikes, open
PB similar to the scape bracts, red, scaled, 4 cm long, much shorter than the sterile bases of the spikes, which are

slender, rod-shaped, up to 30 cm long and up to 9 mm wide
FB upright, 3–5 cm long, 1 cm wide, no keel, naked, smooth, nerved and scaled at the tip
Fl up to 3·5 cm long, dark blue
Se elliptical, blunt, 1·8 cm long, no keel
St and Pi protruding from the flower
Ha As yet, known only from northern Peru (Department of Cajamarca); epiphytic in dry forests at 700 m.
CU Bright, sunny, little watering.

Vriesea rodigasiana Morren, 1882 (Ill. 170)

Syn. *Tillandsia rodigasiana* Baker, 1888; *Tillandsia tweedieana* Baker, 1888; *Tillandsia citrina* Baker, 1889; *Vriesea citrina* Morren, 1889; *Vriesea tweedieana* Mueller, 1893; *Vriesea vitellina* Mueller, 1893

Pl up to 60 cm high including the inflorescence
Fo in a dense rosette 25–30 cm in diameter
LS oval, 7–9 cm long, 5–6 cm wide, dark lavender
LB lineal, rounded, with a thorn tip or long-tapered, up to 15 cm long, 2·5 cm wide, green
Sc almost upright, slender, not longer than the leaves, red or green
SB shorter or as long as the internodes, upright, red
I up to 36 cm long, 17 cm wide, composed loosely of 4 to 8 spikes with few flowers
PB similar to the scape bracts, spread, red, shorter than the almost upright spikes of 3 to 8 flowers
FB yellow, up to 1 cm long, open, distichous, shorter than the elliptically lanceolate sepals, which are up to 1·6 cm long
Fl almost upright, up to 3·5 cm long, yellow; petals flared at the tip
St and Pi protruding from the flower
Ha Brazil.

Vriesea sagasteguii L. B. Smith, 1968 (Ill. 310)

(*sagasteguii* named after A. Sagastegui, collector)
V. sagasteguii belongs to the group containing *V. cylindrica*, *V. harmsiana* and *V. tillandsioides* all of which have a cylindrical inflorescence.

Pl stemless, flowering to 80 cm high
Fo numerous, upright to outspread, forming a rosette 40–60 cm in diameter
LS longish-oval, not clearly distinguished, 8–10 cm long, 8–9 cm wide, dark brown scaled
LB narrow-lanceolate, long pointed, about 5 cm wide above the sheath, 35–45 cm long, both sides densely silvergrey scaled, with curled up margins
Sc upright to upwards, sturdy, to 30 cm long, grey scaled
SB densely overlapping, the sheaths enclosing the scape, grey scaled, longer than the internodes, with longer or shorter upright blade
I cylindrical, compound with 8 to 14 upright, spirally arranged spikes, 30–40 cm long, 4 cm thick
PB upright, 6–9 cm long, 4–6 cm wide with short pointed tip, red, densely grey scaled, as long as or shorter than the spikes, these 8–10 cm long, 2–3 cm

wide, laterally flattened, short stalked 1–2 cm long, 5–8 flowered

FB in two rows, overlapping, upright, red and densely grey scaled, strongly keeled, claw shaped curved tip, 3–4 cm long, longer than the sepals

Fl upright, sessile, 5·5–6 cm long, in two rows

Se free, 2·7 cm long, green, naked, pointed

Pe 5·5 cm long, 1 cm wide, the lower third white, otherwise blue-violet, red after flowering; tips upright or slightly flared, with two 9 mm long white scales at the base

St extending beyond the petals

Ha North Peru (Olmos Valley, 1,500–2,000 m)

CU Very fine with its silver white rosette and primary bracts. It grows well on columnar cactus and flowers regularly.

Vriesea saundersii (Carr.) Morren, 1894 (Ill. 174)

Syn. *Encholirion saundersii* Carr., 1872; *Tillandsia saundersii* K. Koch, 1873; *Vriesea botafogensis* Mez, 1894

Pl 60–70 cm high including the inflorescence

Fo numerous, arched, flared, forming a dense rosette 40 cm in diameter

LS longitudinally oval, 14–15 cm long, 9–10 cm wide, yellowish-brown with irregular reddish-brown spots

LB up to 30 cm long, 6 cm wide above the sheath, lineal, short, thorn-tipped, recurved, grey green, light grey-scaled, irregular, reddish-brown spots especially on the underside

Sc upright, 30–35 cm long, 7 mm thick, naked

SB the basal ones similar to the leaves, the upper ones with a shorter, somewhat spreading blade, longer than the stalk segments, grey-scaled, with reddish-brown spots

I 20–30 cm long, 15 cm wide, loosely compound

PB similar to the upper scape bracts, shorter than the spikes, which are almost upright, short-stemmed, 10–15 cm long, 6–7 cm wide and have 5 to 8 loosely arranged flowers

FB shorter than the sepals, 3–3·5 cm long, 1·5 cm wide, oval-elliptical, rounded, greenish-yellow

Fl yellow, distichous, spread, 5 cm long; distance between the flowers about 1·5 cm

Se 3 cm long, 1·1 cm wide, longitudinally oval, with rounded tips

St and **Pi** somewhat shorter than the petals

Ha Brazil.

CU a very pretty species recommended for cultivation especially because of the colourful leaves.

Vriesea scalaris Morren, 1879 (Fig. 48)

(*scalaris* ladder-like; because of the ladder-like form of the pendulous inflorescence)

Syn. *Tillandsia scalaris* Baker, 1888

Pl 20 cm high

Fo 10 to 15 leaves forming a rosette about 20 cm in diameter

LS broadly oval, 8–9 cm long, 6 cm

wide
LB 10–15 cm long, narrower at the base than in the middle, green, naked, with finely ciliated edges
Sc slender, pendulous, 10–16 cm long, 3 mm thick, green
SB tightly overlapping, longer than the stalk segments, 3 cm long, green
I pendulous, simple-spiked, open, few flowers, 7–13 cm long, 9 cm wide
FB loose, distichous, shorter than the sepals, 3–4 cm long, 2·5 cm wide, keeled, red at the base, yellow toward the tip
Fl green, spreading to recurved, 5·5 cm long, short-stemmed
Se 3–3·5 cm long, 1·3 cm wide
St and **Pi** protruding from the flower
Ha Brazil; epiphytic on trees.
CU especially impressive at bloom time with its colourful inflorescence.

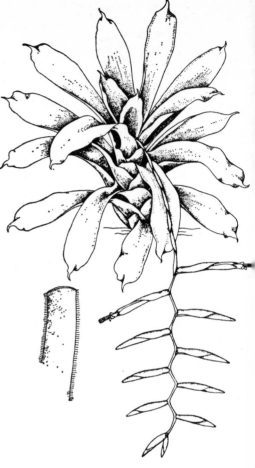

Fig. 48 *Vriesea scalaris* Morren (× $\frac{1}{3}$).

Vriesea splendens (Brongn.) Lem., 1850–1851 (Plate 63)
(*splendens* splendid)
Syn. *Tillandsia splendens* Brongn., 1845; *Vriesea speciosa* Hook., 1848

Pl stemless, up to 1 m high including the inflorescence
Fo 10 to 15 leaves, forming a dense rosette 30–35 cm in diameter
LS longitudinally oval, 10–13 cm long, 6–8 cm wide, brown-scaled, reddish-brown bands
LB 30–60 cm long, 4–6 cm wide, lineal, broadly rounded, thorn-tipped, green, both sides with reddish-brown to dark violet lateral bands
Sc upright, 30–35 cm long, up to 1 cm thick
SB longer than the stalk segments, upright, lying close to the scape, banded with reddish-brown

I a simple sword up to 55 cm long, up to 6 cm wide
FB upright, tightly overlapping, 6–8 cm long, keeled, bright red, narrowly triangular, long-tapered
Se elliptical, blunt, 2·5 cm long, yellow or white, red tips
Pe strap-like, up to 8 cm long, yellow

St protruding from the flower
Ha Surinam and Venezuela.
CU known by the name 'Flaming Sword', it is a trade plant of great importance and is grown in great numbers. In cultivation, usually only select forms are found, while the wild species is very rare.

Vriesea tillandsioides L. B. Smith, 1963 (Ill. 173)

(*tillandsioides* tillandsia-like; refers to the similarity to a grey tillandsia)
In its non-blooming stage *V. tillandsioides* can easily be confused with a grey-leafed *Tillandsia*. It is very closely related to *Vriesea cylindrica* and *V. harmsiana* (see page 233).

Pl up to 60 cm high including the inflorescence
Fo numerous, forming a rosette about 40 cm in diameter
LS broadly oval, 5–6 cm long, 4–5 cm wide, dark brown
LB narrowly lanceolate, long-tapered, up to 30 cm long, 3 cm wide above the sheath, both sides heavily grey-scaled
Sc upright, short, hidden by the leaves
SB tightly overlapping, the sheaths lying close to the scape, the blades narrow, arched away from the scape, heavily grey-scaled
I slender, spindle-shaped, 13 cm long, 2 cm thick, compound
PB upright, elliptical, 5 cm long, multifarious, overlapping, pink, with short, grey scales, the primary bracts hide the spikes, which are short, sessile, elliptical, 3·5 cm long, 1·5 cm wide and have 5 to 7 dense flowers
FB distichous, lanceolate-pointed, up to 2·3 cm long, longer than the sepals, distinctly keeled, heavily grey-scaled, thin, nerved at the tip
Fl pale blue, almost sessile, upright
Se longitudinally elliptical, pointed, 1·5 cm long, not keeled, naked
St protruding from the flower
Ha Northern Peru (in the vicinity of Huancabamba); epiphytic on trees, up to 2,000 m.

Vriesea vanhyningii L. B. Smith, 1960 (Ill. 172) (*vanhyningii* named after the bromeliad enthusiast Van Hyning)

Very closely related to *V. tonduziana* L. B. Smith from Costa Rica and is perhaps even identical with it. Numerous transitional forms between the two would so indicate.

Pl stemless, up to 40 cm high including the inflorescence
Fo about 20 leaves, forming a broad funnel rosette
LS oval, 10–11 cm long, 6–7 cm wide, green
LB tongue-shaped, long-tapered, up to 25 cm long, up to 5 cm wide above the sheath, on the underside short-scaled

Sc upright to arched upward, up to 40 cm long and 4 mm thick, green
SB longer than the stalk segments, tightly overlapping, surrounding the scape
I simple, tightly overlapping, up to 15 cm long, 1·5 cm wide, up to 10 flowers
FB 3·5 cm long, up to 2 cm wide, upright, oval, with a rounded tip, during bloom and thereafter somewhat one-sided, green, soon dry and then showing a distinct wart formation
Fl during and after bloom one-sided, upright, up to 3 cm long, white
Se 2·5 cm long, 1·3 cm wide, green, oval, rounded
Ha Mexico.

Glomeropitcairnia Mez, 1905

This genus comprises only two species
Glomeropitcairnia penduliflora (Griseb.) Mez, 1905
Syn. *Tillandsia penduliflora* Griseb., 1864; *Caraguata penduliflora* Wittm., 1890
Ha The Lesser Antilles, Dominica and Martinique.

Glomeropitcairnia erectiflora Mez, 1905
Ha The island Margerita on the coast of Venezuela.

Both species are reminiscent of *Guzmania zahnii* and the other species related to it and which were formerly included in the genus *Schlumbergeria* Morren. They differ from these by the formation of a semi-posterior ovary, a feature which puts them close to the members of the subfamily *Pitcairnioideae*. But the glomeropitcairnias differ from the *Pitcairnioideae* in the structure of the seed, which, like in *Tillandsia*, bears a tassel of hair at the base. Because of the seed structure, *Glomeropitcairnia* forms a transition between the *Pitcairnioideae* and the actual *Tillandsioideae* (*Tillandsia*, *Vriesea* and *Guzmania*).

Mezobromelia L. B. Smith, 1935

Mezobromelia bicolor L. B. Smith, 1935. The only known species: it resembles *Guzmania zahnii* (see page 226 and Plate 57) but differs from it by its very open inflorescence and the ligulate petals. The petals have two small, triangular scales at their base, as in *Vriesea*. In contrast to *Vriesea*, however, the petals are fused at the base as in *Guzmania*.

According to Smith, it is known only in Colombia and Ecuador at altitudes from 2,100–2,400 m*.

---

* Fig. 70 in L. B. Smith *The Bromeliaceae of Columbia*. In *Contributions from the United States National Herbarium*, Washington, 1957.

Main hybrids of Tillandsioideae

As soon as a group of plants become popular and start to be grown on a large scale for the trade, then it is not long before some cross-breeding work is done. People try to increase the available number of forms, to enhance the beauty of the wild plants, and (especially) to improve certain features such as flower size, flower colour, flower durability, etc. This can be done in two ways—by selection and by hybridization. In the process known as selection, the most beautiful examples are chosen from a population of wild plants and are then propagated, usually vegetatively. After a few years, there usually appears on the scene forms that greatly surpass the beauty of the original wild plants. For example, we hardly ever find species plants of *T. anceps, T. cyanea, T. lindeni, Vriesea splendens*, etc. in cultivation, only select forms.

In the crossing or hybridization of two different species, or even of two different genera (e.g. *Cryptobergia*, a bi-generic hybrid between *Cryptanthus* and *Billbergia*), it is hoped that a new combination of features of the two parents will result, so that the hybrids will surpass the parents in growth form and colour, depending on which one strives for. But not every cross has commercial value. In addition, the hybrids are subject to certain trends in fashion. What is 'stylish' today can be old-fashioned and forgotten tomorrow. Many of the bromeliad cultivars that were popular around the turn of the century no longer exist.

The rationale behind selective breeding and hydridization is really that hybrids appear even in nature—but less in the genus *Tillandsia* than in the genera *Vriesea, Guzmania, Billbergia, Aechmea*, etc.

Since the Belgians were the first to carry out work on the cultivation of bromeliads, it is natural that the first crosses were made in Belgium. Here we must mention the name of E. Morren. His first cross, *V. psittacina* × *V. carinata*, is still in the trade under the name *V.* × *morreniana**. The crossing of this hybrid with *Vriesea scalaris* resulted in *Vriesea* × *retroflexa*. It was with the genera *Vriesea* and *Billbergia* that hybridizers experimented especially. In 1888 came the first bi-generic cross, *Vriesea* × *magnifica*, from the crossing of *Vriesea splendens* with *Guzmania zahnii*.

Some far-reaching work on bromeliads began around the year 1890, a sign of the great interest in the plant group at that time. Belgian, French and German gardeners took up bromeliad cultivation. In France, L. Duval was especially successful. He specialized in the hybridization of vrieseas. From his work, we have still the hybrids *Vriesea* × *rex*—a four-fold cross—and *V.* × *poelmanii*. Also worthy of mention are the

---

* The sign × before a species name indicates that it is a hybrid.

Frenchmen Chantrier, Dutrie and the Belgian De Smet. In Germany, it was W. Richter who took up bromeliad cultivation and produced a number of interesting hybrids and select forms of great value to the trade. In the U.S.A. M. B. Foster enriched the array of bromeliads with his many beautiful cultivars. Especially famous are his *Aechmea* hybrids such as *Aechmea* × '*Bert*' (*Aechmea orlandiana* × *A. fosteriana*) and his famous *Aechmea* × '*Fosters' Favourite*', a plant with special charm; it has dark red leaves and pendulous, bright orange flowers. It came from the cross *Aechmea victoriana* × *A. racinae*.

Today, people are still developing beautiful select forms and hybrids. In Germany, Hans Gülz of Bad Vilbel was successful in the creation of the beautiful hybrid *Guzmania* × *Symphonie*.

Tillandsia

From this genus relatively few hybrids have been developed; they seem to be rare even in nature. In Mez (1932) we find only the hybrid *T. rectangula* Baker × *T. cordobensis* Hicken from the province of Cordoba in Argentina.

Rohweder (1956) cites a possible hybrid between *Tillandsia brachycaulos* and *T. caput-medusae* from El Salvador.

Artificial hybrids are:

Tillandsia × *duvali* Duval et André (*T. cyanea* × *lindeni*)
Tillandsia × *oeseriana* M. B. Foster (*T. flabellata* × *T. tricolor*)
Tillandsia × '*Victoria*' M. B. Foster (*T. ionantha* × *T. brachycaulos*)
W. Richter's nameless bi-generic hybrid *Tillandsia biflora* × *Guzmania minor* has not been of any great importance to the florist or nurseryman.

Guzmania

In this genus, too, relatively few hybrids are known. The splendid *Guzmania lingulata* with its varieties *splendens* and *cardinalis* are favourite 'parents'; *Guzmania* × *intermedia* Richter (*G. lingulata* var. *cardinalis* × *G. splendens*), *Guzmania* × *lingulzahnii* Dutrie (*G. lingulata* var. *cardinalis* × *G. zahnii*).

Under this cross, the selected forms which Dutrie named *Guzmania* × *insignis, G.* × *chevalieri, G.* × *Victrix* are designated as *Guzmania* × *magnifica* Richter (*G. lingulata* var. *cardinalis* × *G. minor*), *Guzmania* × '*Mignon*' Gülz (*G. minor* × *G. lingulata*), *Guzmania* × '*Symphonie*' Gülz (*G. zahnii* × *G. peacockii*).

Vriesea

Within the subfamily of Tillandsioideae, the most frequent hybrids are in this genus.

The preferred parent plants are *V. barilletii* Morren, *V. carinata* Wawra, *V. duvaliana* Morren, *V. incurvata* Gaud., *V. psittacina* (Hook.) Lindl., *V. splendens* (Brongn.) Lem., *V. saundersii* (C. Koch) Morren, *V. rodigasiana* Morren and *V. van geertii* Hort.

Of these, the most important ones have been *V. barilletii*, because of its beautiful sword and good growth, and *V. psittacina* and *V. carinata*, because of their vibrant parrot colours.

Although the following list makes no claims to be complete, Mez (1932) lists the following natural hybrids:

V. carinata × *V. paraïbica* Wawra (Brazil)
V. conferta × *V. erythrodactylon* F. Müller (Brazil)
V. incurvata × *V. ensiformis* Ule (Brazil)
V. incurvata × *V. erythrodactylon* F. Müller (Brazil)
V. poenulata × *V. procera* (Mart.) Wittm. var' *gracilis* (Gaudich.) Mez.

Among the cultivated hybrids are:

V. × *brachystachys splendens* Duval (*V. carinata* × *V. psittacina* var. *rubrobracteata*)
V. × *cardinalis* Duval (*V. carinata* × *V. psittacina* var. *rubrobracteata*)
V. × *'Deutscher Zwerg'* Gülz ((*V.* × *erecta* × *V. carinata*) × *V. corcovadensis*)
V. × *donneai* Makoy ex Witte (*V. barilletii* × *V. guttata*)
V. × *'Favourite'* (*V. ensiformis* × ?)
V. × *'Flamme'* Richter (*V.* × *vigeri* × *V. barilletii*)
V. × *gemma* Duval (*V. barilletii* × (*V. carinata* × *V. psittacina*) × (*V. duvaliana* × *V. incurvata*))
V. × *'Gigant'* Richter (*V.* × *poelmanii* × *V.* × *versaillensis*)
V. × *gloriosa* Duval (*V. barilletii* × *V. incurvata*)
V. × *'Gnom'* Richter (*V. corcovadensis* × *V.* × *poelmanii*)
V. × *imperialis* Duval (*V. mirabilis* × *V.* × *rex*)
V. × *kitteliana* Hort. (*V. barilletii* × *V. saundersii*)
V. × *'Komet'* (*V. corcovadensis* × *V.* × *sceptre d'or*)
V. × *mariae* André (*V. carinata* × *V. barilletii*)
V. × *morreniana* Leod. ex Morren (*V. psittacina* × *V. carinata*)
V. × *poelmanii* Duval (*V.* × *gloriosa* × *V.* × *van geertii*)
V. × *rex* Duval ((*V. barilletii* × *V. morreniana*) × (*V. carinata* × *V. psittacina* var. *rubrobracteata*))
V. × *Septre d'or* Hort. (*V. saundersii* × *V. gloriosa*)
V. × *versaillensis* Truff. (*V. duvaliana* × *V. carinata*)
V. × *vigeri* Duval (*V. rodigasiana* × *V.* × *rex*)

Bi-generic crosses:

V. × *magnifica* Makoy (*V. Splendens* × *Guzmania zahnii*); *Guzvriesea magnifica* M. B. Foster.

8 Sub-family Bromelioideae

Key for the identification of the genera

Identification Stage	Plant Characteristic	Genus
1	a Flowers mostly not forming single headed inflorescence; not with ovary ripening at same time and not developing into eatable fleshy berry-like multiple fruit (syncarpium)	see **4**
	b Flowers forming cone-to-head-shaped inflorescence; ovary ripening at same time and forming (when ripe) eatable fleshy berry-like multiple fruit (syncarpium) (Ill. 204, 205)	see **2**
2	a Syncarpium not larger than 10 cm (Ill. 175)	*Acanthostachys*
	b Syncarpium larger than 10 cm, with sterile leaf-tufts at the tip	see **3**
3	a Syncarpium with sterile leaf tufts at tip; plants without stolons	*Ananas*
	b Syncarpium without sterile leaf-tufts; plants with stolons	*Pseudananas*
4	a Inflorescence branches not with normal leaves; only with small inflorescence bracts	see **6**
	b Inflorescence branches set with normal leaves up to tip (Ill. 235, Fig. 52)	see **5**
5	a Violet flowers, gathered together into head-shaped panicle (Fig. 4)	*Andreá*
	b White flowers; lengthened inflorescence, in loose arrangement; upper part of inflorescence compound (Ill. 235, Fig. 119)	*Orthophytum*
6	a Inflorescence lengthened; simple or compound	see **18**
	b Inflorescence remains shortened; cone or dense headed with very short,	see **7**

rarely extended shaft, supporting
(usually) flat outspread rosette
inflorescence or inflorescence sunk in
centre of rosette (Ill. 50) or elevated
on lengthened shaft; mostly with
brightly coloured tract leaves
surrounding the flowers, forming an
involucre (Ill. 216–218)

7   a   Filaments of the stamens united into a   *Bromelia*
ring at the base, partly fused with the
petals (Fig. 49, a-b). Plants very large,
with stout, robustly spined leaves and
frequently with thick stolons

      b   Bases of filaments not united into ring   see **8**
nor fused with petals

8   a   Bases of filaments (at least inner   see **10**
ones), fused more or less at high level
(Fig. 49, c-d), but unlike *Bromelia* not
formed into a proper ring

      b   All filaments free to the base   see **9**

9   a   Stamens projecting beyond flower   *Ochagavia*
(Fig. 49, d); petals without ligules at
base

      b   Stamens enclosed in flower; petals   *Fascicularia*
with ligules at base (growths at base of
petals)

10   a   Petals at base without ligules (scales)   see **13**

      b   Petals at base with ligules (scales)   see **11**

11   a   Petals united more or less at high level   *Wittrockia*
united in a tube

      b   Petals free to base   see **12**

12   a   Inflorescence surrounded by foliage-   *Orthophytum*
like bracts (Ill. 234)

      b   Inflorescence surrounded by brightly   *Canistrum*
coloured bracts (Ill. 216–218)

13   a   Small, low rosette plants; petals blunt,   *Cryptanthus*
outspread, white, the upper bracts
similar to the foliage (Ill. 220–223)
eastern Brazil

      b   Larger rosette plants   see **14**

14   a   Inflorescence not cone shaped, but   see **16**
frequently gathered together into
heads (glomerate); frequently
surrounded by brightly coloured
'heart leaves'

	b Inflorescence densely cone shaped, sometimes simulating a lateral position; not surrounded by brightly coloured 'heart leaves'	see **15**
15	a Inflorescence simple, spiked; frequently occupying a lateral position (Fig. 61)	*Greigia*
	b Inflorescence composed of up to five rows of compound spikes	*Streptocalyx*
16	a Inflorescence simple, free parts of petals outspread and pointed	*Neoregelia*
	b Inflorescence compound	see **17**
17	a Free parts of petals upright with rounded tips; flowers slightly open	*Nidularium*
	b Free parts of petals horizontally outspread and pointed; flowers wide opened (only in Amazon basin, sub-genus *Amazonicae*)	*Neoregelia*
18	a Inflorescence compound	see **25**
	b Inflorescence simple, racemed or spiked	see **19**
19	a Petals with scales at base (through magnifying glass)	see **22**
	b Petals without scales at base (through magnifying glass)	see **20**
20	a Flowers stalked, in many flowered raceme; eastern Brazil (Fig. 60)	*Fernseea*
	b Flowers sessile	see **21**
21	a Sepals pointed or with moderate prickle tip; petals mostly ligulate	*Aechmea*
	b Sepals neither pointed nor prickle tipped; petals not ligulate; Colombia	*Ronnbergia*
22	a Sepals pointed or with prickle tip	*Aechmea*
	b Sepals not pointed nor prickle tipped	see **23**
23	a Flowers slightly zygomorphic; petals recurved to rolled-up (convoluted) (Fig. 55); inflorescence frequently flexuous, zigzag, pendulous; inflorescence bracts conspicuous and brightly coloured, short-lived	*Billbergia*
	b Flowers radial; petals upright, not rolled-up; inflorescence not pendulous, upper-part curved over	see **24**
24	a Flowers sessile, spiked	*Quesnelia*
	b Flowers stalked, in many-flowered,	*Neoglaziovia*

upright racemes

25   a   Petals with scales at base (through   see **29**
                 magnifying glass)
        b   Petals without scales at base   see **26**

26   a   Bases of filaments fused into short   *Bromelia*
                 tube, and to bases of petals (Fig. 49,
                 a-b); inflorescence dense, frequently
                 surrounded with brightly coloured
                 scape bracts; plants large with firm,
                 robustly spined leaves
        b   Bases of stamens not fused into a   see **27**
                 short tube, but filaments more or less
                 united at high level with petals

27   a   With scale shaped scape bracts in the
                 shoulders of the inflorescence sections
        b   With larger, conspicuously coloured   see **28**
                 scape bracts in the shoulders of the
                 inflorescence sections

28   a   Scape bracts yellow   *Nidularium*
                                                seidelii
        b   Scape bracts not yellow   *Streptocalyx*

29   a   Sepals with prickle tip or prickled,   *Aechmea*
                 seldom blunt, then long-tailed when in
                 seed (interesting to see through a
                 magnifying glass when long or cross
                 sectioned through the ovary, Fig. 49, e)

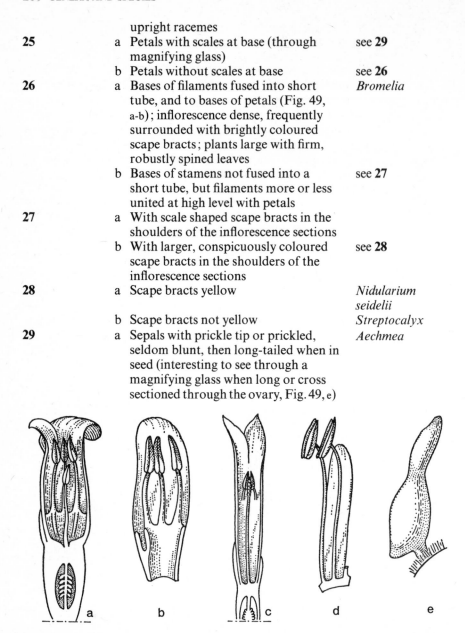

Fig. 49 *Bromelia laciniosa*: a longitudinal section through flower; b single petal with its stamens; c *Neoregelia concentrica* (Vell.) L. B. Smith, longitudinal section through the flower; d *Ochagavia lindleyana* (Lem.) Mez, petal with stamens; e *Aechmea luddemanniana* Brogn. seed pod.

	b Sepals neither pointed nor prickle tipped; seed not long tailed	see **30**
30	a Flowers either zygomorphic or petals convoluted; inflorescence mostly pendulous with brightly coloured scape bracts	*Billbergia*
	b Flowers radial; petals not convoluted	see **31**
31	a Flowers long stalked (1–4 cm), in very open or cylindrical, glomerate panicle	*Portea*
	b Flowers sessile or very short stalked, in convoluted to head-like inflorescence sections	see **32**
32	a Petals above the ovary free or united to a tube	*Hohenbergia*
	b Petals above the ovary united into a distinct bue	*Gravisia*

Acanthostachys Klotsch 1841

The genus *Acanthostachys*, the 'Prickle ear' is represented from Brazil to Paraguay and Argentina by only one species *A. strobilacea* Klotsch. This species has been frequently joined with the genus *Ananas* from which it is distinguished by its flower structure. The cone-shaped multiple fruit head resembles that of the pineapple, but is much smaller, although it is just as edible and equally sweet-tasting.

Acanthostachys strobilacea (Schultes fils) Klotsch 1841 (Fig. 50, Ill. 175) (*strobilacea(us)* cone-shaped; on account of the cone-shaped inflorescence)
Syn. *A. exilis* Bertoni, 1925; *Hohenbergia strobilacea* Schultes fils, 1830

Pl stemless, increasing by short stolons and thus growing in clusters
Fo Few leaves in loose open rosette
LS very long, narrow, not clearly defined from the plant sheath
LB straight narrow, channelled, edges serrated, underside covered with grey scales
Sc slender, upright or pendulous to 50 cm long, scaly, shorter than the leaves
I in all respects cone-shaped to 5 cm
SB similar to the leaves, one or two below the inflorescence, widely spread out
FB widely triangular to ovoid, ending in an acute point, red-orange
Se keeled, ending in a point, to 11 mm long
Pe yellow, to 2·5 cm long
Fr yellow, berry-like, eatable multiple aggregate fruit (syncarpium)
Ha Brazil, Argentina, Paraguay
CU an attractive, widely cultivated species, specially recommended to the beginner. It can be cultivated either in an orchid basket or as a terrestrial plant, but looks better in the former.

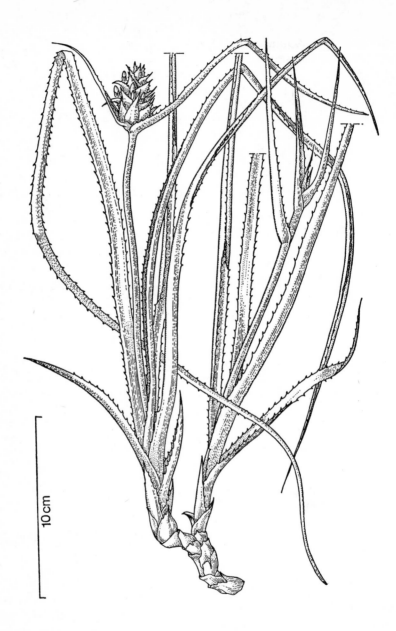

10 cm

Fig. 50 *Acanthostachys strobilacea* Klotsch; flowering plant.

Aechmea Ruiz and Pavon 1794

This genus was named *Aechmea* in 1794 by the Spanish naturalist and South American traveller Pavon.

The name comes from the Greek *aichme* and means 'lance head', a reference to the prickly points of the flower bracts and flower sepals (Fig. 51, 1–2a, D).

The genus comprises some 150–180 species and their large size accounts for the fact that only a small number of them find their way into enthusiast's collections.

You will only find the larger members of Aechmeas in a botanical garden or exhibition, where they make up a colourful display because of their long lasting, prominent flower heads (inflorescence). Several species (*A. fasciata, A. chantinii, A. orlandiana* and others) have in recent times found their way into the ornamental plant business. Today, the nurseryman offers cultivated selected forms or hybrids and mass produced plants. Most flower shops can now at least offer *A. fasciata*, a plant whose month-long flowerhead is most attractive (Plate 68).

Points which distinguish an aechmea from the other members of the Bromeliad family are:

1 Leaves vary to form a wide funnel to a narrow tube slim rosette, and are frequently crossbanded (Ill. 181) or conspicuously dappled (Plate 75) with the margins distinctly spined (Fig. 51, 1b; only a few kinds have the leaf margins all but smooth; their tips are widely rounded, terminating in a more or less long narrowed pointed tip (Fig. 51, 1a).

2 Flower bracts, inflorescence bracts, and sepals are vividly coloured, pointed at the tip, seldom blunt, compact firm and leathery (at least the flower bracts and sepals).

3 The petals (represented in the genus *Vriesea*) frequently carry (on the upper side at their base) two spoon-shaped organs or ligules (Fig. 51, 3a L) fused together into a short tube. The inferior ovary (Fig. 51, 3 O) develops to form a ripe, pulpy (occasionally dried), frequently brilliantly coloured berry (Fig. 51, 4 a-b). Berries are very ornamental and their colour remains for long periods.

A number of species are self-fertile and they fruit without fertilization.

In their native habitat the strong nectar of the flowers attracts hummingbirds who thus pollinate the plants.

The main area of distribution of the *Aechmea* southerly is Brazil, though a few species spread out from there.

The *Aechmeas* live chiefly epiphytically on trees, rarely on the ground, in tropical cloud and rain forests. However, they also occur in dry areas, where they frequently have a very firm leaf construction (xeromorphic). It follows that on account of their distribution most *Aechmeas* require to be cultivated in temperate conditions and half shade. Some kinds—*A. recurvata, A. purpureo-rosea, A. nudicaulis* and others—can be placed in full sunlight, where their leaves take on an intensive bright red colour e.g. *A. recurvata* (Plate 80).

Propagation is mainly by offshoots and, in the trade, by seed sowing. The seeds grow readily and quickly in contrast with those of the *Tillandsia*.

Few pests trouble Aechmeas other than woolly aphids and scale. However the disease known as Aechmea Rot is an attack by fungus which often causes the plant to die in a very short time (see page 72).

A selection of decorative species worth growing and suitable for the small greenhouse owner are given below.

Aechmea angustifolia Poepp and Endl., 1838 (Ill. 176)

(*angustifolia(us)* narrow leaved)
Syn. *Hoplophytum angustifolium* (Poepp. and Endle) Beer, 1857; *Hohenbergia angustifolia* (Poepp. and Endl.) Baker, 1871; *A. cummingii* Bak., 1879; *A. boliviana* Rusby, 1907; *A. cylindrica* Mez, 1913; *A. inconspicua* Harms 1929

Pl stemless, with inflorescence up to 75 cm high
Fo 10–15, 50–70 cm high, forming an ellipsoid based rosette
LS 10–14 cm long, to 10 cm wide, appressed thickly with brown scales
LB strap shaped 3–6 cm wide, green, grey scaled, edged with brown spines to 3 mm long.
Sc upright, at first covered with white, wool, later bare, reddish-brown or green
I 15–35 cm long, firm to loose, with numerous outward spreading spikes
SB elliptical, tapered to a point, brown-red, grey scaled, toothed, soon drying, longer than the internodes
PB at the base of the inflorescence similar to the scape bracts 5–6 cm long, narrow-triangular with spined edges, red-brown, longer than the spikes, reducing towards the top of the in-florescence to 5–10 mm long, turning back against the stem
Sp outward spreading to recurved, each with about 10 closely set flowers 2–4·5 cm long; flexuous axis flattened, red-brown with green centre strip
FB broad-oval, with terminal spine to 5 mm long, brown-red, the base covered with white wool, shorter than the ovary
Fl 1·2–1·6 cm long, distichous
Se 4–5 mm long, spine tipped, free, asymmetrical, yellow
Pe lineal, 8–9 mm long, short pointed tip, yellow
Ov at the time of blooming greenish-yellow, then white; lilac-brown when ripe
Ha Costa Rica to Peru and Bolivia
CU widely popular and frequent plant.

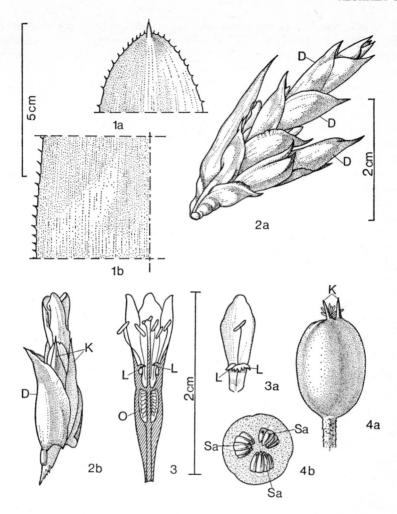

Fig. 51 *Aechmea pubescens* Baker: 1a-b leaf tip and leaf margin; 2a single flowering spike; 2b single flower; D flower bract; K sepal; 3 longitudinal section through a flower; 3a single petal; L liqula; O ovary; 4a ripe fruit; 4b cross section through ripe fruit showing seeds (Sa).

Aechmea bambusoides L. B. Smith, 1961 (Ill. 178)

(*bambusoides* like bamboo; this plant when in bloom has its inflorescence in tiers which bear a distinct resemblance to a small bamboo)

Pl with inflorescence 1–1·20 m high
Fo an upright, bottle-shaped rosette
LS broad oval 10–15 cm long, 8–11 cm wide, upperside dark black-brown, underside yellow-brown; margins not spined

LB wide triangular, long terminal point, upright to spreading, green, to 60 cm long, up to 7–10 cm wide above the sheath (average 4–5 cm wide), margins plain or with only a few robust upward bent green to brown thorns to 9 mm long

Sc upright, slender 50–60 cm long, 3–4 mm in diameter, reddish-brown, bare or sparsely grey scaled

I 40–50 cm long, 6–7 cm wide, loosely compound with tier arrangement, convoluted, spiked for part of the inflorescence

SB very long, narrow, upright, at upper part of the scape shorter than the internodes, reddish-brown veined, lightly grey scaled; the uppermost bracts about 12 cm long and 9 mm wide

PB narrow, lineal-lanceolate, longer than the clusterlike spike, soon drying

FB triangular, 1 cm long, shorter than the sepals, nerved, reddish-brown

Se free, naked, green, reddish-brown prickled, 9 mm long

Pe yellow, 1·5 cm long, 2·5 cm wide, its tip not flared

Ov naked, green, 5 mm long

Ha Brazil

Aechmea bracteata (SW.) Griseb 1894 (Plate 64)
(*bracteata(us)* bractleaved; refers to the shiny red coloured bracts
Syn. *Bromelia bracteata* SW., 1788

Pl with inflorescence 1·70 m high and higher

Fo numerous, to 1·30 m long

LS very large and decorous, ellipsoidal bottle-shaped build, 25–30 cm long, 20–25 cm wide, thickly brown scaled

LB strap shaped, to 1 m long and 10 cm wide, with a long terminal tapered point or rounded and spined tip, pale green, grey scaled; upward curled leaf margins with 1 cm long, bent green spines

Sc upright, beginning white woollen, later bare, about 1 cm thick

I 10–70 cm long, loose compound, pyramidal; with basal compound branches

SB more or less overlapping, lancoeolate, pointed, the lower part of the scape greenish, shiny red toward the tip, the margins not toothed

PB at the base of the inflorescence similar to the upper bracts; shiny red, longer than the basal branches, reducing towards the top and becoming scale like, green

Sp spreading, loose, distichous, 7 to 14 flowers

FB as long as the sepals, green, somewhat white woolly, wide oval, tapered prickled tip, keeled at the upper part of the scape, nerved

Se triangular-oval, assymetric, spine tip 3–4 mm long, green, with membranous edge, beginning somewhat white woolly, soon bare

Pe lineal, with the point widely rounded to 1 cm long, nail-like; nail to 2 mm wide, lamina 5 mm wide, pale yellow

Ov ellipsoid to spherical, green somewhat white woolly, turning to blue with the ripening of the seeds

Ha Mexico to Central America

CU more impressive in fruit than in flower.

Aechmea bromeliifolia (Rudge) Baker 1883 (Ill. 179)
(*bromeliifolia*(*us*) bromelia leaved; the plant is similar to a species in the genus *Bromelia*)
Syn. *Tillandsia bromeliifolia* Rudge, 1807; *Bromelia tinctoria* Mart., 1828; *Macrochordium pulchrum* Beer, 1857; *Aechmea conspicuiarmata* Bak., 1889; *A. macroneottia* Bak., 1889; *A. tinctoria* Mez, 1892; *A. pulchra* Mex 1892; *Hoiriri bromeliifolia* Kuntze, 1898; *A. eriostachya* Ule, 1908; *A. ellipsoidea* Rusby, 1927

Pl with inflorescence 60–90 cm high
Fo 12–20 in a tubular rosette
LS oval to long-elliptic, 10–20 cm long, to 12 cm wide, entire, with the upper part very much spined
LB strapshaped, very variable, long tapering to rounded and spine tipped often recurved, 60–100 cm long, 4–9 cm wide, dark grey, thickly grey scaled and ribbed; margins with brown spines to 1 cm long
Sc upright, sturdy, thickly white woollen
I simple-spiked, ellipsoid or cylindrical, dense, many flowered to 15 cm long, 3–4 cm in diameter, white woollen
SB lanceolate-oval, smooth, thin, red, weak
FB wider than longer, green, white woollen, not toothed, stump-tipped, two-keeled, much shorter than the sepal, enclosing the base of the flower
Fl sessile, nearly horizontal, tight spiralled arrangement, the axis covered, to 1·7 cm long
Se almost circular, to 7 mm long, with 2 mm of its length fused, pale green, grey scaled
Pe 1·5 cm long, greenish yellow, turning to black after blooming
Ov short, densely white woolly
Ha Brazil, Guatemala, Honduras to Paraguay and Argentina.

Aechmea calyculata (Morren) Baker 1879 (Ill. 177)
(*calyculata*(*us*) with an involuere)
Syn. *Hoplophytum calyculatum* Morren, 1865; *Aechmea sellona* Baker, 1889; *Echinostachys pineliana sensu* Wittm, 1891 non Brongn., 1854–58

Pl with inflorescence more or less 60 cm high
Fo 8–12, in a loose rosette
LS long-oval, 13–15 cm long 5–6 cm wide
LB lineal, with rounded but prickled tip, 3–5 cm wide, to 50 cm long, green, grey scaled, violet spot at tip; margins small prickled
Sc upright, circular, brown, white woolly, not exceeding the leaves
I simple-spiked, firm, pyramidal to
cylindrical, to 5 cm long, 3 cm in diameter, many flowered
SB upright, lying against the scape, the lower section shorter than the internodes, thin, red, soon drying
FB narrowly-triangular, lanceolate, as long as the flowers or somewhat longer, thin, red, entire
Fl sessile 1·7 cm long, in a spiral formation
Se to 1·5 mm fused, 4 mm long, with 4 mm longer, brown spine tip,

greenish-yellow, naked, somewhat ver-rucose
Pe to 1·4 cm long, yellow, with rounded tip

Ov 4 mm long, yellowish red, naked or somewhat white woolly
Ha Brazil, Argentina.

Aechmea candida Morren 1889 (Ill. 180)

(*candida*(*us*) white; refers to the white woolliness of the inflorescence and its parts)

Pl with inflorescence 50–70 cm high
Fo an open rigid rosette from 50–60 cm in diameter
LS longish-oval, 16–17 cm long, 8–10 cm wide, green, thickly grey scaled
LB 20–45 cm long, 4·5–6·5 cm wide, lineal, with rounded prickle tip, both sides grey scaled (the underside thicker) and towards the tip banded
Sc upright, round, green, white woolly, 7–8 mm thick, not exceeding the rosette
I 15–20 cm long, 6–8 cm wide, narrow-pyramidal, at the base spirally arranged in open compound spikes, single towards the top, from 5 to 9 flowers, thickly white woollen
SB upright, open spiral, longer than the internodes, thin, pale rose, white woolly, not permanent
FB as long as or somewhat shorter than the sepals, triangular, spined, 4–5 mm wide, thin, pale rose, white woollen
Fl sessile, about 1·5 cm long
Se pale yellow, white woollen, the upper part not fused, with 3–3·5 mm long brown spined tip
Pe white, 1 cm long, 3 mm wide, with the tip rounded and hooded
Ov pale yellow, woolly, up to 5 mm long
Ha Brazil.

Aechmea caudata Lindm., 1891 (Ill. 182)

(*caudata*(*us*) with a tail; refers to the long inflorescence)
Syn. *Aechmea platzmannii* Wittm., 1891; *A. henningsiana* Wittm., 1891

Pl with inflorescence to 1 m high
Fo 10–15, forming a funnel shaped rosette from 50–80 cm in diameter
LS wide-oval, 13–14 cm long, 9–10 cm wide, green, upperside mottled lilac
LB to 80 cm long, 6–8 cm wide, lineal with a wide rounded spined tip, dark green, grey scaled; margined with brown 1 mm long spines
Sc stout, upright, round, white woolly, not longer than the leaves
I 20–25 cm long, to 10 cm wide, lower half branched, upper simple; axis white woollen from the middle to the top on blooming
SB longer than the internodes, upright, lying close to the stem, thin, soon drying
PB as long as or shorter than the outward spreading spikes, brown-lilac-rose grey scaled, the lower to 5 cm long
Sp 5 to 7 flowers, to 5 cm long, 2 cm wide; the axis flexuose, white woollen
FB shorter than the sepals, to 1·5 cm long, triangular, brown-lilac-rose with brown spined tip, grey scaled, becoming yellow after blooming

Fl sessile, 2·5 cm long
Sc orange coloured, grey scaled, base fused to 1 cm long (including the brown spines) becoming yellow after flowering

Pe 1·5 cm long, yellow
Ov 1 cm long, white woolly, yellow after flowering
Ha Brazil.

VARIETIES

var. *variegata* M. B. Foster, 1953 (Plate 65) Leaves with wider white longitudinal stripes
var. *eipperi* Reitz, 1962 Petals with pale blue base and blue tip

Aechmea chantinii (Carr.) Baker, 1889 (Plate 66, Ill. 181)
Syn. *Billbergia chantinii* Carr., 1878
Because of its beautiful leaves and shiny red, striking inflorescence bracts, one of the loveliest bromeliads. Although brought to Europe in 1877, it has always been rare in cultivation until in recent years, when more and more choice forms have been produced following its rediscovery in Amazonia in the region of Iquitos. Even more attractive than the species is the variegated form, its leaves not only white cross banded but with yellowish margins and longitudinal stripes (a fine coloured plate is to be found in the *Journal of the Bromeliad Society*, Vol. XX, No. 4, 1972).

Pl with inflorescence 40–100 cm high
Fo about 10, upright to almost outward spreading, a rosette from 50–60 cm in diameter
LS outward spreading, almost circular, 10–12 cm long, 8–10 cm wide, upper side lilac, grey scaled
LB wide-lineal, 30–40 cm long, 5–6 cm wide, the tip rounded and with terminating spine, green, each side cross-banded grey; margins with 3–4 mm long brown spines
Sc upright, longer than the leaves, 5–7 mm thick, upper part red, grey scaled
I short pyramidal with 5–10 short compound spikes, to 15 cm long, to 8 cm in diameter
SB upright, somewhat longer than the internodes, thin, shiny red, grey scaled, toothed
PB the basal ones similar to the inflorescence bracts, very large, thin, shiny red, grey scaled, toothed, longer than the spikes, flared to folded back, the upper ones small and insignificant
Sp 5 to 8 flowered, spread out, short stalked, 4–5 cm long with flattened 7–10 mm long, red peduncle
FB oval, 1·1 cm long, red with yellowish tip, naked to scattered scales
Fl sessile, to 3 cm long
Se free, 1·3 cm long, red with yellow tip
Pe narrow-lineal, pointed tip, yellow
Ov 8 mm long, greenish red
Ha Brazil (?), Northern Peru (Amazonia).

Aechmea coelestis (C. Koch) Morren, 1875 (Ill. 183)
(*coelestis* bluish; refers to the pale blue flowers)
Syn. *Hoplophytum coeleste* C. Koch, 1856

Pl with inflorescence 50–70 cm high
Fo compact, nearly upright, a rosette from 60–70 cm in diameter
LS narrow-oval, 15–18 cm long, 6–8 cm wide
LB wide-lineal, the tip rounded and prickled, 3·5–6 cm wide, 50 cm long, grey scaled, margins curled outwards, prickles 1 mm long
Sc upright, round, 6 mm thick, not exceeding the leaves, dark green, white woolly
I pyramidal, many flowered, to 20 cm long, 8 cm wide, few flowers at the base, spreading, spirally arranged compound spikes, reducing at the tip to single flowered
SB loose, upright, thin, shorter than the internodes, pale rose, grey scaled, perishable
PB similar to the inflorescence bracts, as long as the spikes or slightly longer
Sp loose, 6 flowered, 1 cm long, stalked, more or less 3·5 cm long, white woollen
FB 1·5 cm long, 5 mm wide, shorter than the sepals, triangular with long terminal spine, reddish-brown, white woollen
Fl sessile, to 2 cm long
Sp 4 mm long, 2 mm high fused, greenish-rose, white woollen; terminal spine 3 mm long, red-brown
Pe pale-blue, to 1·3 cm long with the tip hooded and inclined together
Ov 8 mm long, greenish-rose, white woollen
Ha Brazil.

Aechmea comata (Gaud.) Baker, 1879 (Plate 67)
(*comata(us)* tuft; by reason of the tuft-like inflorescence)
Syn. *Pothuava comata* Gaud., 1851; *Hoplophytum lindenii* Morren, 1865; *Aechmea lindenii* Baker, 1879; *Macrochordium lindenii* Wittm., 1891.

Pl 60–100 cm high
Fo a nearly upright, slim rosette
LS narrow-oval, 18–20 cm long, 7–9 cm wide, upperside lilac
LB lineal-strapshaped, to 60 cm long, 4–5 cm wide, the wide rounded terminal prickle, margins short brown prickled
Sc upright, round, green, white woolly, not longer than the leaves, 50–60 cm long, 5 mm wide
I upright, simple-spiked, thick, many flowered, ellipsoid to 8 cm long and 3–4 cm in diameter
SB lying close to the stem, upright, shorter than the internodes, 3–4 cm long, quickly withering
FB red, thin, triangular, leaflike, 1·4 cm long, at the base 8 mm wide, quickly drying
Fl nearly upright, closely set spiral arrangement, 2 cm, sessile
Se inverted-eggshaped to elliptical, short fused, nearly symmetrical with 1–2 mm long spined terminal point, yellowish red
Pe yellow, the tip hooded and rounded
Ov yellowish red, greying towards the flower, 1 cm long
Ha Brazil
CU Usually as var. *mackoyana* L. B. Smith, which has yellowish-white longitudinally striped leaves.

Aechmea cylindrata Lindm., 1891 (Plate 96)

(*cylindrata(us)* cylindrical; refers to the cylindrical inflorescence)
Syn. *Aechmea cylindrata* var. *micrantha* Lindm., 1891; *A. hyacinthus* F. Mueller, 1893

Pl stemless, 50–60 cm high when in flower
Fo numerous, in a flat flared rosette
LS long-oval, 10–12 cm long, 5–10 cm wide, green, pale lilac blotched or striped, grey scaled
LB lineal, 30–40 cm long, to 6 cm wide, rounded spined tip, dark green, upper side lightly scaled, underside heavier scaled. Margins prickled
Sc Sturdy, upright to angled, 30–40 cm long, 1–1·3 cm wide, round, green, becoming white woollen
I upright, moderately firm, cylindrical, many flowered, 15–20 cm long, 2–3 cm in diameter
SB Those at the base shorter, those above longer than the internodes, upright, close to the stem, thin membraneous, soon drying, red dependent on their position, 3–6 cm long, somewhat white woollen
FB narrow-triangular, long pointed, the basal ones longer than the flowers, to 4 cm long, the upper ones shorter, 5 mm long, red, slightly while woollen, not toothed
Fl sessile, 2 cm long
Se 3 mm high fused, with the tip rounded but with brown terminal spined to 5 mm long rose coloured
Pe 1·2 cm long, light blue, with darker blue, dome shaped widely rounded tip
Ov towards the flower green, somewhat white woolly, especially at the base of the inflorescence, 8 mm long
Ha Brazil
CU very decorative plant when in bloom.

Aechmea dealbata Morren, 1889 (Plate 69)

(*dealbata(us)* becoming white; refers to the dense powdery scaled appearance of the leaves)
Syn. *Hoplophytum dealbatum* Morren, 1889; *Bilbergia glaziovii* Regel, 1885

Pl when in bloom about 80 cm high, the inflorescence hardly taller than the leaves
Fo very few, an almost upright rosette from 60–70 cm in diameter
LS long, narrow-oval, 20–23 cm long, 6–7 cm wide, underside green, upperside red-lilac, both sides densely grey-scaled
LB tongue-shaped lanceolate, pointed tip, 50–70 cm long, 5–7 cm wide, green, underside grey cross-banded; margins at the middle of the blade section curled upwards, dense brown prickles
Sc upright, 30–40 cm long, 8–9 mm thick, dark red, densely white woolly, round, smooth
I single or few (2 to 3) densely compound spikes, cylindrical, 7–8 cm long, 6 cm wide
SB upright, spirally arranged, lying close to the scape, longer than the internodes, 4–5 cm long, 1·5 cm wide, lanceolate, long pointed, red, nerved, grey scaled, the margin prickled
PB similar to the upper inflorescence

bracts, red, densely grey scaled, toothed longer than the spikes
FB densely spirally arranged, upright, 3–4 cm long, oval-triangular, long pointed, dark red, nerved, dense grey white-felted scaled, margins prickled
Fl sessile, 3·5 cm long
Se 1 cm long, fused to the middle, assymetric, red, densely white woolly
Pe 3 cm long, rounded tip, base white, upper half violet, when finished flowering strong red
Ov 1 cm long white woollen
Ha Brazil, closely allied to *Aechmea fasciata.*

Aechmea distichantha Lem., 1853 (Plate 71)

(*dischantha(us)* bipartite flowering; refers to the bipartite arranged flowers). (Arranged in two opposite rows)
Syn. *Tillandsia polystachia* Vell., 1825 non L., 1762; *Aechmea excavata* Baler 1879; *A. brasiliensis* Regel, 1885; *A. myriophylla* Morren, 1887; *Quesnelia distichantha* Lindm., 1891; *A. polystachya* Mez, 1892; *A. polystachya* var. *excavata* Mez 1896; *Hoiriri polystachya* Kuntze 1898; *A. polystachya* var. *myriophylla* Hassler, 1919; *A. platyphylla* Hassler, 1919

Pl with inflorescence 30–70 cm high
Fo many flowered, a rigid rosette
LS elliptical or longish, clearly flared, to 30 cm long much wider than the blades
LB tongue-shaped, short or long pointed, prickly, 2·5–8 cm wide, 1–1·50 m long, dull-green, grey scaled, the margins dense brown spined
SC upright, white woolly
I loose to dense compound, wide-pyramidical
SB lying close to the scape, longer than the internodes, overlapping, elliptical, narrowing to the tip, margins usually not prickled, rose or green
PB wide-oval, prickle tip, usually shorter than the spikes, rose, white woollen
Sp few to many flowered, upright or spreading to outward spreading
FB distichous, 5–7 mm long, nearly as long as and surrounding the ovary, thin, nerved, rose, white woollen
Fl nearly upright, 1·5–2·9 cm long
Se assymetric, longish to nearly square, prickled tip, 0·5–1·3 cm long, free or short fused, thin, nerved, rose and more or less white woolly
Pe 1·8 cm long with blunt tip, blue, white or purple
Ov short-cylindrical or cupshaped, 4–6 mm long, white woolly.

VARIETIES
var. *distichantha* (Plate 71) Inflorescence loose or moderately loose, nearly wide-pyramidical; spikes more or less spreading, many flowered; leaves usually short pointed tip; flowers purple or blue
Ha Brazil, Bolivia, Paraguay, Uruguay, Argentina.
var. *distichantha f. albiflora* L. B. Smith 1943. Differing from the type by its sessile white petals
Ha Brazil.

var. *Schlumbergeri* Morren, 1892
Syn. *Chevaliera grandiceps* Griseb., 1879; *Aechmea grandiceps* Mez, 1892; *A. involucrata* Rusby, 1907; *A. polystachya* var. *longifolia* Cast., 1925; *A. rubra* A. Silveira, 1931; *A. involucrifera* Mez 1934 Inflorescence stout, long, slender, cylindrical or spindle form; spikes upright, few flowered; large flowers; leaves usually with a long pointed tip
Ha Brazil, Bolivia, Paraguay, Argentina.
var. *glaziovii* (Baker) L. B. Smith 1943 (Ill. 185)
Syn. *Aechmea glaziovii* Baker, 1879; *Quesnelia wittmackiana* Regel, 1888
A. jucunda Morren, 1889; *A. regelii* Mez, 1892; *A. wittmackiana* Mez, 1892; *A. pulchella* Morren, 1892.
Short inflorescence, thick, egg-shaped-cylindrical; spikes upright, few flowered; the plant smaller than the type; leaf tip usually rounded with terminal prickle.
Ha Brazil

Aechmea fasciata (Lindl.) Baker, 1879 (Plate 68, Ill. 186) (*fasciata(us)* banded; because of the bands on the leaves)
Syn. *Billbergia fasciata* Lindl., 1828; *Hohenbergia fasciata* Schult. f, 1830; *Hoplophytum fasciatum* Beer, 1857; *Billbergia rhodocyanea* Lem., 1847; *Quesnelia rhodocynea* Wawra, 1883; *Aechmea rhodocynea* Wawra, 1883; *A. leopoldi* hort., 1889

Pl stemless
Fo a tube shaped to funnel shaped rosette
LS oval, 10–13 cm long, 9–10 cm wide, upper side lilac, grey scaled, underside green, grey scaled, margins not prickled
LB to 50 cm long, 6 cm wide, wide-lineal, the tip widely rounded but with a 4 mm long spine, densely consistently grey scaled or distinctly banded; margins with small brown to 4 mm long spines
Sc upright to 1 cm in diameter, reddish-brown, densely white woollen 30–40 cm long
I thick, knob-like-pyramidal, at the base branched, towards the tip simple
SB loose at the base of the scape, greenish-white to delicate rose and greenish tip, 5–6 cm long, 1–1·5 cm wide, prickled margins, lightly white woollen, at the lower part of the inflorescence thick and forming a kind of involucre; 7–9 cm long, the base 1–2 cm wide, narrow-triangular, long pointed, rose, with prickled edges
PB like the upper inflorescence bracts, much longer than the spikes
FB similar to the inflorescence bracts, but smaller, triangular, long pointed, 4–5 cm long, at the base to 1·5 cm wide, rose, lightly white woolly, the margins rose coloured prickled
Fl sessile, to 3·5 cm long
Se to 1 cm long, at the base about 2·5 mm high fused, assymetric, rounded tip, rose, otherwise white, white woollen scaled
Pe 3 cm long, with rounded tip, blue,

fading after flowering to a rose colour and white towards the base
Ov 6–8 mm long, 5–6 mm wide, nearly globular, white woolly

Ha Brazil
CU nurseries have developed forms with beautiful leaves and large inflorescences.

VARIETIES
var. *Purpurea* (Guillon) Mez
Syn. *Billbergia rhodocyanea purpurea* Guilon, 1883; differing from the type only by the reddish-purple leaves; known only in cultivation
var. *variegata* hort. Leaves with creamy-yellow longitudinal stripes.
Mez (1934) established the difference between *A. dealbata* and *A. fasciata*. With *A. dealbata* the backs of the bracts are thick white and with *A. fasciata* only scattered white lepidate.
In *A. dealbata* the inflorescence is usually unbranched, in *A. fasciata* we find the base of the inflorescence branched.

Aechmea ferruginea L. B. Smith, 1932 (Plate 77)

(*ferruginea(us*) rust-brown; refers to the rust-brown scales of the inflorescence

Pl stemless, with inflorescence 70–80 cm high
Fo numerous forming, a nearly upright funnel-shaped rosette from 50–80 cm in diameter
LS to 20 cm long, 10–14 cm wide, distinctly flared, wide oval, dark brown scaled, not toothed
LB 60–70 cm long, lineal with rounded and prickled tip, 4–7 cm wide, green, grey scaled, margins closely set with spines to 4 mm long, brown
Sc slender, upright, shorter than the leaves, to 40 cm long, 7 mm thick, felted dark brown scaled
I pyramidal, to 17 cm long and 7–8 cm wide, at the base loose several times compound, towards the tip simple, the

axis thickly felted dark-brown scaled
SB loose, longer than the internodes, lanceolate, long pointed, thin, soon drying, brown felted scaled
PB the basal ones similar to the inflorescence bracts, longer than the spikes, half way up shorter than the spikes, at the tip absent
Sp loose flowered, spreading to widely spread, 4 cm long, 5 to 10 flowered
FB absent, or existing with tiny scales
Fl 2·8 cm long, about 1 cm long
Se assymetric, 4 mm long, red and brown felted scaled
Pe 1·3 cm long, at the base white changing to blue at the tip
Ha Peru.

Aechmea filicaulis (Griseb.) Mez 1896 (Ill. 184)

(*filicaulis* threadlike branched; refers to the thin, threadlike long hanging inflorescence)
Syn. *Billbergia filicaulis* Griseb., 1864

Pl stemless, epiphytic

Fo numerous, forming a rosette from

40–50 cm in diameter
LS distinct, 10–12 cm long, 6–7 cm wide, longish-oval
LB lineal, somewhat widened in the upper part, pointed spine tip, to 40 cm long, at the base 2 cm wide, in the upper part 3–4 cm wide, upper side dull green or reddish, underside reddish, and grey scaled, margins more or less strongly toothed
Sc long and thin, hanging, 1–2 mm thick, green to red
I hanging, loosely compound, 50–100 cm long
SB at the base 4–5 mm wide, longer or shorter than the internodes, according to their positions on the scape, green-red, quickly drying
PB red, shorter than the spikes, at the base of the inflorescence 6–7 cm long at the tip very much smaller
Sp loose, few flowered, to about 1·5 cm long, thin, green stalked
FB insignificant, scale-shaped, green with darker margins, 2–3 mm long
Se assymetric, 1–3 cm long, 5 mm wide, thin, partly cutaneous, green with small, dark terminal prickle, at fruiting time becoming blue
Pe 5 cm long, white, nail like, 3–4 mm long and wide, with widened 7 mm wide lamina
Ov 1·8 cm long, green, at fruiting time becoming blue
Ha Venezuela
CU the only bromeliad with a very thin and very long pendulous inflorescence. It must be cultivated epiphytically. Requires a tall greenhouse.

Aechmea fosteriana L. B. Smith, 1941 (Ill. 187)

(*fosteriana(us)* named after the American bromeliad collector and explorer M. B. Foster)

Pl stemless, 50–70 cm high
Fo 10–14, a slender tubular rosette
LS wide-oval, 10–14 cm long, 8–10 cm wide, upper side lilac-brown, with smooth margins
LB 40–60 cm long, 4–6 cm wide, rounded and short terminal spine, green, dense grey scaled on both sides, underside irregularly dark flecked
Sc thin, pendant, about 60 cm long, 3–4 mm thick, greenish brown, lightly white woolly, soon becoming bare
I 7–10 cm long, 6–8 cm wide, pendant, with 4 to 8 few flowered loosely compound spikes
SB widely spaced, much shorter than the internodes, upright clinging to the scape, 7–8 cm long, greenish-brown
PB similar to the upper inflorescence bracts, longer than the spikes at the base of the inflorescence reducing towards the top, thin, reddish-brown, grey scaled
Sp flared to spreading, loose 4 to 6 flowered, 4–5 cm long, 2–3 cm wide, with a sessile disc-like (thalamus) thickening at the inflorescence axis
FB 1·2 cm long, somewhat longer than the ovary, 9 mm wide, green with a rounded tip but short terminal point
Fl 3 cm long, sessile
Se free, 1 cm long, green
Pe 2 cm long, yellow
Ov 9 mm long, green, bare
Ha Brazil
CU a very pretty plant, worth growing.

Aechmea fulgens Brongn., 1841 (Plate 72)

(*fulgens* bright; refers to the bright red flower organ (calyx, flower and fruit)

Pl stemless, with inflorescence 40–50 cm high
Fo numerous in a wide funnel-shaped rosette
LS oval, 10–11 cm long, 6–7 cm wide, brown spotted, scaled
LB 30–40 cm long, 5–6 cm wide, ending with a rounded and spined tip, green, underside grey-waxy, limp, margins widely prickled
Sc upright, robust, to 25 cm long, 5–10 mm thick, round, red, bare
I to 20 cm long, the lower half loosely compound with upright to spreading spikes reducing to simple at the top
SB thin, longer than the internodes, withering, drying and hanging and outlasting the flowers
PB the basal ones similar to the inflorescence bracts, perishable, the upper ones reducing to small triangular scales
Sp 3 to 8 flowered, 5–8 cm long with thin, swollen-jointed, red rachis
FB small scales or absent
Fl sessile, to 2 cm long, arranged loose tripartite on spike
Se 5 mm long, red, bare, with the tip rounded
Pe 1·2 cm long, the tip rounded, violet, changing to red with the fading of the flowers
Ov round, bare, becoming red on ripening
Ha Brazil.

VARIETY

var. *discolor* (C. Morren) Brognn., 1841 (Plate 72)
Syn. *Aechmea discolor* C. Morren, 1846
Distinguished from the type by the strong red colour of the undersides of the leaves. Only known in cultivation; a very decorative plant.

Aechmea gamosepala Wittm., 1891 (Ill. 188)

(*gamosepala*(*us*) fused sepals; refers to the fused (united) sepals)
Syn. *A. thyrsigera* Speg., 1917; *Chevalieria thyrsigera* Mez, 1934

Pl with inflorescence to 50 cm high
Fo 15–20, nearly upright, a rosette from 50–60 cm diameter
LS long-oval, 10–12 cm long, 5–7 cm wide, green
LB 25–35 cm long, 3·5–4·5 cm wide, strap shaped-lineal, ending with a rounded spined tip or short point, green, the underside grey scaled; margins only towards the tip somewhat prickled
Sc upright, to 30 cm long, 4 mm thick, round, bare, green
I upright, loose and many flowered, cylindrical, 15–20 cm long, 3 cm diameter; rachis clearly seen
SB triangular, long pointed, loose, spirally arranged, soon drying, longer than the internodes, the upper ones 2 cm long, at the base 4 mm wide, entire
FB narrow-triangular with long prickle, reddish-brown, entire, reaching to about half-way along the ovaries

Fl sessile, to 1·5 cm long, until blooming nearly upright, when flowering and
after horizontal to angled
Se to 4 mm long, assymetric, fused
halfway, with brown prickled terminating spine, rose coloured

Pe pale blue, 9 mm long, the tip rounded
Ov 6 mm long, naked, rose, round,
smooth
Ha Brazil, Argentina.

Aechmea germinyana (Carr.) Baker, 1889 (Ills. 189 and 190)
(*germinyana*(*us*) named after the French Count Alfred de Germiny)
Syn. *Chevalieria germinyana* Carr, 1881
When in flower a beautiful and decorative plant, closely resembling *A.
veitchii* (Plate 79).
The difference between the two (Ills. 189 and 190) is that *A. germinyana*
has flower bracts at the top not reflexed, a little shorter than the developed
flowers, and the margins densely and finely-toothed, whereas *A. veitchii*
has flower bracts at the top reflexed, nearly as long as the developed
flowers, the margins large toothed.

Pl stemless, with inflorescence to 1 m
high
Fo 20–30 in wide funnel shaped rosette
LS distinct, 15–18 cm long, 7–9 cm
wide, elliptical, both sides grey short-
scaled
LB straplike to 80 cm long, lanceolate,
pointed with brown terminating spine,
5–6 cm wide, upper side green, underside green to lilac, both sides grey
appressed scaled; margins red prickled
Sc strong, upright, 30–50 cm long,
1 cm in diameter, round, pale green
naked
I upright, thick, many flowered, egg-
shaped to cylindrical, 8–10 cm long,
6–7 cm wide, with sterile flower bracts
to the tip
SB upright, tightly overlapping and
covering the stem and spirally arranged, longer than the internodes,
lanceolate-elliptical with prickled tip;
margins finely toothed, green to
brown-red, appressed scaled
FB oval-elliptical, wide pointed to
3·5 cm long, both sides grey-scaled,
red, bract-tip horizontal, not reflexed,
a little shorter than the developed
flowers, margins toothed
Fl sessile, to 3 cm long
Se free, assymetrical, 1·4 cm long, with
red spined tip, white, with the margins
thinly cutaneous
Pe to 2·5 cm long, white, pointed tip
Ov 6 mm long, flattened, laterally
winged
Ha Colombia, Panama.

Aechmea gracilis Lindm., 1891 (Ill. 191)
(*gracilis* delicate, distinctive small and delicate plant)

Pl stemless, with inflorescence
30 cm–50 cm high
Fo about 10, forming a nearly upright
rosette from 30 cm in diameter
LS narrow-oval, 8–10 cm long, 4–5 cm
wide, upper side to some extent dark

lilac, grey scaled
LB to 30 cm long, 3–4 cm wide, narrow-lineal, the tip rounded and prickled, dark green, grey scaled, the margins remotely spined
Sc about 20 cm long, shorter than the leaves, upright, 5 mm thick, round, green, at first white woolly, later bare
I simple, loose to 10 cm long, few flowered, rarely compound spikes with fewer than 2 to 3 flowers
SB remote, the basal ones shorter than the internodes, 2–3 cm long, upright, thin, soon drying
FB lanceolate to triangular, small, thin, wide, red, the basal ones to 1·5 cm long 5 mm wide, soon drying, the margins not toothed
Fl spirally arranged, sessile, to 2·5 cm long
Se about 6 mm long, 3 mm fused, with brown spine tip, light red, bare
Pe 1·4 cm long, the tip rounded, pale blue; margins and bases darker blue
Ov bare, light red, 1·2 cm long
Ha Brazil.

Aechmea kertesziae Reitz, 1952 (Plate 70)
(*keresziae* named after the collector Kertesz)

Pl stemless, with inflorescence 70–80 cm high
Fo upright to arching, forming a rosette from 40–50 cm in diameter
LS longish-oval, 17–18 cm long, to 7 cm wide, green, upper side blotched lilac
LB to 60 cm long, 3–5 cm wide, lineal, the tip rounded and spine tipped, dull green, upper and under sides densely grey scaled (towards the tip less heavily scaled), margins curled with remote 0·5 mm long spines
Sc upright, round, 8 mm thick, to 45 cm long, white woolly
I simple, loose, cylindrical, many flowered, 20–30 cm long, 4 cm wide; stem white woolly
Sc upright, round, 8 mm thick, to 45 cm long, white woolly
SB loosely lying close to the scape, upright, thin, 5–6 cm long, 8–9 mm wide, soon drying
FB spirally arranged, the basal ones longer, the upper ones shorter than the flowers, entire, thin, red, soon drying
Fl spreading to outstretched 2·7 cm long, opening upward and downward from the middle of the flower stem, the lower more widely spaced than the middle and upper ones
Se 1 cm long, short fused, red at flowering time, white woolly, soon becoming bare
Pe 1·9 cm long; yellow, hooded with a rounded top
Ov 7 mm long, orange coloured at flowering time, white woolly
Ha Brazil.

Aechmea luddemanniana (K. Koch) Brongn., 1934 (Ill. 192)
(*luddemanniana* named after the collector Luddemann)
Syn. *Pironneava luddemanniana* K. Koch 1866; *Lamprococcus caerulescens* Regel, 1871; *Aechmea caerulescens* Baker, 1879)
The form known as 'Marginata' 'Mend' is especially decorative because

of its wide white leaf edge, and when grown in half shade assumes a rose tint. (A good colour plate can be found in the *Journal of the Bromeliad Society*, Vol. XXI, No. 6, 1971).

Pl with inflorescence 25–70 cm high

Fo several, a nearly upright big rosette, to 70 cm in diameter

LS long-elliptical, to 13 cm long, to 7 cm wide, green, upper side brown scaled

LB tongue-shaped, pointed, 20–50 cm long, 4–7 cm wide, green to reddish brown, grey scaled, margin with 1–2 mm long spines

Sc upright, round, more or less 30 cm long, the leaves not exceeded, 7–10 mm thick, green, grey scaled

I loosely compound at the base, from several few flowered racemes to simple at the top, cylindrical to narrow-pyramidical, 10–30 cm long, white powdered

SB generally overlapping, the basal ones longer, the upper ones shorter than the internodes, elliptical to lineal-lanceolate, not prickled, thin, not permanent

PB much shorter than the racemes, narrow, almost thread-like, not permanent

Racemes short, outward spreading to 10 flowers, spirally arranged, about 2–4 cm long

FB thread-like, insignificant, shorter than the pedicle, not permanent

Fl 6 mm long, stalked, to 1·5 cm long (without stalk)

Se assymetrical, spine tipped, 3·5 mm long, green with brown tip, lightly scaled

Pe 9 mm long, rounded tip, upright, hardly unfolded, blue, changing to red after flowering

Ov 6 mm long, blue berry, scattered with white scales

Ha Guatemala, Honduras.

Aechmea mariae-reginae Wendl., 1863 (Ill. 193 and Plate 4)
(*mariae-reginae* named after the German Queen Maria)
One of the best Aechmeas and of truly queenly appearance, when in flower but seldom seen in cultivation.

Pl stemless, very large

Fo to 15–20, forming a wide funnelled rosette, to 1 m high and 1·6 m wide

LB wide-lineal with rounded and short spined tip, to 80 cm long, 8 cm wide, both sides green, the margins densely toothed

SB concentrated towards the flower head, the upper ones longish-lanceolate, with short terminating point, the margins toothed, angled back and downwards, bright carmine-red

Sc 60–70 cm long, sturdy, white woolly

I simple spiked, cylindrical, to 20 cm long and 6 cm thick

FB very small, prickled, hardly 4 mm long, margin smooth, hidden

Fl sessile, dense massed, to 2 cm long

Se 1·1 cm long, with a thorn, densely powder scaled

Pe white, with a violet blotched rounded tip

Ov to 5 mm long, densely powder scaled

Ha Costa Rica.

Aechmea mexicana Baker, 1879 (Ill. 194)
(*mexicana* named after Mexico where the plant is found)
Syn. *A. bernoulliana* Wittm., 1891

Pl stemless, with inflorescence over 1 m high

Fo numerous, forming a wide funnel shaped rosette over 1 m in diameter

LS oval, 20–25 cm long, 13–17 cm wide, brown, thickly scaled

LB strap-shaped, pointed or rounded with terminating spine, 80–90 cm long, 6–12 cm wide, narrowing above the sheath, green, irregularly blotched dark green, especially on the underside appressed grey scaled. Margins with 2 mm long spines

Sc upright, robust, fuzzy grey scaled, to 2·5 cm thick

I 30–70 cm long, loose paniculate, in all nearly cylindrical to pyramidical, fuzzy scaled

SB to 30 cm long, much longer than the internodes, lineal-lanceolate, long pointed, entire, rose, pale scaled, the upper bracts reflexed

PB lineal-lanceolate, as long as or shorter than the branches, rose coloured. Inflorescence element: spreading, loose 5 to 10 flowers with fuzzy scaled rachis

FB filamentous, much shorter than the flower-stalk

Fl spreading to widely extended, with a fuzzy scaled stalk to 3 cm long

Se free, assymetrical, triangular-oval, spine tipped, 6 mm long, fuzzy scaled

Pe 1–1·5 cm long, upper part 5 mm wide, red or lilac

Ov 6 mm long, round or ellipsoid, greenish, fuzzy scaled

Fr white

Ha Mexico, Guatemala, Costa Rica, Panama, Ecuador.

Aechmea miniata (Beer) hort., 1889 (Plate 73)
(*miniata(us)* small; the plant blooms when only 40 cm high)
Syn. *Lamprococcus miniatus* Beer, 1857

Pl with inflorescence 30–40 cm high

Fo forming a loose funnel shaped rosette to about 40 cm in diameter

LS elliptical, 11–12 cm long, 7–8 cm wide, grey-brown spotted with scales

LB to 45 cm long, at the base 3·5–4 cm wide, towards the middle becoming to 5 cm wide, cusp tipped, both sides green, the margins small prickled

Sc upright, round, bare, dark red, not longer than the leaves

I compound, to 10 cm long and 7 cm wide, with several spreading and widely outstretched spikes

SB tightly overlapping, upright, thin, longer than the internodes, soon drying

PB the basal ones similar to the inflorescence bracts, shorter than the spikes, soon drying, the upper ones greatly reduced, scale shaped

Sp 2 to 10 flowered

FB very small, scale shaped (seen under a magnifing glass), red

Fl sessile to 1·4 cm long, open bipartite (distichous) arranged along the genuflect red rachis

Se free, bare, to 4 mm long, coral-red

Pe to 1 mm long, rounded tip, blue turning to red after flowering

Ov 4 mm long, red, bare

Fr lively red

Ha Brazil.

VARIETY
var. *discolor* (Beer) Beer, 1889 (Plate 73)
Syn. *Lamprococcus miniatus* var. *discolor* Beer, 1857.
Distinguished from the type by the undersides of the leaves being red-violet. Only known in cultivation. Very decorative, small plant, with its red fruit lasting a month or more and similar to *A. fulgeus*.

Aechmea nudicaulis (L.) Griseb., 1864 (Plate 74)

(*nudicaulis* naked-stem; named after the scale-shaped, sometimes absent, flower-bracts)
Syn. *Bromelia nudicaulis* L., 1753; *Billbergia nudicaulis* Lindl., 1827; *Hoplophytum nudicaule* K. Koch, 1857; *Hohenbergia nudicaulis* Baker, 1871; *Pothuava nudicaulis* Regel, 1882

Pl very variable in size, 30–70 cm high
Fo forming a compact, few-leaved, tubular shaped rosette
LS large, elliptical, more or less 15 cm long, upper side brown or lilac, dense brown scaled
Lb wide-tongue-shaped, ending in a cusp, 20–90 cm long, 6–10 cm wide, green, grey cross-banded on the underside; margins with 4 mm long black spines
Sc upright, not longer than the leaves, round, green or red, white flocked-scaled, 8–10 mm thick
I simple spike, loose, cylindrical, 5–25 cm long; rachis whitish scaled, later becoming bare
SB shiny red, upright, longer than the internodes, elliptical, pointed, the margins weak toothed, those on the lower half of the inflorescence dense
FB small, triangular, pointed, shorter than the sepals, red, sometimes scale shaped (under a magnifier) or entirely missing
Fl sessile, 2·2 cm long, sirally arranged
Se assymetric, 5–10 mm long, free, spine tipped, yellowish green, scaled (under a magnifier)
Pe tongue-shaped, pointed, only a little longer than the sepals, yellow, 1·2 cm long
Ov more or less 5 cm long, green to yellowish-green, grey scaled
Ha Mexico, Central America, West Indies, Venezuela, Brazil.
CU *A. nudicaulis* and its varieties are free flowering, easy to grow beginners' plants.

VARIETIES
var. *nudicaulis* (Plate 74)
Flower bracts small, scale -shaped or entirely missing.
Ha Mexico, Central America, West Indies, Venezuela, Brazil.
var. *cuspidata* Baker, 1879
Syn. *Tillandsia unispicata* Vell., 1835; *Pothuava spicata* Gaud., 1851; *Aechmea nudicaulis* var. *sulcata* Mez 1896.
Triangular shaped flower bracts, conspicuous, sepals yellow; petals yellow.

Ha Brazil.
var. *aureo-rosea* (Antoine) L. B. Smith, 1955
Syn. *Hoplophytum aureo-roseum* Antoine, 1881; *Aechmea aureo-rosea*
Baker, 1889.
Triangular shaped flower bracts, conspicuous, sepals yellow mottled red;
petals red.
Ha Brazil.

Aechmea orlandiana L. B. Smith, 1941 (Plate 75)
(*orlandiana* named after the town of Orlando, Florida, U.S.A.)

Pl stemless, with inflorescence to 50 cm high

Fo 15–20 forming a funnel shaped rosette from 40–50 cm in diameter; the outer leaves shorter than the inner ones

LS longish-elliptical, 14–15 cm long, 7–9 cm wide, upper side dark lilac, grey scaled; margins smooth

LB lineal, 30–35 cm long, 4 cm wide, light green with dark violet, cross banding and mottling, grey scaled, leaf tup lanceolate terminating with spine; margins with dark violet narrow edge and 2 to 2·5 long spines

Sc upright, to 6 mm thick and 30 cm long, round, green or reddish

I compound, dense with 5 to 8 more or less, upright spikes, short pyramidal, 7–10 cm long, 5–6 cm in diameter

SB the basal ones shorter, the upper ones longer than the internodes, upright, pressed close to the scape and nearly enfolding it, red, grey scaled

PB somewhat longer than the spikes, to 4 cm long, wide-oval, long pointed, red, grey scaled, the margins spiny to the tip

Sp spirally arranged, short stalked, 4 to 6 flowered, 3–4 cm long, 2 cm wide

FB dense spirally arranged, upright, longer than the sepals, about 2·3 cm long, long pointed, bare, red

Se free, 1 cm long, whitish

Pe yellowish white, 2 cm long, 4 mm wide

Ov 1 cm long, bare, greenish-white

Ha Brazil

CU a very decorative and worthwhile plant, even when not in flower because of its beautiful leaves.

VARIETY

var. *variegata* 'Ensign' leaves with green-white longitudinal stripes. Additional plants are produced only vegetatively by removing offsets or young growths.

Aechmea ornata (Gaud.) Baker, 1879 (Plate 76)
(*ornata(us)* adorned; named for the bright red inflorescence leaves)
Syn. *Chevalieria ornata* Gaud., 1843; *Aechmea hystrix* E. Morren, 1880;
Echinostachys hystrix Wittm., 1891

Pl with inflorescence to 80 cm high

Fo stiff forming an upright rosette

from 80–90 cm in diameter
LS wide-oval, 15–18 cm long, 8–10 cm wide, brown scaled
LB 70–100 cm long, narrowing above the sheath to 7–8 cm wide to more or less 4 cm wide in the middle, strap shaped, pointed with brown terminating spine, green, both sides grey scaled; margins densely edged with small brown spines
Sc 40–50 cm long, 1·5 cm thick, upright round, pale green, grey scaled
I single-spiked, dense, cylindrical, 8–15 cm long and to 4 cm thick
SB like the leaves, stiff with prickle tip, upright, the lower ones green and prickled, the upper ones bright red, with only the tip spined, always longer than the internodes and enclosing the scape
FB asymmetrical, boat shaped, keeled at the base enclosing the base of the inflorescence, 9 mm long, greenish-brown, grey scale spotted; terminal spine thin, brown, 1–6 cm long
Fl sessile, about 2·4 cm long, densely spirally arranged around the rachis
Se 1·2 cm long over terminal spine, 5 mm wide, nearly triangular, green, slightly grey scaled, terminating spine brown, 2 mm long
Pe 1·5 cm long, 6 mm wide, lilac to violet
Ov 7 mm long, white, grey scaled
Ha Brazil.

VARIETIES
var. *hoehneana* L. B. Smith, 1955
Inflorescence about 3 cm in diameter, longer than in the type, flowers slender, blue.
Ha Brazil (São Paulo, Parana)
var. *nationalis* Reitz, 1952
Leaves green and yellow striped.
Ha Brazil (Santa Catarina)

Aechmea penduliflora André, 1888 (Ill. 195)
(*penduliflora(us)* pendulous-flowered; because of the pendant inflorescence)
Syn. *Billbergia paniculata* Mart., 1830 *Aechmea paniculata* R. et P., 1820; *A. schultesiana* Mez, 1892; *A. friedrichsthalii* Mez et Donn.-Smith, 1894; *A. inermis* Mez, 1904

Pl with inflorescence about 60 cm high
Fo forming a loose rosette from 60–80 cm in diameter
LS distinct, asymmetrical, 10–12 cm long, 6–7 cm wide, oval, brown scaled
LB 40–60 cm long, narrow, widening towards the tip, 2–4 cm wide, grey scaled, channelled at the base, then flat, densely spined (1 mm long, the upper part frequently smooth; terminating with a 6 mm long spine
Sc upright or overhanging, not taller than the leaves, 3–5 mm thick, red, commencing white woolly, soon becoming bare, 30–40 cm long
I 7–15 cm long, 6–10 cm wide, with numerous, nearly upright to spreading compound spikes; axis red, com-

mencing white woolly, soon becoming naked

SB upright, pressed close to the stem or enclosing it, cutaneous, red, widely spaced at the base of the stem and shorter at the internodes, higher up closer and longer than the internodes

PB the basal ones nearly as long as the spikes, cutaneous, red, to 1 cm long, narrow, green

Sp spirally arranged, 5–6 cm long, 1·5 cm wide, 5 to 10 flowers

FB 2·5 mm long, spreading, roundish, thin, green, at flowering time nearly reaching the length of the ovary

Fl loosely distichously arranged, sessile, upright, 1·8 cm long

Se not symmetrical, inverted (ovate shaped), free, green, about 5 mm long

Pe 1·3 cm long, with fine, short spine tip, orange coloured

Ov yellow to red, becoming somewhat white woolly, later becoming bare, about 5 mm long

Ha Colombia, Costa Rica, Venezuela, Ecuador, Peru, Brazil.

Aechmea pineliana (Brongn.) Baker, 1879 (Plate 78, Ill. 196)
(*pineliana* named after the grower Charles Pinel)
Syn. *Echinostachys pineliana* Brongn., 1854–58; *Machrochordium pinelianum* Lem., 1862; *Echinostachys rosea* Beer, 1857

Pl with inflorescence to 60 cm high

Fo several forming a funnel shaped rosette

LS wide-oval, 8–11 cm long, 5–8 cm wide, upper side lilac, covered with grey scales

LB 30–70 cm long, to 3·5 cm wide, lineal with a rounded terminal spine, dark green, both sides dense appressed silver-grey scaled, often cross-banded on the underside; margins with narrow reddish-brown edge and 4–5 mm spines

Sc upright, 30–35 cm long

I simple-spiked, dense, cylindrical, many flowered to 7 cm long and 3 cm thick

SB upright, longer than the internodes, bare to few scaled, lacquer red, entire at the base, somewhat prickled at the tip

FB roundish-kidney-shaped (reniform) enclosing the base of the flower head, dark-red, densely white woolly scaled, with a long setaceous (bristly), light-brown terminating spine to 1·1 cm long

Fl sessile, densely spirally arranged, nearly horizontal, to 1·5 cm long

Sc free, with terminating spine to 8 mm long, bare and towards the tip somewhat white woolly, assymmetrical

Pe 8·5 mm long, yellow, after blooming turning black-brown

Ov 3 mm long, white woolly

Ha Brazil to Peru.

VARIETY

var. *minuta* Found by M. B. Foster in the Brazilian state of Esperito Santo; it is smaller in all its parts than the type; the leaves turn rose-coloured and assume a beautiful silver-sheen when subject to intense sunlight.

Aechmea pubescens Baker, 1879 (Ill. 197, Fig. 51)

(*pubescens* downy-woollen; referring to the polos (hairy) appearance of the inflorescence)

Pl 40–120 cm high, in its habitat very variable

Fo 10–15, forming an upright rosette, the outer leaves spreading, the inner ones forming a rube

LS inde-elliptical, to 15 cm long, grey to brown scaled, upper side lilac, margins entire

LB to 1 m long, 3–4 cm wide at the base, widening in the middle to 5–6 cm, terminating in a long point, green, sometimes lilac edged, margins at the base with 2·5 mm long spines, at the middle remotely spaced and smaller, and entirely free of spines at the tip

Sc slender, white-woolly, reddish-green, 4 mm in diameter, to 60 cm long

I upright, 10–35 cm long, to 10 cm wide, generally many flowered loose compound spikes, the basal ones once again branched, red axis, white woolly

SB Densely overlapping (imbricate), longer than the internodes and enclosing them; to 10 cm long, to 1·5 cm wide, green, reddish blotched or red, lanceolate-oval, margins entire

PB red, grey scaled, but quickly becoming bare, the basal ones longer than the spikes, the upper ones shorter, their margins entire

Sp lineal, densely bipartite, 8 to 18 flowered

FB covering the rachis, wide-oval, spine tipped, 1·2 cm long, the tip keeled; green with cutaneous, entire, sometimes white woolly margin

Se 6 mm long, nearly triangular, assymmetrical, keeled, spine tipped, green, cutaneous edged

Pe 1 cm long, narrow strap shaped at the base, wide-oval at the tip, blunt, pale violet

Ov nearly elliptical, triangular, steel-blue

Fr steel-blue

Ha Honduras, Nicaragua, Panama, Costa Rica, Colombia.

Aechmea racinae L. B. Smith, 1941 (Plate 82)

(*racinae* named for Racine Foster, wife of the well known bromeliad grower and collector M. B. Foster)

Pl very small, to 40–50 cm even when in flower

Fo 10–12 in an open rosette

LS distinctly set, long-oval, 10–14 cm long, 6–7 cm wide, whitish-green, each side covered with small brown scales

LB tongue-shaped, more or less 50 cm long, pointed tip, 2 cm wide at the base, 3·5 cm wide above, fresh green; margins entire or hardly prickled

Sc to 45 cm long, pendulous, 3 mm thick, round, delicately longitudinally fluted, reddish-brown, naked

I simple, loose racemes, few flowered, pendulous, about 12 cm long, 5 cm in diameter, with shining red verrucose rachis

SB always longer than the internodes, thin, lineal pointed, about 9–11 cm long, long-nerved, naked, red-brown

FB scale shaped, minutely small (under a magnifier)

Fl laxly scattered, horizontal, short stalked, about 3 cm long with stalk

Se fused at the base, the upper third free 6–7 mm long, 2 mm wide, shiny red, verrucose, with yellowish-green rounded tip
Pe 1·2 cm long, yellow, blackish at the base

Ov fusiform (spindle-shaped) 1 cm long, 5 mm diameter at the middle, shiny red, verrucose
Ha Brazil
CU is especially suited to the small greenhouse and window garden.

VARIETY
var. *erecta* L. B. Smith 1950
Differs from the type by its short, upright inflorescence.
Ha Brazil

Aechmea recurvata (Klotsch) L. B. Smith, 1932 (Plate 80)
(*recurvata(us)* recurved; refers to the recurved leaves) Syn. *Macrochordium recurvatum* Klotsch, 1856; *Hohenbergia legrelliana* Baker, 1871; *Aechmea legrelliana* Baker, 1879; *Portea legrelliana* Benth. et Hook, 1883; *Billbergia legrelliana* hort. ex Baker, 1879; *Ortgiesia legrelliana* Baker, 1889; *Ortgiesia palleolota* Morren, 1871; *Ortgiesia tillandsioides* var. *subexerta* Regel, 1875

Pl when in bloom 15–20 cm high
Fo very few, although forming a tube-like rosette
LS spoon-shaped, to 7 cm long, longish-oval
LB narrow-triangular, long-pointed, longer than the sheath, 1 cm wide at the base, recurved, gradually narrowing to a long sharp point, the margins somewhat outward turned and spined, the inner leaves turning rose at blooming, and when exposed to the sun the entire plant assuming a shiny rose colour
Sc clearly formed
I simple, densely strobilate, 5–6 cm long, 3 cm wide, many flowered, surpassing the leaves
FB red, membraneous, triangular-pointed, the margins clearly entire, about as long as the sepals
Fl sessile, to 3·5 cm long
Se to 1·7 cm long and 5 mm wide, oval, rounded and spined at the tip, fused for a quarter of the length at the base, rose-coloured, bare or few scaled
Pe 2·5 cm long, 4 mm wide, white at the base, rounded at the tip, rose to lilac-red
Ov white, bare, 1·4 cm long
Ha Brazil.

VARIETIES
var. *recurvata* (Plate 80)
Inflorescence longer than the leaf sheaths; the stem clearly visible; leaf sheath green, flower bracts serrated
var. *ortgiesii* (Baker) Reitz, 1962
Syn. *A. ortgiesii* Baker, 1879; *Ortgiesia tillandsioides* Regel, 1867; *Portea tillandsioides* Nichols., 1886

Leaf sheaths green; leaves and bracts robustly serrated
var. *benrathii* (Mez) Reitz (Plate 81)
Syn. *A. benrathii* Mez, 1919; *A. rupestris* F. Mueller, 1899
Leaf sheaths spotted violet; leaves and bracts only sparsely prickled or
without.
In var. *ortgiesii and benrathii*, the scape or inflorescence stem is very short
or missing; the leaf sheaths cover much of the inflorescence. (The
Bromeliad Society *Bulletin*, Vol. XV, No. 3, 1965, carried a colour plate of
var. *ortgiesii* on the cover).

Aechmea tillandsioides (Mart.) Baker, 1879 (Ill. 198)
(*tillandsioides* tillandsia-like)

Pl stemless, very variable

Fo 5–12, forming an upright funnel-shaped rosette

LS elliptical, 12–15 cm long, membraneous at the base, reddish-brown

LB lineal, long pointed, 50–90 cm long, 2–6 cm wide, green, both sides appressed, scaled; margins edged with 3 mm long brown spines

Sc upright, much shorter than the leaves, 30–40 cm long, 3–5 mm thick, round, greenish-brown, white woolly, becoming bare towards the flowers

I mostly loosely compound, seldom simple

SB overlapping (imbricate), somewhat lanceolate, long pointed, thin, toothed margins, red, grey scaled

PB the basal ones similar to the scape bracts, mostly exceeding the spikes, red, grey scaled

Sp 5–15 cm long, 1·5–2 cm wide, extended, flattened, 4 to 12 flowered; rachis visible, genuflected green

FB firm, distichous, wide-elliptical to inverted-ovate (egg-shaped) 1–1·7 cm long, widening to 1 cm, nerved, entire margined, becoming keeled at the tip (mostly separated from the flower), red, often somewhat grey scaled

Se to 7 mm long, 3 mm wide, free, elliptical, spine tipped, somewhat assymmetrical, yellow

Pe 1·3–1·6 cm long, pointed, yellow

Ov 7 mm long, greenish-yellow

Ha Mexico, Colombia, Central America.

VARIETIES
var. *tillandsioides*
Syn. *Billbergia tillandsioides* Mart., 1830; *Aechmea vriesioides* Baker 1879; *Aechmea xiphophylla* Baker, 1889
Inflorescence, 10–30 cm long, mostly interrupted at the base.
Ha Colombia, Brazil, Venezuela, Guyana.
var. *kienastii* E. Morr., 1896; *A. squarrosa* Baker, 1890 non Baker, 1889
Inflorescence with finger-like branches, seldom simple.

Aechmea triangularis L. B. Smith, 1955 (I!l. 199)
(*triangularis* triangular)

Pl with inflorescence to 40 cm high

Fo a loose, stiff, upright to spreading small rosette from about 60 cm in diameter

LS elliptical, 18 cm long, 9 cm wide, green, upper side reddish flecked

LB 30–40 cm long, upper half of the sheath to 6 cm wide, very marked narrow triangle, long pointed, turned out tip, margins with 5 mm long black-brown spines

Sc upright, 25–30 cm long, 8–10 mm thick, red, covered with white woolly scales

I simple, strobilate (cone-shaped), cylindrical, 6 cm long, 2 cm thick, densely white scaled

SB always longer than the internodes, the lower ones closely pressed to the stem, the upper ones spreading, wide-elliptical, thin, 6–7 cm long, red, scattered with grey scales towards the tip; margins of the upper inflorescence bracts weakly toothed

FB almost round, spine tipped, 8 mm long, longer than the ovary, robust, keeled on each side

Se 6 mm long, fused for 2 mm at the base, the upper part free, assymetric, blackish-red, white scaled

Pe 1·2 cm long, lavender to purple, black after flowering, blunt at the tip, two scaled at the base

Ha Brazil (Espirito Santo, Santa Teresa).

Aechmea veitchii Baker, 1877 (Plate 79, Ill. 190)
(*veitchii* named after the English botanist Veitch)
Syn. *Chevalieria veitchii* Morren, 1878
This beautiful plant is closely allied to *Aechmea germinyana*, to which the reader is referred for the principal differences

Pl stoloniferous, 1 m high with inflorescence

Fo 12–17, forming a loose, cyathiform (cup shaped) rosette

LS not distinctly delineated from the blade about 4 cm wide and 5 cm long

LB 25–90 cm long, 3–6 cm wide, tongue-shaped, spine tipped, upper side pale green, bare or hardly scaled, underside dense grey scaled; margins weakly spined

Sc upright, sturdy, green, naked, round, 35–50 cm long, 8 mm thick

I simple, densely cylindrical or conical, 10–14 cm long, 3–5 cm in diameter

SB always longer than the internodes, upright, lying against the scape, 6–8 cm long, long pointed, prickly, underside densely grey scaled

FB dense, spirally arranged, as long or longer than the flowers, to 2·5 cm long, recurved from their base, terminating in a 1·5 cm long pointed tip, red, nerved, grey scaled, margins toothed

Fl sessile, nearly upright, 2 cm long

Se asymmetrical, free, pointed, prickly, greenish-white with rose coloured tip, 1·3 cm long

Pe white, tongue-shaped, blunt

Ov about 7 mm long, 1 cm wide, triangular, greenish-white, naked

Ha Costa Rica, Panama, Colombia.

Aechmea warasii E. Pereira, 1972 (Ill. 200)

(*warasii* named after the botanist Waras)
A very beautiful plant, briefly described here for the first time.

Pl stemless, about 30 cm high

Fo a funnel shaped rosette from about 40–50 cm in diameter

LS distinctly delineated, wide-oval, 10–14 cm long, 8–9 cm wide, brown scaled, somewhat lilac flecked

LB tongue-shaped, somewhat narrowed at the base, widening towards the widely rounded but spined tip, about 20–30 cm long, the upper part to 7 cm wide, green, upper side sparsely grey scaled; margins with lilac brownish edges, with widely spaced and few spines

Sc pendulous, to 50 cm long, 4 mm thick, round, brown, bare, verrucose (warty)

I pendant, mostly simple, occasionally with the lower part having 3 to 5 compound spokes, few flowered; to 30 cm long, 3 cm thick; rachis red, verrucose

SB apart from the two basal ones, always longer than the internodes (stem segments), to 12 cm long, to 2 cm wide, lilac, becoming dry during flowering

FB 9 mm long, asymmetrical, nearly free, verrucose, the lower side red, changing to dark blue to black at the tip

Pe dark-violet-blue with lighter edge, about 1·5 cm long, 13 mm wide, rounded tip short pointed, turning dark red after flowering

Ov globose, strong red, verrucose, about 1 cm long, 1 cm thick

Ha Brazil.

Aechmea weberbaueri Harms, 1939 (Ill. 201)

(*weberbaueri* named after the German botanist A. Weberbauer who spent much time in Peru)

Pl stemless, 60–70 cm high with inflorescence

Fo the smallest plant forming a large rosette 60–80 cm in diameter

LS longish-oval, 15–20 cm long, 8–9 cm wide, dull green, appressed, grey scaled

LB 50–60 cm long, 6–7 cm wide, lineal, dull green, appressed, grey scaled; margined with 2 mm long spines

Sc upright, sturdy, round, 40 cm long, 8 mm thick, green, white felted

I simple, loose, 10–15 cm long, to 8 cm wide

SB longer than the internodes (stem segments), upright, green, grey scaled, 6–8 cm long, to 2 cm wide, pointed at the tip; margins toothed

FB spirally arranged round the felted axle, longer than the ovary, horizontal, the lower ones to 3·5 cm long, at the base 6 mm wide, long pointed, entire margins, drying before the flowers open, sparsely scaled to white felted, soon bare

Sc 2 cm long, rose coloured, white felted, with 1 cm long brown, spined tips

Pe free, gathered into a rube, the tip somewhat flared, 4 cm long, 6 mm wide, cornflower-blue

Ov 8 mm long, rose

Ha Central Peru.

Aechmea weilbachii Didr., 1854 (Plate 83)
(*weilbachii* named after the German collector Weilbach)
Syn. *A. subinermis* Baker, 1879; *Quesnelia glaziovii* Baker, 1889

Pl about 60–70 cm high with inflorescence

Fo several, forming a loose funnel shaped rosette from 50–70 cm in diameter

LS long-oval, 14–18 cm long, 7–8 cm wide, fresh green, brown scaled

LB 40–60 cm long, upper half of the blade about 2 cm wide, widening to 4 cm with rounded, spined tip, dark green, shiny, bare, margin at the base sparsely spined

Sc upright, 40–50 cm long, 5 mm thick, round bare, soft rose

I red, 10–15 cm long, about 5 cm wide with 5 to 10 spirally and loosely arranged spikes

SB upright, lying against the stem, longer than the internodes, 5–6 cm long, the sheath shiny red merging into a green tip, naked, only the tip becoming scattered with scales

PB spreading to flared, according to their position on the inflorescence longer than shorter than the spikes, triangular-lanceolate pointed, thin, naked, shiny red

Sp spreading or widening, loose, 3 to 5 flowered, 2–3 cm long with red rachis

FB shorter than the sepals, red

Se 8 mm long, 2–3 mm fused, with small spine tip, light lilac

Pe 1·5 cm long, 6 mm wide, rounded at the tip, lilac, changing to brown-black after flowering

Ov somewhat verrucose, lilac-red

Ha Brazil.

Ananas Adans, 1763

The plants in this genus are stemless or with short stems and leaves up to 80 cm long; they are lineal, very stiff, arched, with spined leaf margins. The leaves are nearly naked and forming a dense rosette; the inflorescence stem set with spiny scape bracts; inflorescence variable from globose to long strobilate (cone-shaped) from the crown of which a tuft of leaves form. The primary bracts are fused with the ovary, triangular in section, pointed, parchment coloured, yellowish and with toothed margins.

Flowers, lilac, white or purple.

The fruit is berried, fleshy and juicy (Ill. 204) in which the ovaries and the lower part of the primary bracts are fused (syncarpium).

The seeds are small, thick, ovate (egg-shaped) brown or blackish. The raising of the pineapple by seed is useless (kenocarp). Propagation is achieved by removing young plants growing at the base of the leaf rosette and also from the base of the inflorescence bracts and from the crown (Ill. 204 and 205).

The name *Ananas* is derived from *anana*, a word from the language of the Guarini Indians in Brazil.

Ill. 240 *Quesnelia quesneliana* (Brongn.) L. B. Smith.

Ill. 241 *Streptocalyx angustifolius* Mez.

Ill. 242 *Wittrockia superba* Lindm.

Ill. 243 *Navia heliophila* L. B. Smith.

Ill. 244 *Abromeitiella brevifolia* (Griseb.) Cast.

Ill. 245 *Abromeitiella lorentziana* (Mez) Cast. Resembling *A. brevifolia*; but larger; (*right*) upper corner, flowering shoot.

Ill. 246 *Deuterocohnia longipetala* (Baker) Mez. Part of a natural stand in the valley of the River Jequetepeque, 600 m, North Peru.

Ill. 247 *Deuterocohnia longipetala* (Baker) Mez flowering plant.

Ill. 248 *Deuterocohnia schreiteri* Cast. Rosette.

Ill. 249 *Deuterocohnia schreiteri* Cast. Part of the inflorescence.

Ill. 250 *Puya caerulea* Lindl. Part of the inflorescence.

Ill. 251 *Puya medica* L. B. Smith at 3100 m, near Cajamarca northern Peru. In the background (top right) is a Matucana cactus.

Ill. 252 *Dyckia cineria* Mez.

Ill. 253 *Dyckia choristaminea* Mez.

Ill. 254 *Dyckia fosteriana* L. B. Smith; rosettes.

Ill. 255 *Dyckia fosteriana* L. B. Smith; inflorescence.

Ill. 256 *Dyckia hebdingii* L. B. Smith; habit.

Ill. 257 *Dyckia hebdingii* L. B. Smith; inflorescence.

Ill. 258 *Dyckia hebdingii* L. B. Smith; spike.

Ill. 259 *Fosterella penduliflora* (C. H. Wright) L. B. Smith.

Ill. 260 *Fosterella rusbyi* (Mez) L. B. Smith.

Ill.261 *Hechtia glomerata* Zucc.

Ill. 262 *Hechtia marnier-lapostollei* L. B. Smith; flowering plant.

Ill. 263 *Hechtia marnier-lapostollie* L. B. Smith; inflorescence of male plant.

Ill. 264 *Pitcairnia andreana* Linden.

Ill. 265 *Pitcairnia atrorubens* (Beer) Baker.

Ill. 266 *Pitcairnia integrifolia* Ker.–Gaw.

Ill. 268 *Pitcairnia riparia* Mez.

Ill. 267 *Pitcairnia nigra* (Carr.) Andre.

Ill. 269 *Pitcairnia xanthocalyx* Mart.

Ill. 270 *Puya fastuosa* Mez, showing the thick branching stem of the plant.

Ill. 271 *Puya fastuosa* Mez.; flowering plant near Ocongate, 3600 m, South Peru.

Ill. 272 *Puya raimondii* Harms; flowering plants.

Ill. 273 *Puya raimondii* Harms. Part of the inflorescence from the Cordillera Blanca, Quebrada Pachacola, 4500 m, central Peru.

Ill. 274 *Puya rauhii* at the ice-line in the Cordillera Salcantay, 4800 m, Peru.

Ill. 275 *Puya densiflora* (?) in the deeply cut dry valley of the River Apurimac, southern Peru.

Ill. 276 *Puya laxa* L. B. Smith.

Ill. 277 *Puya mirabilis* (Mez) L. B. Smith, habit.

Ill. 278 *Puya mirabilis* (Mez) L. B. Smith; part of the inflorescence.

Ills. 279–280 *Tillandsia acostae* Mez and Tonduz; habit and spike.

Ill. 281 *Tillandsia albertiana* V. Vervoost.

Ill. 282 *Tillandsia capitata* Griseb.

Ills. 283–284 *Tillandsia califani* Rauh; habit and inflorescence.

Ill. 285 *Tillandsia cryptopoda* L. B. Smith; b r a n c h e d inflorescence.

Ill. 286 *T. cryptopoda* type form with simple inflorescence.

Ill. 287 *Tillandsia ehrenbergiana* Klotsch.

Ill. 288 *Tillandsia esseriana* Rauh and L. B. Smith.

Ill. 289 *Tillandsia friesii* Mez.

Ills. 290–291 *Tillandsia fraseri* Baker; habit (*left*) and part of inflorescence (*right*).

Ill. 292 *Tillandsia gymnobotrya* Baker.

Ill. 293 *Tillandsia hildae* Rauh; type plant growing wild.

Ill. 294 *T. hildae* in cultivation.

Ills. 295–296 *Tillandsia heterophylla* E. Morren; habit and inflorescence.

Ills. 297–298 *Tillandsia ignesiae* Mez; habit (*left*) and inflorescence (*right*).

Ill. 299 *Tillandsia karwinskyana* Schult. f.

Ill. 300 *Tillandsia polita* L. B. Smith.

Ill. 301 *Tillandsia propagulifera* Rauh; the type plant.

Ill. 304 *Tillandsia subconcolor*; typical growth near Quillabamba, southern Peru.

Ill. 302 *Tillandsia spiculosa* Griseb. var. *spiculosa*, habit and part of the inflorescence (Ill. 303 as inset).

Ill. 305 *T. subconcolor*; inflorescence.

Ill. 306 *Tillandsia viridif* Baker.

Ill. 308 *Tillandsia werdermannii* Harms; on the desert beach near Tacan.

Ill. 310 *Vriesea sagasteguii* L. B. Smith, in cultivation growing on Careus peruvianus; (*inset*) part of the inflorescence.

Ill. 307 *Tillandsia micans* L. B. Smith.

Ill. 309 *Vriesea hitchcockiana* L. B. Smith, in the valley of the Olmos, 1200 m. North Peru.

Ill. 311 *Vriesea barclayana* L. B. Smith.

Ill. 312 *Vriesea crenulipetala* L. B. Smith.

Ill. 313 *Vriesea fosteriana* L. B. Smith; habit.

Ill. 314 *V. fosteriana*; inflorescence.

Ill. 315 *V. fosteriana* L. Smith f. 'Red Chestnut.'

L. B. Smith includes 5 species in the genus, their habitat ranging from Brazil to Paraguay and Guyana.

Ananas comosus (L.) Merril, 1971 (Ill. 204 and 205)
(*comosus* crown-like; refers to the terminal leaf tuft)
Syn. *A, sativus* (Lindl.) Schult. f., 1891; *A. sativa* Lindl., 1827; *A. ananas* Voss, 1895; *Bromelia ananas* L., 1753
The pineapple is the only economically useful plant of the bromeliad family and it was the first bromeliad to be brought to Europe, in the year 1690. Since the fruit is likened in shape to that of a pine cone, it is known in South America as *Pina*. Syncarpium over 15 cm, inflorescence stem thin and short; floral bracts (at maturity of ovary) barely covered, margins with saw-like teeth

Pl stemless or short-stemmed, without stolons
Fo 30–50 in a dense rosette
LB lineal-lanceolate, to 1·5 m long, the margins spined terminating in a spined tip, grey-green to reddish (when in a sunny position)
Sc 30–50 cm long, thick, reddish

I Dense, globular shape, to 30 cm long
SB dense and upright
FB fused with the ovary, reddish-yellow
Fr to 30 cm long and to 15 cm thick, fleshy, sweet, eatable
Ha Brazil; now cultivated in most tropical countries.

VARIETIES
var. *porteanus* Koch
Leaves strongly spined, olive-green with yellow longitudinal stripe down the middle.
var. *variegatus* hort. (Plate 84)
Syn. *Ananas striatifolius* hort.
Leaves green and yellowish-white striped, known only in cultivation.
Recent cultivation has produced leaves striped in three colours, green, white-yellow and reddish.

In addition to *A. comosus* the following species are known.

Ananas ananassoides (Baker) L. B. Smith, 1939
Syn. *Acanthostachys ananassoides* Baker, 1889; *Ananas microstachys* Lindm., 1891; *A. savitus* var. *microstachys* Mez, 1892; *A. guarinituds* Bertoni, 1919; *A. comosus* var. *microstachys* L. B. Smith, 1934.
Syncarpium smaller than 15 cm—stem thin, long; leaf margins, spined, basal spines bent forward; inflorescence many flowered, up to 15 cm long.
Ananas bracteatus (Lindl.) Schult., 1830.
Syn. *Ananassa bracteata* Lindl., 1827; *Ananas sagenaria sensu* Mez, 1934

non Schult. 1830.
Syncarpium over 15 cm, inflorescence stem thick and short; floral bracts covering ovary and highly coloured. In the trade, a pygmy variety has been developed with a small inflorescence and vivid red bracts.
Ananas fritzmuelleri Camargo, 1943.
Syn. *A. silvestris* Fritz Mueller, 1896; *A. bracteatus* var. *ablus* L. B. Smith, 1939.
Syncarpium over 15 cm, inflorescence stem thick and short; floral bracts covering ovary and not coloured.
A. lucidus Mill., 1768.
Syn. *A. erectifolius* L. B. Smith, 1939.
Syncarpium under 15 cm; stem thin, long; leaf margins not spined.
Ananas nanus (L. B. Smith) L. B. Smith, 1962
Syn. *A. ananassiides* (Baker) L. B. Smith var. *nanus* L. B. Smith, 1939.
Syncarpium under 15 cm—stem thin, long; leaf margins spined, basal spines bent forward; inflorescence poorly flowered, 4 cm long.
Ananas parguazensis Camargo and Smith, 1968
Syncarpium under 15 cm; stem thin, long; leaf margins spined; basal spines bent backwards.

Andrea Mez, 1896

This is a monocarpic, endemic Brazilian genera with only a single species.

Andrea sellowiana (Baker) 1889 (Fig. 52)
Syn. *Quesnelia solloana* Baker, 1889.
Named after the French botanist E. André (1840–1911)
A small stoloniferous plant, grass-like, lineal-lanceolate leaves in a bushy rosette. Above the densely brown scaled sheath, the blade is long and narrow with the underside thickly scaled and only spined at the base.
The strong stem is similar to that of *Orthophytum* (Ill. 234) with normal leaves; the violet flowers are arranged in a compound panicled head, surrounded by the inflorescence bracts in the form of an involucre. This plant is rarely found in cultivation in Europe.

Androlepis Brongn., 1870

This genera contains two species from Guatemala. Each has a funnel-shaped rosette and is stoloniferous. The leaves are ovate, wide-lineal, terminating in a spine tip; the margins spine edged, upperside bare, underside scaled. The inflorescence is clearly formed, flowers sessile, yellow and arranged in a cylindrical panicled spike. The flower spikes contain 1 to 5 flowers and are very short, and therefore the inflorescence appears to be simple.

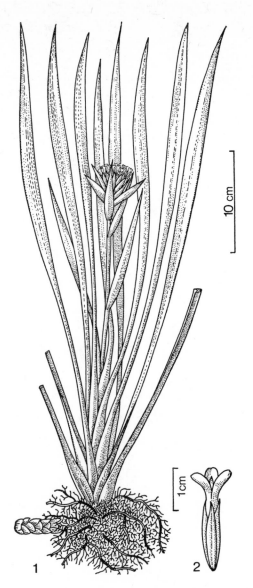

Fig. 52 *Andrea   sellowiana* (Baker) Mez.: 1 flowering plant; 2 single flower.

Androlepis skinneri (C. Koch) Brongn., 1870 (Fig. 53)
(*skinneri* named after the botanist Skinner)
Syn. *Aechmea skinneri* Baker, 1889; *Aechmea leucostachys* Baker, 1889;

Billbergia skinneri hort. Linden ex Morr., 1871

Pl 1–1·2 m high when in bloom

Fo numerous, forming a funnel-shaped rosette, to 80 cm high

LS only a little widened and not delineated from the blade

LB including the sheath, up to 80 cm long, 5 cm wide, wide-lineal, gradually becoming pointed, the margins densely edged with small saw teeth, underside grey scaled

Sc prominent, shorter than the leaves

I to 20 cm long, cylindrical, to 3 cm thick, appearing to be simple, thickly mealy scaled

SB the basal ones upright, the upper ones turned down, the edges toothed

Fl usually to 3 on a very small axle, scale shaped bract (Fig. 53, 2)

FB very small, often missing

Pe yellow

Ov inferior, scaled (Fig. 53, 2–3a)

Ha Guatemala, Costa Rica.

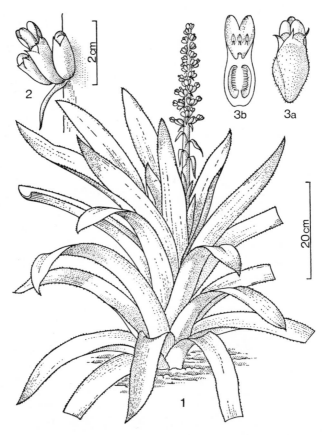

Fig. 53 *Androlepis skinneri* (C. Koch) Brongn.: 1 flowering plant; 2 flowering spike; 3a flower; 3b longitudinal section through flower.

Adrolepis donell-smithii (Bak.) Mex, 1896
Syn. *Aechmea donell-smithii* Baker, 1896
This is distinguished from *A. skinneri* by the tri-pinnate inflorescence and bare ovaries.
These species are seldom found in (European) cultivation.

Araeococcus Brongn., 1841

Plants with grass-like growth; their narrow lineal leaves form rosettes or bushes and have scale covered leaves and smooth or strongly-spined leaf margins.

The double-to-multiple racemose compound inflorescence carries insignificant green or violet stalked flowers.

The genus contains four species and their habitat extends from Brazil, through Venezuela to Costa Rica.

Araeococcus flagellifolius Harms, 1929 (Ill. 206)
(*flagellifolius* whip-like leaves)

Pl 70–100 cm high or long
Fo very few, in a tube-like rosette, the outer ones much shorter than the inner
LS the inner leaves 6–10 cm long with smooth margins
LB 60–90 cm long, narrow, whip-like, upper half of the sheath toothed, becoming smooth edged towards the tip, more or less scaled
Sc thin, much exceeded by the leaves
I 10–13 cm long, loose flowered, amply branched, the basal branches horizontal, 5–7 cm long, shortening at the tip; rachis very thin
SB few, grasping the scape, wider than the stem
FB much shorter than the branches, lanceolate, pointed tip, cutaneous
Fl 6–8 mm long, thin, horizontally stalked, with a small, bare, scale-shaped flower bract at the axil
Se oval-longish, 3 mm long
Pe oval-longish, blunt, whitish, 6 mm long and 2 mm wide
Ha Brazil, Amazonia.

The other species are:
Araeococcus micranthus Brogn., 1841
(*micranthus* small flowered)
Syn. *Aechmea micrantha* Brongn. ex Brongn., 1841; *Bromelia acanga* Roem. et Schult., 1821; *Bromelia lindleyana* Lem., 1852–53
Leaves lineal-lanceolate; inflorescence few branched, with strong flexuose 3 to 8 flowered branches
Ha Tobago, Trinidad, Guyana, Brazil.
Araeococcus pectinatus L. B. Smith, 1931
(*pectinatus* comb-like; refers to the floral bracts)
Similar to the previous species, stoloniferous; inflorescence 15–20 cm long, few branched, with red axil.
Ha Costa Rica

Araeococcus parviflorus (Mart.) Lindm., 1891 (Fig. 54)
(*parviflorus* small flowered)
Syn. *Aechmea parviflora* Baker, 1879; *Lamprococcus chlorocarpus* Wawra, 1866.
Leaves 6–10, forming a tube-like rosette; leaves 30 cm long and 1·8 cm wide; inflorescence ample-flowered; very small flowers (Fig. 6, 2)
Ha Brazil

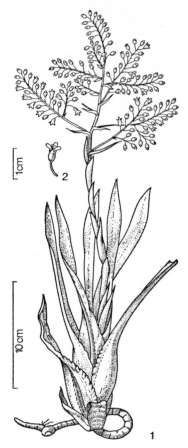

Fig. 54 *Araeococcus parviflorus* (Mart.) Lindm.: 1 flowering plant; 2 flower.

Billbergia Thunb., 1821

There are today about 60 known species in this genus, which is named after the Swedish botanist Gustav Johannes Billberg (1772–1844). Their range extends from South Mexico to Bolivia, North Argentina, and Brazil, where a great number of species have been collected; most are epiphytic and rarely terrestrial.

In contrast to many other bromeliad groups, the colourful inflorescence and flowers of the billbergias do not last very long. It is a disadvantage which decreases their popularity, so that relatively few varieties are available from nurseries; only *Billbergia nutans* seems to be readily available in an assortment of forms.*

Because of their partly compact and hard upright leaves, many billbergias are very suited as house-plants and adapt well to the dry air associated with the environment of a room. They are best kept in half-shade, in a moderately humid place and at a temperature of from 15–18°C.

The leaf margins are usually 'armed' with spines and the blades are more or less densely scaled with the scales frequently arranged as white crossbanding (e.g. *B. zebrina*, *B. vittata*). In general, the pendant inflorescence is in the form of racemes (simple or compound), spicate or panicled, frequently with a flexuose axis.

The scape carries large, brightly coloured (red or pink) scape bracts; but these soon loose their colour.

The flowers are radially symmetrical or slightly zygomorphic (equally bisected in any one plane) and, in many species, are rolled back to the base of the anthers in a spiral (Fig. 55). With few exceptions (e.g. *B. chlorantha*), no ligules are present at the base. Stamens and styles (pistils) extend very clear beyond the petals; the inferior ovary when fully developed becomes a many seeded berry (Ill. 64).

Billbergia chlorantha L. B. Smith, 1945 (Ill. 207)
(*chlorantha(us)* greenish flowering)

Pl to 35 cm when in flower
Fo several, forming a dense funnel shaped rosette from more or less 35 cm in diameter
LS elliptical, 10–12 cm long, 7 cm wide, underside pale green, grey scaled, upper side lilac
LB to 20 cm long, 3 cm wide above the sheath, tongue shaped with rounded short spined tip, pale green, grey scaled; margins black-brown spined; spines to 2 mm long
Sc to 20 cm long, pendant, 4 mm thick, round, green, naked

I loose compound, pendulous, about 17 cm long, with green, bare rachis
SB arranged densely overlapping about 6–7 cm long, upright, much longer than the internodes, pale rose to white, densely white scaled
PB the upper ones similar to the inflorescence bracts, wide-oval with 2 mm long tip, pale rose to white, densely white scaled, thin, more or less 4–6 cm long
Sp short, 1 to 2 flowered
FB small, scaly, green, 2–3 mm long
Fl sessile, 5·5 cm long

* Many billbergia hybrids have been produced by numerous collectors and horticulturists; specialist nurseries usually have a selection of these hybrids in their lists. Several are suitable as house or window plants.—*Ed.*

Se 1·6 cm long, assymmetrical, blunt, bare, pale green
Pe lineal, blunt at the tip, 4 cm long, with split scales at the base
Ov 1·5 cm long, green, bare, ribbed lengthwise
Ha Brazil.

Billbergia decora Poepp and Endl., 1838

(*decora(us)*) ornate, tasty, beautiful; refers to the large carmine red inflorescence bracts
Syn. *Helicodea baraquiniana* Lem., 1864; *Billbergia boliviensis* Baker, 1889
Von Metz placed all billbergias with the petals spirally rolled back to the anthesis in the sub-family Helicodea. In this group are *B. rosea* (Ill. 211) and *B. zebrina* (Plate 91 and Fig. 55)

Pl stemless, to 80 cm high
Fo 8–10 in a narrow tubelike rosette
LS narrow-elliptical, about 10 cm long
LB strap-shaped, pointed, 5–6 cm wide, 50–60 cm long, recurved for about half its length, dark green, grey banded and mottled; margins edged densely with 1–1·5 mm long spines
Sc slender, pendant, white powdered
I simple-spiked, about 10 cm long, 10 cm wide, pendant, many flowered; rachis thickly covered with white powder
SB large, elliptical, pointed, carmine-red, densely placed at the lower part of the scape
FB scale-shaped, wide-oval, 4 mm long
Fl densely spirally arranged, sessile, to 8 cm long
Se assymetrical, unequal in length, the longest to 1·2 cm, the others to about 7 mm long, pointed or rounded with prickle tip, white powdered
Pe green; spirally rolled back at flowering, lineal, pointed, to 6·8 cm long, with 2 naked scales at the base
Ov nearly globose, 8 mm long, white powdered
Ha Brazil, Peru, Bolivia
CU rarely found in cultivation.

Billbergia distachia (Vell.) Mez, 1892 (Ill. 208)

(*distachi(y)a(us)* in two rows (of ears))
Syn. *Tillandsia distachia* Vell., 1825; *Tillandsia distaceai* Vell., 1835; *Billbergia ensifolia* Baker, 1889; *B. burchellii* Baker, 1889; *B. Bakeri sensu* Lindm., 1891; *B. caespitosa* Lindm., 1891; *B. regeliana* Mez 1916

Pl 35–45 cm high with inflorescence
Fo few, forming a narrow, upright, tubelike rosette
LS oval-triangular, to 12 cm long and 6 cm wide, green, grey scaled, upper side somewhat lilac-violet coloured, margins not spined
LB narrow-lineal to lanceolate, to 50 cm long, 3–4 cm wide, green, grey scaled; margin with widely spaced prickles
Sc upright and curved, 30–35 cm long, bare, whitish-green, 5 mm in diameter
I to 10 cm long, simple, panicled,

pendant, open, few flowered, with genuflect rachis
SB densely overlapping, upright enclosing the stem, longer than the internodes, rose with greenish tip, grey scaled, becoming larger towards the top
FB scale-shaped (through a magnifier), bare
Fl short stalked, to 6·5 cm long with stalk
Se thin-cutaneous, green with blue tip, 2 cm long, bare
Pe 5 cm long, green with blue tip, flared for the upper third
Ov 1 cm long, green, bare, ribbed
Ha Brazil; both the type and the varieties
CU very variable in growth, leaf form and flower colour.

VARIETIES
var. *distachia* (as described above)
var. *straussiana* (Wittm) L. B. Smith, 1950 (Ill. 208)
Syn. *B. pallescens* sensu Baker, 1878; *B. bakeri* Morren, 1880; *B. bakeri* var. *straussiana* Wittm., 1885
Fl green, wider than the type (to 5·5 cm) not spined
Se green, blue, tipped
Pe green
var. *concolor* Reitz, 1952
Leaves, sepals and petals green.
var. *maculata* Reitz, 1952
Leaves yellow blotched.

Billbergia iridifolia (Nees and Mart.) Lindl., 1827 (Ill. 209)
(*iridifolia(us)* iris leaved)
Syn. *Bromelia iridifolia* Nees and Mart., 1823

Pl to 40 cm high
Fo few, forming a short, upright, tubelike rosette
LS long-oval, 10–15 cm long, 4–5 cm wide, green, grey or brown, appressed, scaled
LB lanceolate, the tip curled, about 30 cm long, 3–5 cm wide, upper side dark green, underside densely grey scaled; margins with widely spaced prickles
Sc arched-overhanging, 15–20 cm long, 5 mm thick, round, rose, bare, somewhat grey
I pendant, simple, loose, to 20 cm long, 13 cm wide, bare rachis, rose
SB loose, arranged spirally, longer than the internodes, lying against or spreading from the stem, more or less 5 cm long, 1·5–2 cm wide, sparsely scaled, rose, soon withering, thin
FB similar to the inflorescence bracts, as long as the flowers at the base, becoming shorter as they ascend, thin, rose, bare, somewhat grey
Fl arranged spirally, horizontal, 5–6 cm long, widely spaced
Se to 2·3 cm long, short spined tip, yellowish green, often delicate-rose and pale-blue tipped
Pe to 4·3 cm long, 6 mm wide, the lower portion yellowish-green, the

upper light-blue, flared or reflexed
Ov l cm long, longitudinally ribbed,
rose coloured
Ha Brazil.

VARIETY
var. *concolor* L. B. Smith, 1955
The petals are pale yellowish-green and not blue tipped.

Billbergia leptopoda L. B. Smith, 1945 (Plate 86)
(*leptopoda*(*us*) thin stalked; refers to the thin flower scape

Pl 30–40 cm high with inflorescence
Fo few, in a short, tube-like rosette
LS narrow, nearly triangular, 7–8 cm long, 3–4 cm wide at the base; upper side somewhat white to pale violet; margins not spined
LB 30–35 cm long, narrowing above the sheath, 1·5 cm wide, 2·5–3·5 cm wide halfway up the blade, lanceolate, pointed, green, irregularly yellowish blotched, underside grey scaled; margins with brown spines 2 mm long and widely spaced
Sc upright, slender, bare, to 25 cm long, 5–6 mm in diamter becoming rose-coloured towards the top
I upright, few flowered, simple-panicled, 10–15 cm long, with bare, genuflect, rose-coloured rachis
SB lanceolate, pointed, 4–5 cm long, upright, thin rose, soon withering
Fl 6 cm long, flared, on thin to 2 cm long red stalk
FB similar to the upper inflorescence bracts, lanceolate, pointed, thin, rose, soon withering, longer than the ovary
Se narrow-elliptical, short pointed, 18–21 mm long, rose coloured with blue tip, bare
Pe 3·4–4·5 cm long, 5 mm wide, lineal, greenish-yellow with blue reflexed tip
Ov ellipsoid, 13–17 mm long, furrowed, bare, red
Ha Brazil.

Billbergia macrocalyx Hook., 1859 (Plate 87)
(*macrocalyx* large sepals) (calyx)
Syn. *Billbergia quintusiana* Wittm., 1890

Pl forming a tube like rosette
LS long-oval, 16–20 cm long, 7–10 cm wide, upper side lilac, grey scaled
LB wide-strapshaped, pointed, upright, the upper part recurved, rolled back at the tip; 35–45 cm long, 4–6 cm wide, dark green, marked by yellow blotches, underside grey crossbanded; margins with 1·5 mm long spines
Sc upright to overhanging, to 25 cm long, round, 8–9 mm thick, white felted
I simple-panicled, to 15 cm long and 10 cm wide; axis densely white woolly
SB oval-lanceolate, 12 cm long, much longer than the internodes, 3·5 cm wide, thin, red, white woolly scaled
FB the basal ones similar to the upper inflorescence bracts, red, the upper ones narrow-lanceolate, scale shaped, insignificant
Fl arranged spirally, nearly upright,

often secundat flowering, stalked, to 7·5 cm long (including stalk)
Se 2·4 cm long, 7 mm wide, thin, pale blue, white powdered (farinose)
Pe to 5·3 cm long, yellowish-green, light blue bordered, the light-blue tip flared
Ov 1·3 cm long, greenish-red, farinose
Ha Brazil.

Billbergia morelii Brongn, 1848 (Ill. 210)

(*morelii* named after the plant collector Morel who first introduced the plant to Europe)

Pl forming a tube like rosette
LS longish-narrow-oval, 15–16 cm long, 5–7 cm wide, green, grey scaled, upperside lilac; margins not spined
LB lineal, pointed, to 40 cm long and to 5·5 cm wide, green, grey-scaled; margins spined; brown overhanging, 30–40 cm long, 5 mm thick, rose coloured
I pendant, single-spiked, few to many flowered, to 15 cm long and to 10 cm in diameter; rachis white woolly, rose
SB overlapping, upright, lying against and enclosing the scape, with shorter, greenish tip, rose to red, grey scaled, thin, longer than the internodes
FB the basal ones similar to the upper inflorescence bracts, longer than the flowers, red, grey scaled, thin, and only scale-shaped (under a magnifing glass) as the uppermost part of the inflorescence
Fl nearly sessile, flared, 5–5·5 cm long
Se 1·7 cm long, 6 mm wide, thin, red with blue rounded tip, elliptical-lineal, grey-woolly.
Pe to 4·5 cm long, 6 mm wide, rounded at the tip, greenish-yellow at the base, becoming dark blue, flared to rolled back
Ov to 7 mm long, furrowed
Ha Brazil.

Billbergia nutans H. Wendl., 1869 var. nutans (Plate 88)

(*nutans* pendant, nodding; refers to the pendant inflorescence)
Syn. *B. linearifolia* Baker, 1889; *B. bonplandiana* Gaud., 1892; *B. minuta* Mez, 1916

Pl stemless, 40 cm high with inflorescence, increasing by several short stolons and growing into a thick bushy leaf growth
Fo 12–15, in a narrow funnel-like rosette
LS not clearly developed, indistinguishable from the blade, imperceptible, nearly triangular, 4–5 cm long, 2–2·5 cm wide at the base; margins not spined
LB narrow-lineal, long pointed to 60 cm long, 1 cm wide, appressed scaled; margins at lower part spined but not the upper part for about 10–20 cm
Se arched, to 40 cm long, 3–4 mm thick, green, bare, round
I pendant, simple-inflorescence, few flowers with genuflect rachis
SB similar to the leaves, the basal ones green, longer and narrower, the upper ones red with shorter spread, thin, grey scaled, upright, longer than the in-

ternodes, covering the scape

FB small and insignificant, not longer than the flower stalks

Fl with short 2–4 mm long stalks, 4·8–5 cm long, overlapping

Se oval-lanceolate, pointed, 1·7 cm long, cutaneous, red, changing to green with blue mottling towards the tip, bare

Pe 3·4–4 cm long, green, blue edged, the tip plain green and flared to reflexed

Ov green, bare, 1 cm long

Ha Brazil, Uruguay, Paraguay, Argentina

CU hardy enough to be placed outside in summer.

VARIETY

var. *schimperiana* (Wittm. ex Baker) Mez, 1896
Differing from the type by the entire leaf margins (not spined) and blue edged petals to the tip.
Ha Brazil, Paraguay.

Billbergia pyramidalis (Sims) Lindl., 1827 var. *pyramidalis* (Plate 89)
(*pyramidalis* pyramid shaped; refers to the shape of the inflorescence)
Syn. *Bromelia pyramidalis* Sims, 1815; *Billbergia pyramidalis* var. *bicolor* Lindl., 1828; *B. thyrsoidea* Mart., 1830; *B. longifolia* C. Koch and Bouche, 1856; *B. longifolia* var. 'B' *longifolia* Baker, 1889

Pl to 50 cm high with inflorescence

Fo upright, about 12 in a wide funnel-shaped rosette

LS wide-oval, 11–12 cm long, 7–8 cm wide, upperside lilac, grey scaled

LB to 50 cm long, 3–4 cm wide, narrow-lineal, pointed, grey scaled on both sides, with underside somewhat grey crossbanded; margins with brown spines

Sc upright, round, 8 mm thick, white felted, to 25 cm long, not surpassing the leaves

I upright, simple, many flowered, short inverted pyramid shape, to 12 cm long and 6 cm wide

SB thin, upright, longer than the internodes, pale rose, grey scaled, arranged in open spiral on the lower part of the scape, and densely packed immediately below the inflorescence

FB small scales, reduced or missing

Se to 5 cm long, 6 mm wide, red at the base, merging into violet at the tip, white powdered

Pe upright and reflexed at the tip, bright carmine-red, violet edged, to 7·5 cm long

Ov 1·7 cm long, white powdered

Ha Brazil

CU very fine plants (both type and variety), if a little fleeting in bloom; they should be grouped for preference.

VARIETY

var. *concolor* L. B. Smith, 1954 (Ill. 203)
Syn. *Billbergia thyrsoidea sensu* Lindl., 1852 non Mart., 1830; *B. paxtonii* Beer, 1857
The leaves are wider, fresh green and not scaled; the inflorescence is

entirely red and the petals are not violet edged.
Ha Brazil.

Billbergia rosea Beer, 1857 (Ill. 211)
(*rosea* rose coloured; refers to the colour of the inflorescence bracts)
Syn. *B. granulosa* Brongn. ex Baker 1889; *B. decora* Baker, 1889 non
Poepp and Endl.; *B. porteana* Baker non Brongn., 1889.
B. rosea is distinguished from *B. venezuelana*, to which it is closely related,
by the longer sepals. There is also a difference between *B. rosea* and *B. zebrina*; in *B. rosea* the ovary is grooved whereas in *B. zebrina* (after L. B. Smith) when old it shows black, shiny bumps or protuberances (Pl. 91).

Pl stemless
Fo few in an upright, narrow funnel, to 1 m long, 5 cm wide, with a short flared tip, the margins with 1 mm long spines, frequently reddish overlaid and mottled to cross-banded
Sc pendant, white felted
I pendant, simple, many flowered with white woolly axis, to 20 cm long
SB longer than the internodes, massed on the lower part of the inflorescence
FB small, spreading, to 4 mm long
Fl sessile, to 8·5 cm long, half upright
Se varying lengths, blunt, persisting and verrucose (Rükken)
Pe yellow-green, to 6·5 cm long, spirally rolled back to the anthesis
Ov thick white woolly, grooved
Ha Trinidad.

Billbergia sanderiana Morren, 1884 (Ill. 212)
(*sanderiana* named after the English horticulturist H. F. Sander)
Syn. *Billbergia amoena* sensu L. B. Smith, 1943

Pl stemless, 50–60 cm high
Fo many forming a funnel-shaped rosette from 50–60 cm in diameter
LS wide-oval, 13–14 cm long, 9–10 cm wide, upper side somewhat dark lilac, grey scaled
LB 30–40 cm long, to 6 cm wide, grey scaled, widely rounded tip with spine; margins with nearly black to 1 cm long spines
Sc upright to overhanging, bare, not longer than the leaves
I pendant to overhanging; up to the middle of the inflorescence 2 to 3 compound flowered spikes, simple above to the top, 20–30 cm long, 10–15 cm wide
SB lying against the scape, longer than the internodes, upright, delicate rose
PB standing off from the axis, thin, delicate rose, grey scaled, 7–8 cm long at the base of the inflorescence becoming smaller towards the tip
Sp outspread, to 6 cm long, level, naked, pale green, stalked 2–3 cm long
FB as long as or somewhat longer than the ovary, pale rose with pale blue grey scaled tips
Fl sessile, upright, 6–7 cm long
Se to 2 cm long, green, rounded and scaled at the tip, otherwise bare, blue
Pe 4·5 cm long, green with blue outspread, sparsely scaled tip
Ov 1·5 cm long, bare, greenish-white
Ha Brazil.

Billbergia saundersii Bull. ex C. Koch, 1869
(*saundersii* named after the plant collector Saunders, who introduced the plant to Europe)
Syn. *B. chlorosticta* hort. Saund., 1871

Pl stemless, with many offsets at the base
Fo to 5–6 in a short tube-shaped rosette
LS not clearly defined
LB to 40 cm long and 3·5 cm wide, wide-lineal, with a short-flared tip, margins brown spined, upper side green, underside red-brown, yellowish-reddish blotched or lightly crossbanded
Sc slender, flexuose, overhanging, reddish, whitish scaled
I nodding to pendant, simple, loose-raceme, to 12 cm long, with straight or gently flexuose, white-powered axis, few to many flowered
SB the upper ones to 7 cm long, bright carmine-red
FB diminishing in size towards the top of the inflorescence
Fl to 5–10 cm long stalks; flower to 6 cm long
Se to 2 cm long, short pointed, underside powdered
Pe narrow-tongue-shaped, pointed, yellow up the middle, reflexed with blue violet tip
Ov thick woolly haired
Ha Brazil (Bahia)
CU has been used to make several beautiful hybrids.

Billbergia viridiflora H. Wendl., 1854 (Ill. 213)
(*viridiflora(us)* green flowered; refers to the green flowers)

Pl to 60 cm high
Fo about 10 forming a nearly upright tube-like rosette
LS longish-elliptical, narrow, about 10–13 cm long, 7 cm wide, green, grey scaled with entire margins
LB tongue-shaped, long pointed, 4–9 cm wide above the sheath, to 60 cm long, dark green, grey scaled; margins with 1–4 mm long, green spines
Sc upright or overhanging, about 5 cm thick, to 50 cm long, round, dark-red, white woolly
I simple, loose-raceme, pendant, 20–50 cm long
SB much longer than the internodes, upright, overlapping, to 14 cm long, 2–3 cm wide, dark red, grey scaled, margins toothed
FB narrow, to 2 cm long, 2 mm wide at the base, soon drying, white woolly
Fl arranged spirally, 5–6 cm long stalked
Se 2·1 cm long, long pointed, bare, green
Pe 4·5 cm long, green, pointed, hardly spreading at the tip, 6 mm wide, with 2 scales at the base
Ov green, naked, ellipsoid, 1·3 cm long, becoming yellowish-orange at fruiting
Ha Honduras, Guatemala
CU a very decorative green flowered species.

Billbergia vittata Brongn., 1848 (Plate 90)
(*vittata(us)* banded, striped; refers to the banded leaves)

Syn. *B. amabilis* Beer, 1857; *B. vittata* var. *amabilis* Morren, 1847; *B. leopoldi* C. Koch non Morren, 1856; *B. zonata* hort., 1850; *B. rohaniana* De Vries, 1853; *Bromelia rohaniana* Walp., 1861; *Billbergia moreliana* Lem. non Brongn., 1852; *Tillandsia moreliana* hort., 1851

Pl stemless, to 90 cm high

Fo 8–10 forming an upright, tube-like rosette

LS wide-elliptical, to 20 cm long to 10 cm wide, grey scaled, upper side lilac blotched

LB hard, compact, to over 1 m long, 6–7 cm wide, dark grey, upperside grey scaled, underside grey crossbanded, lanceolate with reflexed spine tips, margins with 5 mm long, brown-black spines

Sc commencing upright, then becoming overhanging or pendant, naked to 50 cm long

I pendant, to 25 cm long, to 10 cm wide, compound at the base becoming simple at the tip, many flowered

SB conspicuous, much longer than the internodes, narrow, to 1·5 cm wide, the middle ones to 20 cm long, thin, red, grey scaled, becoming rolled up, brown and dry

PB longer than the spikes, narrow, 1–2 cm wide, 10–12 cm long, thin, light red, grey scaled, weak, similar to the upper inflorescence bracts.

FB small, scale shaped, 5–6 mm long, light red

Fl sessile, to 6 cm long

Se 2·5–3 cm long, narrow-lineal with cetaceous (bristly) curved back tip, bare, thin, light red and blue tip

Pe 5 cm long, deeper coloured, the upper cm blue, then reddish for 1 cm, the lower ones 2 cm greenish-white, rolled back to the anthesis

Ov bare, ribbed, rose, 1·5 cm long

Ha Brazil

CU highly decorative, even when not in flower, because of its beautifully banded leaves, comparable to *B. zebrina* below.

Billbergia zebrina (Herb.) Landl., 1827 (Plate 91, Fig. 55)

(*zebrina*(*us*) zebra-striped; refers to the zebra-striped leaves)
Syn. *Bromelia zebrina* Herb., 1826; *Billbergia canterae* André, 1897

Fo compact, rigid, few, in a large tube-like rosette to 1·20 m high

LS not clearly defined

LB wide-lineal, to 1·2 m long and 8 cm wide, the tip rounded and short-reflexed, dark-green, grey-scaled, the underside most clearly grey cross-banded, margins with 3·5 mm long spines

Sc to 40 cm long, pendant, white woolly scaled, 8–9 mm thick

I simple-spike, pendant, to 40 cm long and 8 cm wide, densely flowered, with white woollen covered rachis

SB oval-lanceolate with curved edges, long striad, to 15 cm long, 3–4 cm wide, bright rose

FB scaled (under a magnifying glass)

Fl sessile, to 7·8 cm long, arranged spirally

Sc 8–10 mm long, 5–6 mm wide, widely rounded, thick white woolly

Pe to 6·3 mm long, 7 mm wide, pointed, spirally rolled back, green, after flowering greenish-yellow

Ov to 10 mm long, wide cone-shaped

(strobilate), 1·3 cm wide, white woolly with conspicuous black, shiny protuberances when old.

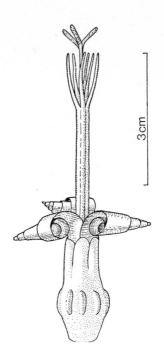

Fig. 55 *Billbergia zebrina* (Herb.) Landl. Flower at the time of anthesis.

Bromelia L., 1753

We have to thank the Swedish botanist Olaf Bromelius (1639–1705) for the family name Bromeliaceae. This came about through the efforts of an explorer and botanist Charles Plumier who named the genus *Bromelia* after Bromelius. The genus had previously been named *Karatas*, an Indian name. Linnaeus confirmed the genus in 1753.

All the known species (about 50) are conspicuous plants and with their long underground stolons form great impenetrable areas in their native habitat. The long rigid leaves, armed with compact, strong and hooked spines, form a loose, upright or spreading rosette. The inflorescence scape is thick and set with multiple, bright, rose-coloured spiny bracts.

The inflorescence is felted and is capitular to cylindrical, with white, reddish or violet flowers, large fruit and yellow berries.

In spite of their very beautiful and decorative inflorescence, bromelia species are of no commercial interest; they require a lot of sunlight and too much space if they are to become fully developed. Therefore they are found only in large plant exhibitions sometimes in association with cactus.

In their habitat (from Mexico and the West Indies to Paraguay and Argentina) they live in company with cactus and are xerophytic, growing in low rainfall areas of thorn forest, where they make up a substantial part of the undergrowth. Indeed, these plants form living fences.

Woe betide travellers who find themselves in these surroundings and who attempt to penetrate or try to cross the area; the hooked spines will cause deep and painful wounds.

Bromelia balansae Mex, 1891 (Plate 95, Fig. 56)
(*balansae* named after the collector Balansa)
Syn. *B. guyanensis* hort. ex Baker, 1889; *B. laciniosa* Baker, 1889

Pl forming a large group to 1·5 m high
Fo above the sheath hardly narrowed, long spine tipped, to 1·5 m long and 3 cm wide, the margin armed with strong to 5 mm long spines
Sc short, thick, white woolly
I compound, elongate, sub-pyramidal to cylindrical, 20 cm long and 9 cm thick, with 10 flowering branches
SB leaf-like, the upper one with bent-back, bright cinnabar red
PB similar to the inflorescence bracts; folded back, the margins armed with stout spines and wider set, redder to rose coloured, with white felted sheath, the upper ones with shorter blades or bladeless, shell-like; underside thickly white felted, shorter than the flowers
FB to 3 cm long and 8 mm wide, lineal-elliptical, nearly bare
Fl short stalked, to 4·5 cm long
Se free, upright, keeled-concave, rounded at the tip, to 2 cm long and 7 mm wide, hairy to bare
Pe to 2·5 cm long and 8 mm wide, upright, violet
Ha Paraguay, Argentina
CU cultivated and grown in the open in the mediterranean area.

Bromelia pinguin L., 1753 (Plate 93)
(*pinguin* named after Pinguin in the West Indies)
Syn. *B. penguin* L., 1771; *Karate penguin* Mill., 1768; *Agallostachys pinguin* Beer, 1857; *Karatas pinguin* Baker 1889; *Ananas pinguin* Trew., 1750–53; *B. ignea* Beer, 1857; *B. paraguayensis* hort. ex Baker, 1889; *Karatas plumieri* Devans ex Baker, 1889; *B. sepiaria* hort. Lovax ex Roem. and Schult., 1830

Pl stemless forming a more or less large grass-like clump
Fo numerous, over 2 m long the upper half bent and curved over, forming a rosette
LS wide, dense and coarse felted with scales
LB lineal, long pointed, 4 cm wide, upperside dark green, underside pale-green and with a few appressed scaled; margins with 1-cm long spines
Sc stout, white woolly
I to about 12 lowered compound panicles, narrow-pyramidal, white woolly
SB like the leaves but with rose-

coloured, somewhat swollen sheath and red, spiny blade

PB the basal ones similar to the inflorescence bracts, with larger sheath and narrower blade; upper ones blade-less, shorter than the flowers

FB lineal-subulate (awl-shaped) with shorter, wider base, 3 cm long

Fl to 6 cm long, clearly stalked

Se upright, narrow-triangular to sub-ulate, pale

Pe lineal-elliptic, 3 cm long, rose col-oured with whiter base and margins, dense white-felted at the tip

Ov narrow ellipsoid, 2 cm long, dense white felted

Ha Panama, West Indies, Guyana, Mexico, Central America.

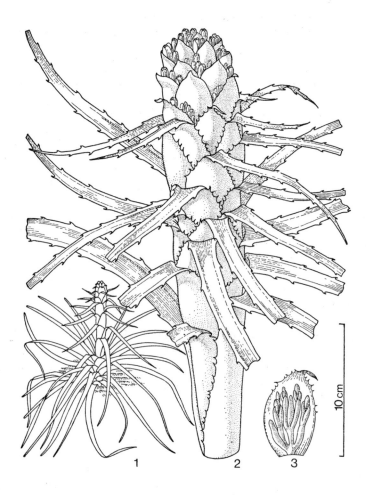

10 cm

1   2   3

Fig. 56 *Bromelia balansae* Mez.: 1 habit; 2 inflorescence (upper scape bracts and basal primary bracts are bright red); 3 part of inflorescence, showing shell-like primary bract.

Bromelia scarlatina E. Morren, 1881 (Fig. 57)

(*scarlatina(us)* scarlet red; refers to the scarlet-red inflorescence bracts)
Syn. *Disteganthus scarlatinus* Nicholson, 1884–1888; *Distiacanthus scarlatinus* Baker, 1894

Pl stemless
Fo 15–20 in a dense rosette
LB extending from the sheath gathered into a petiole (stalk) with toothed margins, 25 cm long and 1·2 cm wide, the lamina (blade) ovate, long pointed, to 50 cm long and 8 cm wide with weakly toothed margins
Sc very short, not exceeding the rosette

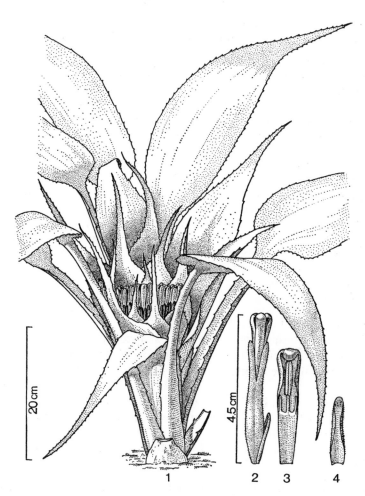

Fig. 57 *Bromelia scarlatina* E. Morren: 1 flowering plant; 2 flower with bract and the three petals; 3 flower-bract.

I gathered together into a nest
SB like the leaf blades but smaller, forming an involucre, scarlet-red
FB small, much shorter than the ovary (Fig. 57, 2) narrow-lineal, underside strongly scaled to bare (Fig. 57, 4)
Fl to 4·5 cm long, 5 mm long stalked
Se 1·5 cm long, 3 mm wide, free,

densely brown scaled
Pe violet, white edged, to 2·7 cm long, with rounded tip
Ov densely brown scaled
Ha Brazil
CU not usually found in cultivation in Europe.

Bromelia serra Griseb., 1879 (Fig. 58)
(*serra* saw; because of the saw-edged leaves)
Syn. *Rhodostachys argentina* Baker, 1889, *Bromelia argentina* Baker, 1892

Pl long, with low outspread leaves, stoloniferous, forming a large plantation (Fig. 58, 1)
Fo to 1·50 m long, pointed, 4 cm wide, scaled, upperside green, underside grey; margins lax with curved 5 mm long spines
Sc short, sturdy, densely scaled
I compound, dense-headed to globose, 6 cm in diameter (Fig. 58, 2)
SB similar to the leaves, densely overlapping, the upper ones bright red with longer, wider, spined blade
PB similar to the upper inflorescence bracts, with shiny-red blade and larger spoon-shaped sheath, the upper ones

nearly bladeless and about 4 cm long, the flowers nearly covered by the bracts
FB strap-shaped, blunt, gathered together into a hood at the tip, 3 cm long, keeled, densely grey scaled
Se longish, hooded tip, 1·5 cm long, keeled, white scaled
Pe eliptical, blunt, fused for 5 mm from the base, blue-white edged with white base
St and **Pi** locked within the flower
Fr ovate, 4 cm long
Ha Argentina, Bolivia, Paraguay, Brazil.

VARIETY
var. *variegata* M. B. Foster, 1955, 2961
Distinguished from the type by the white striped leaves.
Very tough, at blooming time this is one of the best decorative plants, however should not be planted out until about to bloom.

Bromelia urbaniana L. B. Smith, 1967 (Ill. 214 and 215)
(*urbaniana* after the type-habitat, the environs of the State of Cordoba, Argentina)
Syn. *Rhodostachys urbaniana* Mez, 1891; *Deinacanthon urbanianum* (Mez) Mez, 1896

Pl stemless, with long stolons, in flowering condition more or less to 20 cm high

Fo few, forming a loose low rosette
LS 2–3 cm long, about 2 cm wide, brown, wide-oval and with cutaneous

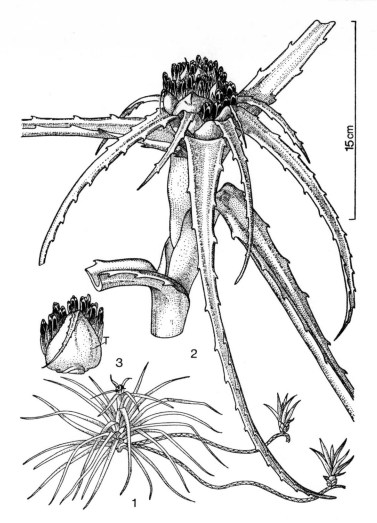

Fig. 58 *Bromelia serra* Griseb. 1 habit; 2 infloresce (scape bracts are bright red); 3 part of inflorescence, showing prinary bract from below.

margins.
LB to 30 cm long, above the sheath to 1·5 cm wide, narrow-lineal, pointed, involucrate; tough, stiff; margins with 3–4 mm long hooked spines, reddish-brown, both sides grey scaled

I simple, few flowered, nestling in the rosette, about 3 cm in diameter
PB to 3·5 cm long, 1·5 cm wide, lanceolate, long pointed, thin-cutaneous, pale brown; margins in the tip region weakly toothed, nerved, grey scaled

Fl upright, sessile to 4 cm long
Se to 1·7 cm long, 8 mm wide at the base, reversed-ovate to oval, spine-tipped, nerved, reddish-brown, dense grey scaled
Pe 1·9 cm long, 8 mm wide at the base, wide lineal, the upper part oval, white to pale rose, brownish nerved at the tip
Ov to 2 cm long, laterally flattened, white felted scaled
Ha Argentina, salt bearing places near Charga de Merced, in the environs of the State of Cordoba and other places. In fact this remarkable bromeliad is only found in salt bearing areas.

Canistrum E. Morren, 1873

This genus is based upon the plant *Canistrum aurantiacum* introduced by E. Morren, an explorer and plant collector. The name *Canistrum* is from the Greek *kanos*, 'a basket', relating the form of the inflorescence, a composite flattish flower head surrounded by bracts, likened to flowers in a basket.

There is a relationship with the genus *Nidularium* in that the flower-head is also sunk in the rosette or raised on a scape.

There are seven known species, six in Brazil and one in Trinidad where they are epiphytic or terrestrial. They are similar to *Nidularium* (page 338) and should be cultivated in the same way, in half shade and humidity—the flower basket should always be filled with water.

The leaves are mostly tongue-shaped, short or long pointed and green, with the underside appressed scaled. They are marginal spined and form a funnel shaped rosette with their large sheaths. The inflorescence scape is short or elongated, whereby the inflorescence bracts form a cup around the inflorescence. This is an exaggerated panicle and takes on the appearance of a composite-capitulum. The short stalked flowers are greenish or white (seldom any other colour); the fruit is a juicy berry, containing many spindle-shaped brown seeds.

Canistrum aurantiacum Morren, 1873 (Ill. 216 and 217)

(*aurantiacum(us)* orange-coloured; refers to the colour of the inflorescence bracts and the flowers
Syn. *Aechmea aurantiaca* Baker, 1879; *Cryptanthus calvatus* Brongn., 1879

Pl stemless, in a wide funnel shaped rosette
LS long-oval, 18–20 cm long, 10–11 cm wide, lilac blotched, the margins entire
LB lineal-strap-shaped, rounded or pointed at the tip with a 3–4 mm brown terminating spine, 50–100 cm long, 3–5 cm wide, green with darker blotches; margins with small 1 mm spines
Sc upright, elongated, 50–60 cm long,

7–8 mm thick, reddish-white
I dense-headed, 4–5 cm in diameter in short compound spikes
SB at the base of the scape longer than the internodes, covering the greater part of the scape, upright with out-spread tip, greenish-red, 7–8 cm long, 2·5–3·5 cm wide, densely placed around the lower part of the in-florescence and forming a kind of cup, wide-triangular with small spined tip, 4–5 cm long, at the base 5 cm wide, orange-red, bare
PB similar to the upper scape-bracts, orange-red
Sp dense, upright, 5 flowered, 3–4 cm long, 2 cm wide
FB 3 cm long, keeled cutaneous, greenish-yellow with red tip, shorter than the sepals, bare, at the tip sometimes simulating dried up mucous through the scaling
Se 1·7 cm long, free, bare, keeled, with 1 mm long spine tip, yellow
Pe 2·4 cm long, 5 mm wide, short spined tip, white at the base, orange-yellow above
Ov to 1·8 cm long, greenish-white, bare
Ha Brazil
CU worthy of cultivation; very dec-orative at flowering time.

Canistrum fosterianum L. B. Smith, 1952 (Ill. 218)
(*fosterianum* named after the bromeliad collector M. B. Foster)

Pl with inflorescence 40–50 cm high
Fo several, forming a small, dense, upright rosette from 20–25 cm in dia-meter
LS long-oval, 7–9 cm long, 5–6 cm wide, brown scaled; margins not spined
LB tongue-shaped to nearly tri-angular, pointed with short spined tip, underside densely grey scaled, green; margins with brown 2·5 mm long spines
Sc upright, 20–25 cm long, 7–8 mm thick, round, the upper part especially densely and coarsely brown scaled
I dense-headed, up to 8 to 10 compound spikes, about 4–5 cm in diameter
SB at the base of the scape shorter than those at the upper part of the scape longer than the internodes, upright, brown scaled, the uppermost bracts forming a kind of cup and surrounding the inflorescence, 6–7 cm long, 4–5 cm wide, orange coloured; margins not toothed
PB similar to the inflorescence bracts of the outer spikes, orange coloured, about 4 cm long, 2·5 cm wide
Sp short, 6 to 10 flowered, upright
FB nearly as long or a little shorter than the sepals, 3 cm long, 1·3 cm wide, cutaneous, white, red at the tip; margins not toothed
Se assymetric, widely rounded at the tip but with a terminating spine, 1·6 cm long, white, rose coloured at the tip
Pe white, 2·5 cm long
Ov white, angular 1 cm long
Ha Brazil.

Canistrum lindenii (Regel) Mez var. *roseum* (Morren) L. B. Smith (Plate 94)
Syn. *Canistrum roseum* Morren, 1879; *Aechmea rosea* Baker, 1889;

?Aechmea fusca Baker, 1889; *?Canistrum fuscum* Morren 1891; *?Canustrum binotii* Mez, 1919; *Canistrum lindenii* var. *roseum* forma *Elatum* Reitz, 1950

The plant shown in Plate 31 is the form *procerum* Reitz, 1952, and is distinguished by the longer inflorescence scape, whereby the inflorescence with its surrounding bracts is raised above the leaf rosette.

C. roseum Morren was placed by Mez as a separate species, but L. B. Smith considered it to be a variatal form of *C. lindenii* (Regal) Mez.

Fo several, forming a funnel shaped rosette from 80–100 cm in diameter
LS large, widened round to oval, about 20 cm long 15–17 cm wide, brown scaled
LB 40–60 cm long, 8–9 cm wide, wide-lineal, sometimes somewhat constructed above the sheath, rounded to pointed at the tip, green with darker blotches; margins with 3 mm long spines
Sc sturdy, upright, to 1·5 cm thick, dense brown felted, to 15 cm long
I dense-headed, 6–6 cm wide, 80–90 flowered
SB the basal ones a little longer than the internodes, surrounding but not entirely covering the scape, upright with outspread, rose-coloured tip, the uppermost forming a cup, 10–11 cm long, 5–6 cm wide, with spined edges, shiny rose-coloured, brown scaled
FB cutaneous, white, 2·5 cm long, 1 cm wide, somewhat shorter than the sepals, dense brown felted
Se 1·2 cm long, 8 mm wide, free, white, dense brown felted
Pe white at the base, with a green tip, 1·5 cm long, 5 mm wide, as long or somewhat shorter than the sepals
Ov white brown felted, 1 cm long
Ha Brazil.

Cryptanthus Otto and Diedr., 1836

About 20 species form the genus and are represented by small stemless, rosette shaped plants lying upon the ground, which on the basis of their vegetative appearance deserves more attention in spite of their insignificant small flowers.

The generic name *Cryptanthus* is derived from the Greek *cryptos*, 'hidden or covered', and *anthos*, 'a flower'.

The general characteristics of the genus are that the sheath, narrowed and stalk-like, widens into strap-shaped leaves, their margins waved, densely and finely toothed, underside scaled, in all forming a flat outstretched rosette. From the axil of the rosette leaves, young plants grow and eventually fall off the mother plant and then take root; this procedure could be followed in cultivation. The compound inflorescence is exceedingly low and sunk in the middle of the rosette. The sessile flowers are insignificant and are either white or greenish. Cryptanthus are native of the dry forests of East Brazil.

These small plants form ideal house plants for window gardens, terrariums, bottles, etc. and do not take up much room. Because of the lovely and differing leaf forms and colours, these plants are favourites with bromeliad collectors and nurseries. They require a rich soil, a relatively high temperature of 20–25°C and half-shade, but can stand full sunlight. Their leaves become coloured and many hybrids have been developed giving even more colourful leaves.

Cryptanthus acaulis (Lindl.) Beer, 1857 (Fig. 59)
(*acaulis* stemless)
Syn. *Tillandsia acaulis* Lindl., 1838; *Madvigia densiflora* Liebm., 1854

Pl stemless or with short but clearly developed stem, forming clumps, to 20 cm high
Fo narrow stalk-like above the short sheath
LB to 13 cm long and to 3 cm wide, narrow-lanceolate, the margins wavy and spined, recurved, upper side pale green, undersides dense white scaled
I few flowered
FB wide-oval, pointed, keeled, bare
Fl to 4 cm long, white scented
Ha Brazil
CU especially fine is the var. *genuinus* Mez, 1896, with the upper sides of the leaves also scaled.

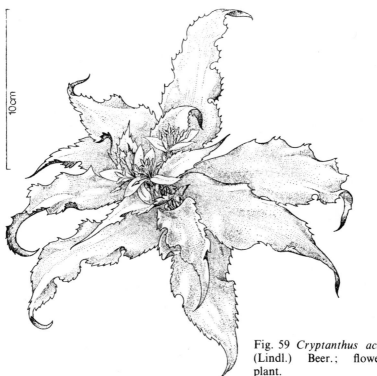

10 cm

Fig. 59 *Cryptanthus acaulis* (Lindl.) Beer.; flowering plant.

Cryptanthus beuckeri Morren, 1880 (Ill. 220)

(*beuckeri* named after Beuker, a plant collector who imported the plant into Belgium)

Syn. *Nidularium beukeri* hort., 1888

Pl stemless or forming a short stem, small, to 15 cm high

Fo to 10, a loose, small rosette

LS short-triangular, 1·5 cm long, 2 cm wide, enclosing the stem base, underside reddish-brown, grey scaled

LB above the sheath narrowing to a long stalk 3–4 cm long; oval flat leaf, long pointed, 8–10 cm long, 3–4 cm wide, the margins toothed and wavy; green, white marbled, rose tinted, underside densely grey scaled

I dense, sessile, few flowered

FB narrow-lineal, pointed, entire

Fl white to 3 cm long

Se to 1·1 cm long, fused to beyond halfway, long pointed, nearly asymmetrical

Pe lineal-lanceolate, pointed, about 2·2 cm long, short fused, flared from the anthesis

Ov to 8 mm long

Ha Brazil

CU very attractive and widely cultivated.

Cryptanthus bivittatus (Hook.) Regel, 1864 (Ill. 221)

(*bi* two, *vittatus* striped; after the two yellowish longitudinal striped leaves)

Syn. *Billbergia bivittata* Hook., 1861; *Tillandsia bivittata* hort. ex Regel, 1864; *Nidularium bivittatum* Lem. ex Baker, 1889; *Cryptanthus moensii* hort., ex Gentil, 1907.

The var. *atropurpureus* Mez is distinguished from the type by the colour of the leaves which turns to purple in the sun and the stripes become less pronounced. Only known in cultivation

Pl small, rosette like

Fo the outer leaves narrowed above the sheath

LB to 18 cm long and 3·5 cm wide, with wavy and densely toothed edges,

upperside bare, green with 2 lighter long stripes, underside weakly scaled

I few flowered

Fl to 2·7 cm long, white

Ha East Brazil.

Cryptanthus bromelioides Otto and Didr., 1836

(*bromelioides* Bromelia-like)

Syn. *C. acaulis* var. *bromelioides* Baker, 1889; *C. acaulis* var. longifolius hort. ex Baker, 1889; *C. carnosus* Mez, 1919.

The plant breeder M. B. Foster introduced var. *tricolor*, one of the most colourful of Cryptanthus; the blades are longitudinally striped creamy-white and overall reddish blushed (Plate 98). Only known in cultivation.

Pl 30–40 cm high, short stemmed and   stoloniferous

Fo several, narrowing above the sheath, to 20 cm long and 4 cm wide
LB wavy margined, small and densely toothed, upperside bare, green, underside white scaled

I ample-flowered, compound, each cluster 4 to 6 flowered
Fl white, to 4 cm long with wide-lineal, blunt petals
Ha Brazil.

Cryptanthus fosterianus L. B. Smith, 1952 (Plate 99)
(*fosterianus* named after the breeder M. B. Foster)

Pl to 12 in a flat, star shaped wide spreading rosette
LS roundish, spoon shaped
LB lineal-lanceolate, long pointed, narrow at the base (but not stalked) to 30 cm long and 4 cm wide, thick, upperside reddish-brown-grey crossbanded, underside thick grey scaled; margins wavy and densely spined
I sitting on the rosette, with few flowered compound spikes
PB similar to the leaves but smaller, with longer tip and heart shaped sheath

Sp the outer ones 3 to 4 flowered, the inner ones mostly 2 flowered
FB wide-oval, thin cutaneous, nearly as long as the sepals
Se 8 mm long, unequally fused in height, wide oval, pointed with toothed margins
Pe white, narrow strap shaped
Ha Brazil
CU one of the best species and with adequate care develops into a show plant. It can be mistaken for *C. zonatus fuscus* (Ill. 220).

Cryptanthus lacerdae Antoine, 1882 (Ill. 222)
(*lacerdae* ragged; because of the strongly toothed leaves)

Pl stemless, hardly 5 cm high
Fo not more than 10–15, in a flat outstretched rosette
LB to 8 cm long and 3 cm wide, hardly narrowed above the sheath, hardly waved margins, densely toothed, dark green with silver-white margins and

centre stripes, underside dense white scaled
I few flowered
Pe white to 15 mm long
Ha East Brazil (?); only known in culture, but not often seen
CU very pretty, small species.

Cryptanthus zonatus (*Vis.*) Beer, 1857 (Ill. 223)
(*zonatus* zoned, banded; because of the banded leaves)
Syn. *Pholidophyllum zonatum* Vis., 1847

Fo several, forming an outstretched, star shaped rosette
LS indistinct, indistinguishable from the leaf blade, green
LB strapshaped, long pointed, 15–

20 cm long to 4·5 cm wide, upperside green with grey-brown cross-banding, underside densely whitish-grey scaled, hence the underside appears to be white.

Sc absent

I dense-headed, compound, nestling in the rosette, few flowered

PB similar to the leaves but shorter, shielding the few flowered spikes

Sp mostly 3 flowered, 2 cm long, arranged spirally

FB oval-lanceolate, long pointed, shorter than the sepals, green, toothed tip

Se to 1·9 cm long, fused for about $\frac{3}{4}$ of their length, asymmetric, long pointed, green, grey scaled, somewhat toothed at the tip

Pe to 3 cm long, the upper ones 1·5 cm flared, 5 mm wide, white, narrow-lanceolate, pointed

Ov to 6 mm long, white

Ha Brazil; similar to *C. fosterianus* but with shorter leaves.

VARIETIES

var. *viridis* hort., 1934

Underside of leaf bare, not scaled, green. Only known in cultivation.

var. *fuscus* Mez, 1896 (Ill. 223)

Syn. *Pholidophyllum zonatum* var. *fuscum* Vis.

Differs from the type by the brown-red colour of the leaves and the light silver-grey banding. Only known in cultivation.

Cryptanthus Hybrids

Cryptanthus x *lubbersianus* (*C. beuckeri* x *C. bivittatus*)

Cryptanthus x *mackoyanus* (*C. acaulis* x *C. bivittatus*)

Cryptanthus x *osyanus* (*C. lacerdae* x *C. beuckeri*)

Bi-generic Hybrids called Cryptbergia

x *Cryptbergia meadii* (*Cryptanthus beuckeri* x *Billbergia nutans*)

x *Cryptbergia rubra* (*Cryptanthus bohianus* x *Billbergia nutans*)

Fascicularia Mez, 1891

The genus is so named (*fascicularia*, forming bushes) because of the crowded arrangement of the flowers.

The plants are terrestrial or epiphytic growing rosettes and the leaves are narrow-lineal, stiff, terminating in a prickled tip with the margins spined. The inflorescence scape is short, glomerate headed, simple-spiked inflorescence nestling in the centre of the leaf rosette which turns red or whitish. Flowers are sessile to very short stalk, blue or blue violet; berry fruits.

The genus is endemic to Chile and comprises 5 species; it is closely allied to *Bromelia* and *Cryptanthus*. Although extremely xerophytic, in culture they should be subject to long periods of dry conditions. Occurring at latitudes of 44° south, they are at the southernmost boundary of the bromeliad family.

Fascicularia bicolor (Ruiz and Pav.) Mez, 1896 (Plate 92)
(*bicolor* two coloured; because of the green and red coloured inner leaves of the rosette)
Syb. *Bromelia bicolor* Ruiz and Pav., 1802; *Billbergia bicolor* Roem. and Schult, 1830; *Rhodostachys bicolor* Benth. and Hook, 1883; *Bromelia albobracteata* Steud., 1857; Hechtia gracilis hort. Haage and Schmidt, 1910

Pl stemless to short stemmed, growing into a large clump or mass
Fo several, in a large, outspread rosette
LB to 45 cm long and 15 mm wide, lanceolate, terminating in a sharp point, stiff, upperside bare, underside brownish scaled, spined margins; the inner leaves of the rosette shorter, lively carmine-red and forming an involucre
I a capitulatum, sessile, nestling, the outer bracts longer than the flowers
Fl to 4 cm long, pale blue
Se wide-lineal blunt, the backs keeled and winged, strongly woolly scaled towards the tip
Pe narrow-elliptical, to 20 mm long
Ov somewhat compressed, scaled
Ha South Chile; on rocky ground
CU in bloom, one of the most decorative and most beautiful of Chilean plants.

Fascicularia pitcairnifolia Mex, 1896
(*pitcairnifloia* Pitcairnia-like leaves)
Syn. *Bromelia pitcairnifolia* C. Koch, 1868; *Rhodostachys pitcairnifolia* Benth and Hook, 1883; *Pourretia joinvillei* hort., 1871; *Billbergia* hort., 1871; *Bromelia joinvillei* Morren, 1876; *Rhodostachys joinvillei* Benth. and Hook, 1883; *Hechtia joinvillei* Riv., 1871; *Pourretia flexilis* hort., 1876; *P. africana* hort., 1890; *P. maxicana* hort., 1876; *Hechtia carnea* hort., 1890.
Distinguished from *F. bicolor* by the flower bracts, much shorter than the 4 cm long flowers. The inner rosette leaves at blooming are red coloured. Of the two species, *F. pitcairnifolia* is better in cultivation.

Fernseea Baker, 1889
This is a monotype genus (only one species), endemic to Brazil.

Fernseea itatiaiae (Wawra) Baker, 2889 (Fig. 60)
(*itatiaiae* named after the mountain Itatiaia)
Syn. *Bromelia itatiaiae* Wawra, 1880; *Aechmea stenophylla* Baker, 1889

Pl growing in grass-like form
Fo several, in a clump-like rosette with short, naked sheath and narrow-lineal blades, to 40 cm long, underside weak scaled, pointed, the margins widely spined
Sc thin, with ovate to lanceolate red bracts

I simple, panicled, many flowered

PB red, ovate, lanceolate, entire

Fl stalked, to 1·7 cm long, grape-purple

Fr somewhat setacous berry with large seeds

Ha until now, known only in the highest regions of Itatiaia, in the province of Rio de Janeiro; seldom found in cultivation

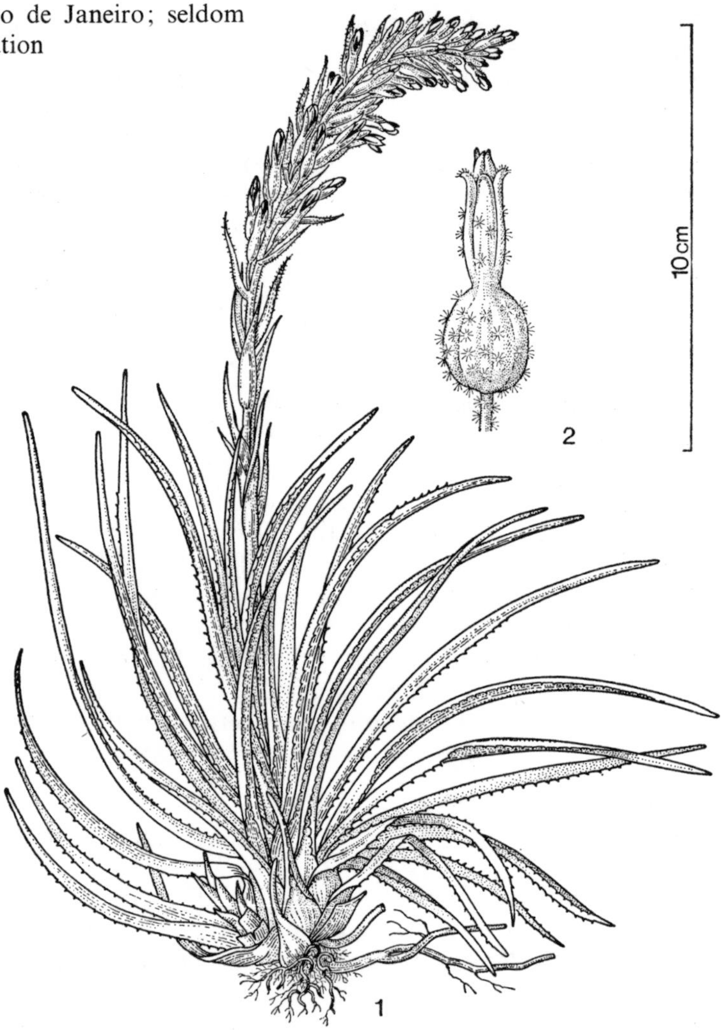

Fig. 60 *Fernseea itatiaiae* (Wawra) Baker: 1 flowering plant; 2 flower and ovary.

Gravisia Mez, 1891

This genus with 6 species is named after a professor of botany A. Gravis; the habitat ranges from Brazil and Venezuela to Costa Rica.

A large stemless rosette with lineal, spine-tipped leaves, the margins strongly spined. The sturdy inflorescence scape carries entire, red or rose-coloured scape bracts. The many-flowered, panicled inflorescence is a sessile or a stalked capitulum-like compound; flowers yellow or orange. In consequence of a strong mucous secretion the individual flower clusters are frequently enclosed by the mucous.

Gravisia aquilega (Salisb.) Mez, 1896 (Plate 108)

(*aquilega* to gather water; probably because of the strong mucous secretion)

Syn. *Bromelia aquilega* Salisb.; *B. exsudans* Lodd., 1824; *Aechmea aquilega* Griseb., 1964; *A. eysudans* Baker, 1889; *A. chrysocoma* Baker, 1889; *A. aquilegioides* Kunthe, 1891; *Gravisia eysudans* Mez, 1892; *G. chrysocoma* Mez, 1892

Pl with inflorescence 80–100 cm high
Fo 15–30 in a funnel shaped rosette
LS large, wide-oval, both sides densely brown scaled, the margins dark brown spined towards the tip
LB narrow strap-shaped, 30–150 cm wide, green, both sides appressed, white scaled; margins thicker at the base, above with widely spaced 1–5 mm long, dark-brown spines
Sc upright, stout, at first white woolly, soon becoming bare
I to 40 cm long, with at first a white woolly axis later becoming bare, interrupted cylindrical, to 10 to 20 compound capitulum-like flower clusters conglomerated, usually enclosed by a mucous secretion

SB long elliptical, pointed, entire, overlapping; longer than the internodes, upper ones red
PB the basal ones similar to the leaves, longer than the individual flower cluster
FB wide-oval, pointed spine tip, bare, keeled, green, shorter than the sepals, about 3–3·5 cm long and 1·5 cm wide
Fl sessile, bare
Se 1·5 cm long, asymmetrical, spine tip, yellowish-orange
Pe 2·5 mm long, 3 mm wide, yellowish-orange, pointed
Ov 1·5 cm long, whitish green, bare
Ha Brazil, Costa Rica, Venezuela, Trinidad, Tobago, Guyana.

Greigia Regel, 1865

The 10 known species which form the genus, named in honour of a Russian D. Greig (1827–1887), occur in mountainous areas from Costa Rica to Chile. On the edge of the forests they frequently form complete thickets.

The form is a fine, mostly stemless terrestrial rosette plant, the leaves long-lineal, stiff, spine-tipped and with strongly spined margins, forming a large rosette.

The inflorescence takes either an apparently lateral or a terminal position and is of a mass-spiked or capitulate form. The sessile or short-stalked flowers are flesh coloured, reddish or whitish-brown near the anthesis; the fruit is a fleshy berry.

The hard, large spines lining the leaves make it very difficult to handle the plants and they are rarely seen even in botanical gardens.

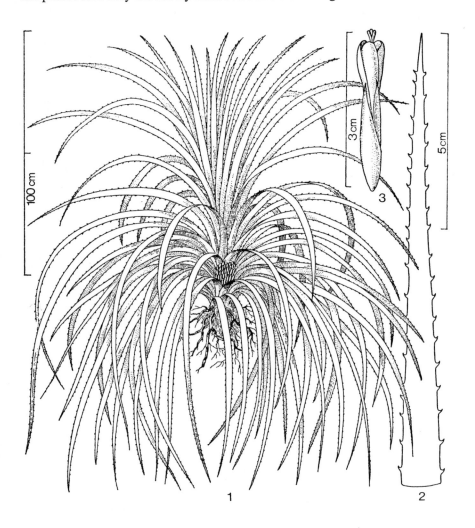

Fig. 61 *Greigia sphacelata* (Ruiz and Pav.) Regel: 1 habit; 2 leaf; 3 flower.

Greigia sphacelata (Ruix et Pav.) Regel, 1865 (Fig. 61)
(*sphacelata(us)* red spotted, torchlike)
The lateral position of the inflorescence is similar to that of *Quesnelia lateralis* (see page 351) and more clearly defined.
Syn. *Bromelia sphacelata* Ruiz et Pav., 1802; *Billbergia sphacelata* Roem. et Schult., 1830; *Bromelia discolor* Lindl., 1838; *Bromelia crassa* Stend., 1857; *Bromelia clandestina* hort., 1880.
The lateral position of the inflorescence is similar to that of *Quesnelia lateralis* and more clearly defined.

Pl stemless or with short thick stem
Fo dense rosette to 1 m high, sometimes to 4 m in diameter
LB to 2 m long, narrow-lineal to 3 cm wide, terminating in a sharp spine tip, margins very stoutly spined, underside moderately brown scaled
Sc very short or missing
I simple-racemose, capitate to globular, in the axil at the base of the rosette leaves
SB outer ones (basal) clearly spined, inner ones (upper) only weakly spined
Fl short and thin stalked, to 7 cm long.
Se lanceolate, long pointed to 4 cm long, fused from the base to 5 cm high
Pe flesh coloured, fused to half way, with upright tips
Ha Chile and perhaps in Peru.

Hohenbergia Schult. f., 1830
The genus is named in honour of Hohenberg, a Württemberg king and a patron of botanical literature. It comprises about 30 epiphytical species, ranging from Brazil and Venezuela to the West Indies and Guatemala.

The form is a rosette of sizeable dimensions with wide lineal, spine-tipped leaves, the margins carrying the spines, with the upperside bare and the underside scaled. The very sturdy inflorescence scape carries white or reddish cutaneous scape bracts; the elongated inflorescence is compound of loose, clawed, strobilate-like, stalked or sessile, spiked clusters. The flowers are laterally strongly compressed, yellowish, whitish or bright blue-violet; the outer stamens are free and between the petals (Fig. 62, 2, aSt), inner stamens are fused with the petals (Fig. 62, 2, iSt). The epigynous petal tube is short or absent. They have an inferior ovary, ripening to a juicy berry.

Hohenbergia augusta (Vell.) E. Morren, 1873 (Ill. 219)
(*augusta(us)* elevated, majestic)
Syn. *Tillandsia augusta* Vell., 1835; *Pironneava glomerata* Gaud., 1843; *Aechmea glomerata* Hook., 1867; *A. augusta* Baker, 1879; *A. multiceps* Baker, 1880; *Hohenbergia ferruginea* Carr., 1881

Pl stemless, 70 cm high with inflorescence

LS wide-oval, to 20 cm long, 15–18 cm wide, brown scaled; margins not spined

LB wide-lineal, blunt terminating with a spine, to 1·20 m long, 8–12 cm wide, green with dark-green blotches, each side grey scaled; margins with 1 mm long strong spines

Sc sturdy, upright, to 2·5 cm thick, round, green, strongly brown felted, 40–50 cm long

I 40–50 cm long and brown felted

SB densely overlapping, upright, longer than the internodes, enveloping the scape completely, the basal ones green, the upper ones drying early and brown felted

PB always shorter than the inflorescence clusters; these globose, short stalked, 1·5 cm long, 6–7 mm wide compound spikes

FB densely overlapping, spirally arranged, keeled, 1 cm long, with 1–2 mm long brown terminating spine, soon drying at the tip, brown felted

Fl to 1·1 cm long, strongly scented

Se keeled, whitish-green

Pe blue-violet, 8 mm long, the tip flared

Ha Brazil.

Hohenbergia stellata Schult, f., 1830 (Plate 109, Fig. 62)

(*stellata(us)* star shaped; because of the star-shaped inflorescence clusters)

Syn. *Aechmea glomerata* Hook. f., 1867; *A. oligosphaera* Baker, 1889; *A. longisepala* Baker, 1889; *Hohenbergia oligosphaera* Mez, 1896

Pl with inflorescence over 1 m high

Fo forming a broad large rosette, to 1 m in diameter

LS wide-oval, 15–20 cm long, 12–15 cm wide, both sides dark brown

LB 60–90 cm long, 7–9 cm wide, dark green, grey scaled, widely pointed, with 1 cm long spined tip, the margins edged with 1 mm long brown spines

Sc sturdy, upright, round, 1·5 cm in diameter, 40–60 cm long, green, white woolly when young

I loose, double compound, to 60 cm long and 15 cm wide clawed inflorescence clusters

SB thin, upright, longer than the internodes, 10–13 cm long, drying during flowering

FB at the lower part of the scape, as long as or somewhat longer than the spikes, the upper ones shorter, drying during flowering

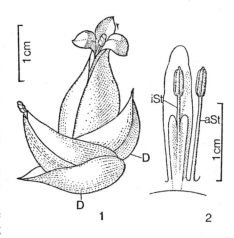

Fig. 62 *Hohenbergia stellata* Schult. f.:
1 two flowers in axis of flower bracts (D);
2 petal with different inner (iSt) and outer (aSt) stamens.

Sp 8 to 12 flowered, to 3 to 6 clawed, each about 3 cm long, flattened, green, sitting on a dense white woollen stalk
FB triangular, long pointed, 2–4 cm long, 1–2 cm wide, partially keeled, bright red, white at the base
Fl laterally strongly compressed

Se 1·5 cm long, white at the base, then bright red, blue at the tip, laterally keeled
Pe 2 cm long, blue
Ha Tobago, Trinidad, Venezuela, Brazil.

Neoglaziovia Mez, 1891

Only two species are included in this genus; both grow as terrestrials or on rocks in East Brazil in the notorious Catinga (thornbush areas); their rigid leaves are utilized in the production of fibre.

The plant is a stemless rosette more than 1 m high, with rigid, narrow-lineal leaves, tapering off to a point and with the margins short spined. The several violet or purple coloured flowers form an upright raceme.

Neoglaziova concolor C. H. Wright, 1910 (Fig. 63)

(*concolor* self-coloured; in contrast to *N. variegata*, it has self-coloured leaves)

Pl stemless, stoloniferons
Fo 8–10, in a rigid compact rosette, leaves 2·5 cm wide, 40–60 cm long, tapering off to a point, both sides densely white scaled
Sc upright, white woolly, much shorter than the leaves, with few leaf-like scape bracts
I about 20 cm long about level with the leaves, simple-raceme with white woolly axel

FB the basal ones longer than the flowers, the upper ones much shorter, white woolly and remaining so
Fl on a short 5 mm long stalk, horizontally extending from the axil
Se brownish, bare, to 1·5 cm long with short, blunt tip
Pe 2 cm long, longish-lanceolate, blunt, violet, with 2 slit scales at the base
Ha Brazil (North Bahia).

Neoglaziovia variegata Mez, 1894

(*variegata(us)* varicoloured, banded)
Syn. *Bromelia variegata* Arruda de Camara, 1810; *Billbergia variegata* Schult., 1830; *Agallostachys variegata* Beer, 1857; *Dyckia glaziovii* Baker, 1889; *Bromelia linifera* hort., 1857
Distinguished from *N. concolor* by the leaves, the upper side green, the underside with wide, white, cross banding.
Ha Brazil.

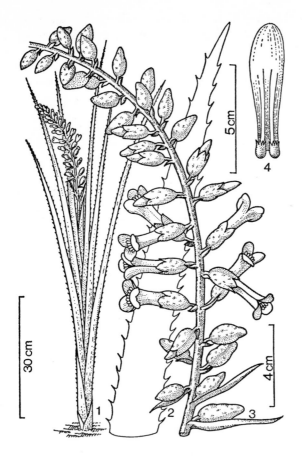

Fig. 63 *Neoglaziovia concolor*
C. G. Wright. 1 habit; 2 leaf;
3 inflorescence; 4 petal with
basal scales (from Bot. Mag.).

Neoregelia L. B. Smith, 1934

The genus is named after a German botanist A. von Regel, who was superintendent of the Imperial Botanic Gardens, St. Petersburg, Russia. Since the name Regelia was previously given to 3 species in the Myrtaceae family, the botanical authority L. B. Smith has given the genus the name *Neoregelia*. (Also, *Regelia* Lindm., 1890; *Aregelia* Maz, 1896 non Kuntze, 1891.)

Usually epiphytic and seldom terrestial, their leaves form a wide funnel shape or a tubular rosette.

The leaf margins are usually spine-edged, the tip rounded with short spine (see Fig. 64, 1); the leaves are bare or scaled.

The inflorescence forms a nest on a short scape, the axel of the simple-raceme is extremely shortened, and all the flowers remain level (Fig. 64, 2), sunk in the centre of the rosette (Ill. 224). Its outstanding appearance is enhanced by the vividly coloured scape bracts called 'heart leaves', which form the involucrum. At the base, the sepals are fused, upright, pointed. The base of the petals form a more or less tall fused tube, seldom free, pointed and with the upper parts flared (Fig. 64, 3). The stamens are enclosed within the tube and their filaments are fused with the petals (Fig. 49, c); the ovary is inferior, developing when ripe into a berry.

The principle area for the 40 or so species which make up the genus is in the rain forests of eastern Brazil; a few come also from eastern Colombia and eastern Peru. With their brightly coloured 'heart leaves', the colour lasts for a long time and many *Neoregelias* are grown as house plants; their leaves are relatively firm. The funnel must be kept filled with water which includes the inflorescence.

It is wise to remove the water from the inflorescence after flowering to prevent development of an evil smelling mess.

Neoregelia ampullacea (Morren) L. B. Smith, 1934 (Ill. 226)

(*ampullacea(us)* ampulla shaped; on account of the ampulla shaped rosette)
Syn. *Aregelia ampullacea* (E. Morren) Mez, 1896; *Nidularium ampullaceum* E. Morren, 1880; *Karatas ampullacea* Baker, 1889; *Regelia ampullacea* Lindm., 1890

Pl stemless, stoloniferons
Fo few, with spoon-shaped wide sheaths forming an ampulla shaped plant, 10–15 cm long and 2–3 cm in diameter
LB lineal-lanceolate, to 15 cm long and 1·5 cm wide, recurved, with rounded spined tip, toothed margins, upper side bare, green, underside reddish brown blotched or banded
I compound head, few flowered, deeply sunk in the base of the rosette and therefore not visible
FB long-pointed, lightly scaled, shorter than the sepals
Pe to 2 cm long, pointed, fused for $\frac{3}{4}$ of their length, blue changing to white at the base
Ha Brazil (State of Rio de Janeiro)
CU with its small ampulla-shaped rosette, a very attractive plant forming a cluster group. Suitable for a small greenhouse.

Neoregelia carolinae (Mez) L. B. Smith, 1934

Syn. *Aregelia carolinae* (Beer) Mez, 1896; *Bromelia carolinae* Beer, 1857; *Billbergia carolinae* hort. von Houlte ex Beer, 1857; *Nidularium carolinae* Lem. ex Baker, 1889; *Karatas carolinae* Antoine, 1884; *Billbergia meyendorffii* Regel, 1857; *Nidularium meyendorffii* Regel, 1859; *Guzmania picta* hort. ex Beer, 1857; *Bromelia rhodocincta* Brongn. ex Baker, 1889

Pl stemless
Fo 12–15, a wide funnel shaped rosette with a diameter to 60 cm
LB to 40 cm long and 3 cm wide, lineal, with rounded short spined tip, the margins densely spined, both sides shiny green, bare, the 'heart leaves' shiny red, with light bluish sheen
Sc very short
I dense compound head, nestling, surrounded by the shiny red inflorescence bracts

FB wide-lineal, with rounded tip, bare, a little shorter than the sepals
Se short fused at the base with 1·6 cm long, free, rounded tip
Pe upright, pointed to 3 cm long, fused to the middle
Ha Brazil (State of Rio de Janeiro)
CU a favourite species, but even more attractive is var. *tricolor* L. B. Smith, 1934 (Plate 101). The leaves toned red and longitudinally striped yellowish-white.

Neoregelia concentrica (Vell.) L. B. Smith, 1934 (Plate 102, Ill. 224, Fig. 64)
(*concentrica(us)* pressed together)
Syn. *Tillandsia concentrica* Vell., 1825; *Bromelia concentrica* (Vell.) Beer, 1857; *Nidularium laurentii* Regel, 1867; *Billbergia aurantiaca* hort. Laurent ex Regel, 1867; *Nidularium acanthocrater* Morren, 1884; *Karatas laurentii* (Regel) Antoine, 1848; *Karatas acanthocrater* (Morren) Antoine, 1848; *Nidularium laurentii* Regel var. *typica* Regel, 1885; *Regelia acanthocrater* (Morren) Lindm., 1890; *R. laurentii* (Regel) Lindm., 1890; *Nidularium concentricum* (Vell.) Mez, 1891; *Aregelia laurentii* (Regel) Mez, 1896; *A. concentrica* (Vell.) Mez, 1896

Pl stemless
Fo a flat outstretched funnel shaped rosette from 70–90 cm in diameter
LS wide, oval, 13–14 cm long, 9–10 cm wide, greenish-brown; the margins not spined
LB 5–10 cm wide, to 30 cm long, wide-lineal with widely rounded spined tip, the margins with 4 mm long black spines, dark green with lilac blotches, underside appressed grey scaled; 'heart leaves' at blooming time lilac

coloured
I simple, many flowered, dense head, nestling, to 7 cm in diameter
FB lineal, to 5 cm long, thin, green, bare
Pe to 4 cm long, about 1·2 cm of the upper part free, pale blue, narrow, long pointed, upright to spreading
Ov to 1·5 cm long, white, bare
Ha Brazil (State of Rio de Janeiro)
CU a very beautiful large species, with its very long, black spines.

Neoregelia coriacea (Antoine) L. B. Smith, 1955 (Plate 100)
(*coriacea(us)* leathery; because of the firm leaves)
Syn. *Karatas coriacea* Antoine, 1884; *Nidularium coriaceum* hort., 1884; *Regelia coriacea* (Antoine) Lindm., 1890

A very fine decorative species, sometimes confused with *N. concentrica*, but distinguished from it by the missing black spines of the leaves

Pl stemless

Fo outspread, a flat rosette with a diameter to 60 cm

LS nearly round, more or less 10 cm long, 8–10 cm wide, pale lilac, grey scaled, the margins not spined

LB tongue shaped, rounded at the tip but not spined, about 30 cm long, 4–9 cm wide, green, appressed grey scaled, the 'heart leaves' turning to lilac at flowering time; margins feebly spined to becoming smooth

Sc 4–5 cm long, 1·8 cm thick, the inflorescent not raised above the rosette

SB dense, scale shaped, 4·5 cm wide, 5 cm long, white with lilac coloured tip

I nestling, simple densely headed, many flowered, to 6 cm wide

FB more or less 4·5 cm long, narrow, long pointed, about half as long as the sepals, green, weakly grey scaled

Fl 4·5 cm long, stemmed

Se 2·4 cm long, asymmetric, long pointed with hooked tip, green, bare

Pe pale blue to white, fused for much of its height, long pointed

Ov 1 cm long, white, bare

Ha Brazil (?).

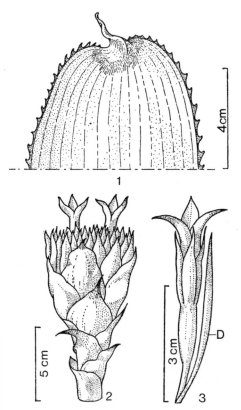

Fig. 64 *Neoregelia concentrica* (Vell.) L. B. Smith: 1 leaf tip; 2 inflorescence; 3 flower with bract (D).

Neoregelia cruenta (R. Graham) L. B. Smith, 1939 (Ill. 228)

(*cruenta*(*us*) black-red; because of the dark red spines on the leaves

Syn. *Bromelia cruenta* R. Graham, 1828; *Nidularium cruentum* Regel, 1859; *N. laurentii* var. *immaculatum* Regel, 1885; *Karatas cruenta* Nichols, 1885; *Regelia cruenta* Lindm., 1890; *Aregelia cruenta* Mez, 1896; *Nidularium longebracteatum* Mez, 1891; *Aregelia longebracteata* Mez, 1896; *A. rubrospinosa* Mez, 1913; *Neoregelia rubrospinosa* L. B. Smith, 1939; *N. longebracteata* L. B. Smith, 1939

Pl stemless

Fo forming a large funnel shaped rosette to 1 m in diameter

LS longish-oval, 17–19 cm long, 10–11 cm wide, green, dark brown scaled and with smooth margins

LB to 80 cm long, 6 to 7 wide with rounded and spined red tip, green, underside appressed grey scaled; margins dark red spined

I simple, densely headed, many flowered, nestling, with 5–6 cm long scape

FB wide-oval to lineal, spine tipped, thin, white to green, bare, shorter than the sepals, 5·5 cm long, 1·5 cm wide at the base

Fl to 6 cm long, short (6–7 mm long) stalked

Se 2·5 cm long, 1·5 cm wide, short fused at the base, asymmetric, green, bare

Pe 3·2 cm long, nail shaped; nail 5 mm wide, white; top part (plate) flared to bent back, 7 mm wide, blue

Ha Brazil (State of Rio de Janeiro, São Paolo)

CU a large plant, outstanding because of its red leaf spines.

Neoregelia cyanea (Beer) L. B. Smith, 1939 (Ill. 227)

(*cyanea*(*us*) blue, cornflower blue; because of the colour of the flower)

Syn. *Hoplophytum cyaneum* Beer, 1857; *Bromelia denticulata* C. Koch, 1859; *Nidularium denticulatum* Regel, 1870; *Karatas denticulata* Baker, 1889; *Regelia denticulata* Lindm., 1890; *Aregelia cyanea* Mez, 1896

Pl stemless, with short stolons

Fo very stiff, to 20 in a loose rosette

LS longish-oval, about 4–5 cm long, upperside partly reddish-brown forming a short tube-like funnelled rosette

LB narrow, strap shaped, about 25 cm long, 7 mm to 2 cm wide, curled backwards, upper side dark grey, underside grooved and grey scaled, widely spaced toothed

Sc short, 3–4 cm long

I nestling, simple, many flowers

FB thinly cutaneous, 2 cm long, 5–7 mm wide, reddish-brown blotched

Fl with 1 cm long, laterally flattened, white stalk

Se reddish-brown, 1·5 cm long with small tip, 2·5 mm fused at the base, asymmetric, bare

Pe 1·7 cm long, fused for a short length at the base, pointed, blue

Ov about 8 mm long

Ha southern Brazil (Minas Gerais)

CU an attractive small species, well worth growing.

Neoregelia eleutheropetala (Ule) L. B. Smith, 1934 (Fig. 65)
(*eleutheropetala(us)* separated corolla free petals)

Pl stoloniferous

Fo to 30, forming a large rosette 50–70 cm in diameter

LS to 12 cm long and 6 cm wide, dark brown, underside shiny

LB 6–8 cm wide, tapering off to a point, green or the inner rosette leaves red, the margin with firm brown spines

I nestling, many flowered, 6–8 cm in diameter

PB the outer ones 4–5 cm long, 2·4–3·2 cm wide, elliptical, with purple coloured spined tip, cutaneous, scaled, the inner ones smaller

FB 4 cm long, oval-lanceolate, raised above the sepals, scaled towards the tip

Fl to 15 mm long stalked

Se free, oval, pointed, 2·1–3·6 cm long, early scaled, later bare

Pe 3·5 cm long, white, free

Ov 1·5 cm long, bare

Ha from northern Peru, through Colombia to Brazil (Amazonia).

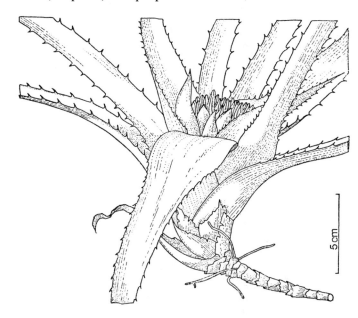

Fig. 65 *Neoregelia eleutheropetala* (Ule) L. B. Smith; flowering plant.

Neoregelia farinosa (Ule) L. B. Smith (Plate 103)
(*farinosa(us)* mealy; because of the underside of the leaves are densely white lepidote

Syn. *Aregelia farinosa* (Ule) Mez, 1934; *Nidularium farinosum* Ule, 1900

Pl stemless, growing in rocky terrain

Fo about 20 in a dense, outspread rosette

LS wide-oval to elliptical, upperside

purple brown, densely scaled, the margins remotely toothed

LB wide-oval, to 60 cm long and 4 cm wide, rounded but with a spine tip, upperside nearly bare, dark green, underside densely white scaled, the marginal spines small, the inner rosette leaves ('heart leaves') shiny blood-red

Sc very short, densely purple coloured or green, pointed, the margins densely toothed, surrounding the inflorescence

I simple, many flowered, surrounded by the red inflorescence bracts

Fl to 16 mm long at the base, fused for 4 mm of its height, greenish, tapering off to a short spine

FB lanceolate-lineal, twisted inwards, to 4·5 cm long and 8 mm wide, whitish green

Pe to 4 cm long, with flared upper part (plate)

Ov white, to 12 mm long

Ha Brazil, Serra do Macahe

CU the underside of the leaves white mealy scaled and the shiny red inner 'heart leaves' makes it an outstanding plant for growing

Neoregelia laevis (Mez) L. B. Smith, 1934 (Ill. 229)

(*laevis* smooth; because of the frequently smooth leaf edges)
One of the few neoregelias with smooth, unarmed leaf margins. On account of the compound inflorescence, this species has been placed in the section Amazonicae.

Pl stemless, short (more or less 10 cm long) flattened, stoloniferous

Fo about 15 in a funnel shaped rosette from 80 cm to 1 m in diameter

LS nearly triangular, 5–7 cm long, 4 cm wide at the base, brownish, both sides dark brown scaled

LB 50–70 cm long, narrow-lineal, long pointed, 2 cm wide above the sheath, upper side green, bare, underside green to reddish-green, scattered grey scaled; margins smooth

Sc short, 4–5 cm long, 1 cm thick, round, bare, whitish

I nestling, dense-headed, set with few flowered compound spikes

PB cutaneous, greenish-white, brown scaled, as long as the spikes, longish-triangular, 3–4 cm long, 1·5 cm wide at the base, with green hardly prickled tip

FB 2 cm long, 3 mm wide, cutaneous, greenish-white; the tip brown and scattered brown scaled

Fl 2·7 cm long, short stalked; stalk 3 mm long, white

Se free, 1·2 cm long, flared, 3 mm wide, long pointed, green, scattered brown scaled, laterally keeled

Pe narrow, white, pointed, spreading, 1·5 cm long, 2 mm wide

Ov 5 mm long, white, naked, laterally keeled

Ha Brazil to Peru (Amazonia).

Neoregelia marmorata (Baker) L. B. Smith, 1939 (Plate 105)

(*marmorata(us)* marbled; because of the blotched leaves)
Syn. *Nidularium laurentii* var. *elatius* Regel, 1885; *Karatas marmorata* Baker, 1889; *Aregelia marmorata* Mez, 1896

Pl to 30 cm high, forming a flat rosette from 50–60 cm in diameter

LS wide-oval, 12–14 cm long, 9–10 cm wide, both sides blotched brown-red, smooth
LB to 40 cm long, 6–8 cm wide, strap-shaped, widely rounded at the tip with terminating spine red blotch, green, irregularly brown-red blotched, both sides grey scaled, the margins black spined; spines 2 mm long
I head-shaped, nestling, many flowered, to 5 cm in diameter
FB 3·2 cm long, 8 mm wide at the base, thin cutaneous, green upperpart, hooded tip, weakly grey scaled
Fl 1 cm stalked, 4·5 cm long
Se 2 cm long, green, nearly bare, asymmetrical, fused for 2 mm above the base, pointed
Pe 2·7 cm long, pale violet to whitish, flared to reflexed tip
Ov white, bare, 1 cm long
Ha Brazil
CU a decorative species, standing light and sun which brings out the marbling. In dull conditions, the leaves remain green.

Neoregelia myrmecophila (Ule) L. B. Smith, 1955 (Ill. 230)
(*myrmecophila* liked by ants)
Syn. *Nidularium myrmecophilum* Ule, 1906; *Aregelia myrmecophila* Mez, 1934

Pl stemless, short stolons
Fo 5–10, forming a flat rosette from 60–70 cm in diameter
LS short, wide-oval, 6–7 cm long, 4–5 cm wide, dark brown, grey scaled
LB narrow-strap shape, long pointed, 35–40 cm long, more or less 2 cm wide, green, appressed grey scaled; margins with brown 2 mm long spines
Sc short, not exceeding the rosette, 2–3 cm long, 5–6 mm thick
I nestling in the rosette, to about 5 densely compound spikes, 4–5 cm in diameter
PB 2–3 cm long, cutaneous, 8–9 mm wide, widely rounded at the tip terminating with a short point
Sp short stalked, the flower bracts united at the base, 4–5 cm long, 1·5 cm in diameter, 10 to 15 flowered
FB boat-shaped lanceolate with spined tip, shorter than the sepals, 2 cm long, flared for 1 cm, bare, only at the tip few scaled, green
Fl short stalked, 3–4 cm long, upright
Se free, 1·3 cm long, flared 5 mm, cutaneous, oval-lanceolate; the tip somewhat crooked and reflexed
Pe 2 cm long, 2 mm wide, pointed, flared tip, white, 2 fringed scales near the middle of the claw (the vertical part of the petal)
St 1·6 cm long, fused with the petals for 9 mm of its height from the base
Ov to 1·8 cm long, flattened
Ha in the rain-forests of Amazonia from Brazil to Peru. Ants make their nests among the flowers, hence the name *myrmecophila*.

Neoregelia olens (Hook. f.) L. B. Smith, 1934 (Plate 104)
(*olens* having a bad smell)
Syn. *Aregelia olens* (Hook. f.) Mez, 1934; *Billbergia olens* Hook. f., 1865
Pl stemless
Fo several, forming a deep, tube like rosette, in which the inflorescence is sunk; the inner rosette leaves are shiny red

LB a little narrowed above the sheath, to 30 cm long and 3 cm wide, nearly bare, the outer ones deep green, tapering to a point, the margins spined
Sc very short or missing
I simple, head shaped, deeply sunk, few flowered
FB elliptical, rounded or pointed at the tip, entire, nearly as long as the sepals
Fl short thick stalked, blue-violet, to 3·8 cm long
Se bare, nearly asymmetrical, fused for a short distance up from the base, wide pointed
Pe at the base fused for a short distance, rounded tip
Ha probably eastern Brazil.

Neoregelia pauciflora L. B. Smith, 1955 (Ill. 225)
(*pauciflora*(*us*) poor flowering; because of the few flowered inflorescence)

Pl epiphytic, to 20 cm high with long, thin outstretched stolons
Fo about 12, small upright, somewhat ampulla like rosette with a diameter from 10–12 cm
LS 7–9 cm long, 4–5 cm wide, wide oval, dark purple on the inner side, margins entire
LB wide-tongue shaped, rounded and spined tip, 10–13 cm long, 3 cm wide, green; young leaves blotched dark purple, when older with grey crossbanding; the margins with brown, to 1 mm long spines
Sc very short, not visible, sunk in the leaf-rosette
I few flowered, nearly 2 cm in diameter
FB shorter than the flower stalk, oval, pointed, thin cutaneous
Fl white, stalked; stalk 2·5 cm long
Se slightly asymmetric, narrow-lanceolate, long pointed, 2 cm long, fused for 1 mm of its height from the base
Pe 3·5 cm long, narrow, long pointed
Ov slender, ellipsoid, 7 mm long
Ha Brazil (Espirito Santo, Santo Teresa).

Neoregelia pendula L. B. Smith, 1963
(*pendula*(*us*) pendant; because of the long, pendant stolons)

Pl epiphytic, rosette with metre-long, thin, pendant stolons
Fo about 15, forming a flared rosette from 30 cm in diameter
LS wide-oval, 3–5 cm long, 2–3 cm wide, entire, towards the tip somewhat toothed, both sides brown scaled
LB lineal, narrowing towards the tip, long pointed, 5–10 mm wide, to 30 cm long, green, bare, black-brown spines, the inner leaf blades turning partly or fully red at flowering time, very much reflexed
I simple, few flowered, head shaped, nestling, about 1 cm in diameter
FB elliptical, pointed, whitish, thin-cutaneous, to 2·5 cm long, lightly scaled
Se wide, pointed, asymmetric, to 2 cm long, fused for a short distance at the base, bare
Ov white, bare, about 1 cm long
Ha northern Peru (Amazonica, Alto Marañon).

VARIETY
var. *brevifolia* L. B. Smith (Ill. 231)
Leaves smaller and shorter, forming an ampulla shaped rosette. The inner leaf blades do not turn red when the plant flowers.

Neoregelia pineliana (Lem.) L. B. Smith, 1936
(*pineliana* after a plant collector named Pinel)
Syn. *Nidularium pinelianum* Lem., 1860; *Karatas morreniana* Antoine, 1884; *Regelia morreniana* Lindm., 1890; *Aregelia morreniana* Mez, 1896; *Neoregelia morreniana* L. B. Smith, 1934; *Aregelia pineliana* Mez, 1934

Pl stemless to short stemmed with short stolons
Fo a flat large rosette from 50–60 cm in diameter
LS clearly distingished, long-oval, 10–12 cm long, 6–7 cm wide, reddish-green, underside grey scaled
LB narrow-lineal, to 50 cm long, 2 cm wide above the sheath, green, underside dense covered with large grey scales; tip rounded with terminating spine; 'heart leaves' red at time of flowering
I simple, dense, head shaped, many flowered, nestling and with short 2–3 cm long scape
FB thin-cutaneous, not exceeding the sepals, green, bare, weakly toothed margins 4·5 cm long, 1 cm wide
Fl 1·3 cm long, stalked, 6·5 cm long
Se 2·5 cm long, asymmetrical, thin, green, bare, fused for a short distance at the base
Pe upright, 4 cm long, the upper 1·2 cm blue and hardly flared
Ov white, bare, 1·2 cm long
Ha Brazil (?).

Neoregelia princeps (Baker) L. B. Smith, 1936 (Plate 110)
(*princeps* princely; because of the plant's shape and its colour)
Syn. *Karatas meyendorffii* Antoine, 1884; *Nidularium marechali* hort., 1889; *Karatas princeps* Baker, 1889; *Nidularium princeps* E. Morren, 1889; *N. spectabile* hort., 1889 non Moore; *Regelia princeps* (Baker) Lindm., 1890; *R. marechali* Lindm, 1890; *Aregelia princeps* (Baker) Mez, 1896; *Nidularium meyendorffii* var. *pruinosum* E. Morren, 1896; *Aregelia marechali* Mez, 1934

Pl stemless
Fo a conspicuous flat rosette from about 80 cm in diameter
LS wide, nearly round, 10–12 cm long, 9–11 cm wide, grey scaled, weakly blotched reddish
LB lanceolate, 30–40 cm long, 4–5 cm wide, green, upperside scattered grey (nearly mealy) scaled, underside dense grey (nearly mealy) scaled, margins with 5 mm long spines; rounded tip terminating with a spine; 'heart leaves' shiny red at flowering time
I many flowered, head shaped, simple, nestling, 4 cm in diameter
FB thin, cutaneous, delicate rose with woolly scaled tip, the innermost sterile, 5–10 mm long

Fl stalked, to 6 cm long with stalk
Se bare, shiny-red, 2·4 cm long, fused for 2 mm of its height from the base, asymmetrical, pointed
Pe 3·5 cm long, fused for 2 cm of its height; tube white, tip violet and pointed
Ov bare, white, 1·1 cm long
Ha Brazil.

VARIETY

var. *phyllanthidea* (Mez) L. B. Smith, 1955
Syn. *Aregelia princeps* var. *phyllanthidea* Mez, 1896
Distinguished from the type by its wider, leaf-like, bright pink 'heart leaves'. Only found in cultivation.

Neoregelia sarmentosa (Regel) L. B. Smith, 1934
(*sarmentosa(us)* root tendrils, to reproduce quickly; probably because of the vigorous formation of young plants)
Syn. *Nidularium sarmentosum* Regel, 1870; *N. denticulatum* var. *simplex* Wawra, 1880; *Karatas sarmentosa* Baker, 1889; *Regelia sarmentosa* Lindm., 1890; *Aregelia sarmentosa* Mez, 1896

Pl stemless, forming a funnel shaped rosette
Fo few, wide-lineal, tapering to a sharp point, the margins hard or lax, edged with 1 mm long spines; the inner rosette leaves greenish-brown or red, upperside nearly bare, underside appressed scaled, to 25 cm long and 2·5 cm wide
I simple, many flowered, clearly headed, sunk deeply in the rosette
FB nearly as long as the sepals, wide-lineal, rounded, purple, entire or with few spines at the tip, scaled on the back
Fl stalked to 5 mm long overall to 2·7 cm long, white
Se bare, about 1·4 cm long, 1 mm fused at the base, short pointed, weakly asymmetrical
Pe to 1·8 cm long, pointed, mostly fused
Ov to 1 cm long
Ha Brazil.

VARIETY

var. *chlorosticta* (Baker) L. B. Smith, 1934 (Plate 106)
(*chlorosticta(us)* green blotched; because of the green blotched leaves)
Syn. *Karatas chlorosticta* Baker, 1889; *Regelia chlorosticta* Lindm., 1890; *Aregelia chlorosticta* Mez, 1896
Distinguished from the species by its strong green-lilac-brown leaf

blotching and somewhat likened to *N. marmorata* (page 332)
Ha Brazil (Rio de Janeiro, Itatiaia, Monte Serrat, Teresopolis)

Neoregelia spectabilis (Moore) L. B. Smith, 1934 (Plate 107)
(*spectabilis* conspicuous; having regard to the large size and to the beautiful leaf colour)
Syn. *Nidularium spectabile* Moore, 1873; *Karatas spectabilis* Antoine, 1884; *Regelia spectabilis* Lindm., 1890; *Aregelia spectabilis* Mez, 1896

Pl stemless
Fo forming a flat rosette from 60–70 cm in diameter
LS wide-oval, 13–14 cm long, 10–11 cm wide, greenish-brown, underside lilac coloured with white cross-banding
LB strap shaped, 4–6 cm wide, to 60 cm long, rounded and vividly blood-red coloured at the tip, spined; upperside dark green, underside lilac with white cross-banding; margins short spined
I simple, dense, head shaped, many flowered, nestling, 5–6 cm in diameter
FB red, thin, bare, finely fringed at the margin, to 3·5 cm long, not exceeding the sepals
Fl 5 mm long stalk, 5 cm long
Se 2 cm long, fused for a short distance at the base, red, thin, bare, asymmetrical, with hooked tip
Pe 3 cm long, free for about 1·2 cm at the top, flared, blue
Ov 1·5 cm long, white, bare
Ha Brazil
CU beautiful plant, very worthwhile having in cultivation.

Neoregelia wurdackii (R. Graham) L. B. Smith, 1963 (Ill. 232)
(*wurdackii* after J. J. Wurdack, who collected in Peru)

Pl epiphyte, to 11 mm wide
Fo about 15, forming a funnel shaped rosette about 80 cm in diameter, densely appressed scaled, the inner rosette leaves rose coloured towards the base
LS nearly circular shaped, about 8 cm long, the margins nearly smooth
LB lineal, narrowed, 1·5 cm wide; margins with black, to 1·5 mm long spines
Sc very short, not exceeding the leaf-rosette
I nestling in the rosette, compound, umbelliferous-raceme, many flowered, 3 cm in diameter
PB elliptic, spine-tipped, shorter than the sepals
FB longish, widely pointed, thin, 2·5 cm long, weakly scaled
Se free, asymmetric, 1·4 cm long inclusive of the 2 mm long spined tip, pale green, hardly scaled, the back keeled
Pe free, white, without scales, long pointed
Ov nearly cylindrical
Ha Peru, Amazonia (Province of Bagna)

Nidularium Lem., 1854

The name is derived from the latin *nidulus*, 'a little nest'. There is some similarity with the genus *Neoregelia*; the differences are summarised in the table below:

Plant Characteristic	*Neoregelia*	*Nidularium*
Leaves	Tongue or strap shaped, seldom narrowed above the middle.	Lineal to sword shaped, i.e. most clearly narrowed above the sheath.
Leaf tip	Mostly rounded, terminating in a short bent spine (a cusp) (Fig. 64, 1).	Mostly tapering to a point (Fig. 66, 1–2) seldom as in Neoregelia with rounded tip terminating in a cusp (Fig. 66, 3)
Inflorescence	Mostly simple, forming a raceme (Fig. 64, 2); flowers clearly stalked (Fig. 64, 3); in the Amazonian species (subgenus Amazonicae, *N. wurdackii, N. pendula*) the inflorescence is frequently compound.	Always compound; flowers sessile, mostly to 3–7 side by side with a wide primary bract in the axel (Fig. 66, 4).
Flowers	The free section of the petals tip flared horizontally (Fig. 64, 3).	The free section of the petal tube formed by the united petals has rounded tips (Fig. 66, 5–6); flowers open slightly (Fig. 66, 5).

Both genera have the filaments of the stamens more or less fused with the upper part of the petals (Fig. 66, 6, Fig. 49, b).

The ovaries are inferior and develop when ripe into berries.

There are today about 25 species known inhabiting the rain forest areas of eastern Brazil.

Cultivation is similar to that for *Neoregelia*.

Nidularium apiculatum L. B. Smith, 1955
(*apiculatum(us*) pointed; on account of the pointed leaves)

Pl stemless
Fo several, forming a rosette from 30–40 cm in diameter

LS long-oval, 9–11 cm long, 5–6 cm wide, pale green, both sides brown scaled

LB narrow-sword shaped, 1–2 cm wide at the base, 3 cm wide for the upper third of the blade, pointed, flat, pale green, irregularly blotched dark green; margins from the base to the middle formed into a channel, with small about 1 mm long spined edges

Sc about 10 cm long, 5 mm in diameter, round, white, bare, the inflorescence raised above the leaf funnel

I dense compound few flowered spikes, about 5 cm in diameter

SB few, pale, upright, about 4 cm long, spine tipped, toothed margin

PB nearly leaf like with flared blades, the outer ones about 8–9 cm long, 4–5 cm wide at the base, green, brown scaled, smooth, red towards the tip, grey scaled, spined, fully covering the spikes and star shaped

FB about 2 cm long, flared for 1 cm, pointed, sparsely spined at the tip, brown scaled

Se 1·8 cm long, fused for 7 mm at its height from the base, rounded, cutaneous margined, red keeled in the middle, sparsely brown scaled

Pe 3·5 cm long, white at the base, the rounded hooded tip dark blue with narrow white edges

Ov 1 cm long, white, bare

Ha Brazil.

Nidularium billbergioides (Schult. f.) L. B. Smith, 1931 (Plate 111, Fig. 66, 1)

(*billbergioides* billbergia-like)

Syn. *Tillandsia terminalis* Vell., 1825; *Hohenbergia billbergioides* Schult. f., 1830; *Tillandsia citrina* Burchell, 1879; *Aechmea billbergioides* Baker, 1889; *Nidularium parviflorum* Lindm., 1891; *N. bracteatum* Mez, 1891; *N. citrinum* Mez, 1921

Pl stemless, with inflorescence 30–40 cm high

Fo several, forming a rosette from about 50 cm in diameter

LS oval, 8–10 cm long, 5–6 cm wide, brown scaled

LB sword shaped, 2–3 cm wide at the base, widening to the middle to about 4 cm, 20–40 cm long, fresh green, bare, pointed, the margins edged with short green spines

Sc upright, to 25 cm long, round, green, bare, 4–5 mm thick

I extended above the rosette, to 5 to 10 few flowered, upright spikes, densely compound, to 8 cm long and 8 cm wide

SB 1–2 similar to the leaves, shorter than the internodes, green, bare

PB forming an involucre around the inflorescence, triangular, pointed, completely covering the spike bases, 6–7 cm long, their tips flared, minutely spined, bare, citron-yellow green towards the tips

Sp sessile, 5 to 6 flowered

FB shorter than the sepals, thin, bare, 1·5 cm long

Se 1·2–1·5 cm long, greenish, bare, fused for 5 mm of its height

Pe upright, 2 cm long, white, rounded at the tip

Ov white, 5 mm long

Ha Brazil

CU a very decorative plant when in bloom.

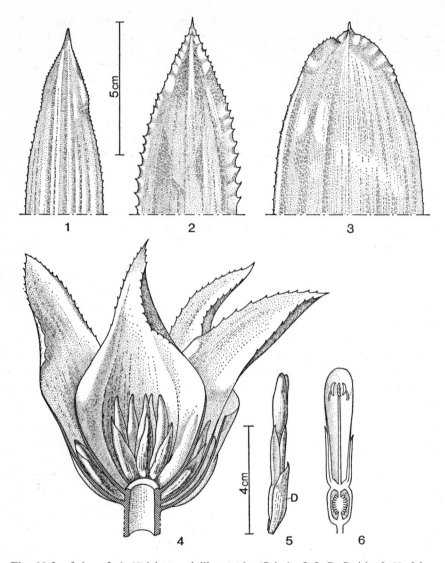

Fig. 66 Leaf tips of: 1 *Nidularium billbergioides* (Schult. f) L. B. Smith; 2 *N. fulgens* Lem.; 3 *N. rutilans* Morr. 4–6 *Nidularium fulgens* Lem.; 4 part of inflorescence; 5 flower and bract (D); 6 longitudinal section through flower.

Nidularium fulgens Lem., 1854 (Plate 112)
(*fulgens* bright; because of the bright red inflorescence leaves)
Syn. *Karatas fulgens* Antoine, 1884; *Guzmania picta* Lem., 1854;

Encholirion pictum hort., 1912; *Bromelia nitens* hort., 1894; *Nidularium rosulatum* Ule, 1900

Pl stemless
Fo 10–15 in a dense, flared rosette
LS clearly set
LB to 30 cm long, narrowing above the sheath and then widening to about 4–5 cm, pointed, the margins sharply toothed, rich green, dark green blotched
Sc very short, on which the inflorescence sits
SB ('heart leaves') similar to the leaves but smaller, outstretched, serrated, bright scarlet-red, nearly enveloping the spikes
FB narrow-oval, pointed, shorter than the sepals
Se reddish, fused for 2 mm in height from the base, the free part lanceolate, pointed tip, to 17 mm long
Pe fused for $\frac{2}{3}$ of its length, 30–40 mm long, with white tube and violet tip
Ha East Brazil
CU a beautiful and decorative species.

Nidularium innocentii Lem., 1855
Syn. *Karatas innocentii* Antoine, 1884; *Regelia innocentii* Ind. Kew., 1895

Pl stemless
Fo several, in a dense, spreading large funnel shaped rosette to 60 cm in diameter
LS wide-oval, 10–13 cm long, 7–9 cm wide, brown scaled
LB sword-shaped, 20–25 cm long, about 3 cm wide at the base, about 5 cm wide in the upperpart, upperside dull-dark green to reddish, underside dark-lilac red, bare, the margins dense minutely prickled
Sc short, 7–8 cm long
I nestling, to 5 cm in diameter with many upright sessile 4 to 6 flowered spikes
PB much longer than the spikes, the upper part bright red, brown scaled, the margins, toothed
FB thin-cutaneous, about 2 cm long, 1 cm wide, broad-oval, the tip rounded, bare, the margins smooth
Fl to 6 cm long
Se 2–2·9 cm long, fused for 7 mm of its height from the base, cutaneous, bare, pale red, keeled
Pe 5·2 cm long, blunt, upright, white, green at the base
Ov to 9 mm long
Ha Brazil
CU they make good houseplants.

VARIETIES
var. *wittmackianum* (Harms) L. B. Smith, 1952
Syn. *N. wittmackianum* Harms, 1928
Leaf blades green with white longitudinal stripes; primary bracts reddish-purple.
var. *lineatum* (Mez) L. B. Smith, 1955 (Plate 113)
Syn. *N. lineatum* Mez, 1913
Leaves green with many white, narrow longitudinal stripes; primary bracts green, changing to red towards the tip

var. *paxianum* (Mez) L. B. Smith, 1950
Syn. *N. paxianum* Mez, 1895
Leaves green with wide, white centre longitudinal stripes; primary bracts green, towards the tip red.
var. *erubescens* hort.
Distinguished from the type by the purple undersides of the leaves.

Nidularium purpureum Beer, 1857 (Ill. 233)
(*purpurum(us)* purple coloured; on account of the grape purple leaves

Pl stemless
Fo several, a flat outspread rosette from 70–80 cm in diameter
LS longish-oval, 12–15 cm long, 6–7 cm wide, greenish-brown, pale brown scaled
LB sword-shaped, pointed, 40–50 cm long, 3 cm wide at the base, 4–5 cm wide for the upper third, grey scaled, the margins small spined, upperside dull green to purple-brown, underside purple-brown
Sc very short, 5–8 cm long, 1 cm in diameter, round, white, bare
I nestling, several few flowered spikes, compound in a dense head
SB similar to the leaves

PB similar to the leaves, purple coloured, arranged spirally and star shaped and completely covering the spikes
Sp flat, few flowered (4–8) short stalked
FB shorter than the sepals, cutaneous, delicate rose, 2–3 cm long, weakly toothed tip
Fl sessile, 5·5 cm long, red or white
Se 2 cm long, flared for 9 mm, brown-red, weakly keeled, cutaneous, bare
Pe 4 cm long, united, the tips rounded and hooded
Ov 1·3 cm long, white, bare
Ha Brazil.

VARIETIES
var. *purpureum*
Syn. *Karatas purpurea* Antoine, 1884
The base of the flowers white, rose pink for the upper third.
var. *albiflorum* L. B. Smith, 1939
Plain white flowers.

Nidularium rutilans Modden, 1885 (Plate 114)
(*rutilans* yellowish-red, red; gleaming)
Syn. *Karatas rutilans* Baker, 1889

Pl stemless
Fo several, in a flattish, large rosette, 50–70 cm in diameter
LS 7–8 cm wide, 10–15 cm long, both sides brown scaled

LB lineal-strap shaped, to 30 cm long, 5–7 cm wide, the tip rounded and terminating with a short spine (Fig. 18, 3) shiny green with dark green blotches, the margins remotely spined

Sc very short, to 8 cm long, not exceeding the rosette
I nestling
PB similar to the leaves but smaller, 8 cm long, 5–6 cm wide at the base, red, scattered brown scaled
Sp densely 5 to 6 flowered
FB thin-cutaneous, keeled, white with red tip, 2·3 cm long, 1·5 cm wide, broad-oval, pointed, entire
Se 1·8 cm long, fused at the base for 2 mm of its height, red towards the tip
Pe 3·8 cm long, about 3 cm fused, with white tube and red tips
Ov 1 cm long, white
Ha Brazil.

Nidularium scheremetiewii Regel, 1857 (Plate 115)

(*scheremetiewii* after the collector Scheremetiew)
Syn. *Karatas scheremetiewii* Antoine, 1884; *Nidularium corcovadense* Ule, 1900

Pl stemless
Fo forming a rosette from 50–60 in diameter
LS 10–11 cm long, 6–7 cm wide, oval, pale brown scaled
LB 30–35 cm long, 2–3 cm wide, upper part of blade, narrowing towards the base, long pointed at the tip, green, the short spined margins to the base often wavy
Sc 7–8 cm long, 5 mm thick, the inflorescence somewhat extended above the leaf blades, white, bare
I compound, dense, somewhat cup-shaped (cyathiform)
SB similar to the leaves
PB much longer than the spikes, densely set and in all forming a cup-shaped involucrum, 10–12 cm long, triangular, long pointed, pale green at the base, brick-red at the tip
Sp mostly 3 flowered, short stalked (2–3 mm) nearly crosswise on the axel
FB thin-cutaneous, 2–2·2 cm long, triangular-oval, pointed, entire
Fl to 5·5 cm long, short stalked
Se 1·5 cm long, fused at the base for 5 mm of its height, white, green towards the tip
Pe 4·5 cm long, fused to form a white tube 3·5 cm long, tip upright, blunt, dark blue with lighter edge
Ov white, 8–10 cm long
Ha Brazil
CU widely cultivated.

Nidularium seidelii L. B. Smith, 1963 (Plate 116)

(*seidelii* named after A. Seidel, a bromeliad grower and collector)

An extraordinary *Nidularium* when in bloom; found by A. Seidel in boggy, swampy ground near the sea in the Brazilian State of São Paulo.

It forms a large funnel-shaped rosette from the centre of which, at flowering time, arises an unusual form of inflorescence for this genus. It is up to 30 cm high, the scape set with leaf-like green scape bracts, above which there forms several tiers of large thin, boat-shaped, citron-yellow primary bracts. Similarly, the flower-bracts and the petals are also citron-yellow in colour.

In cultivation, the plant is easy, increasing by means of offsets.

Nidularium terminale (Vell.) Ule, 1898 (Plate 117)

(*terminale* boundary, limit, doubtless on account of the lengthened scape)
Syn. *Tillandsia terminalis* Vell, 1827; *Hohenbergia terminalis* Beer, 1857

Pl stemless
Fo several, in a dense rosette
LB lineal to wide sword-shaped, to 1 m long, 5–6 cm wide, upperside rich green, underside lighter and frequently blotched, short tip, the margins densely edged with small, brown teeth
I many flowered, simple like *N. billbergioides*, raised on a scape to 0·5 m long
SB forming an involucre; pale green at the base, rose to purple-red towards the tip, the margins densely hooked toothed
FB oval-lanceolate, toothed towards the tip, as long as the sepals
Fl to 6 cm long, pale blue
Se to 2·3 cm long, about 8 mm fused at the base
Pe to 4 cm long
Ha Brazil (on Mount Tiju, 600–800 m in the State of Rio de Janeiro).

Ochagavia Philippi, 1856

(Syn. *Rhodostachys* Philippi, 1857–58; *Ruckie* Regel, 1868)
Plants stemless or short stemmed; many leaved, narrow-lineal, margins strongly spined. Inflorescence simple-spiked, glomerate or short cylindrical, nestling, raised above the rosette leaves; bracts narrow-lanceolate. Flowers red or yellow; ovary inferior; berry fruit.

There are 5 species in the genus, 4 in Chile and 1 in Juan Fernandez. They are seldom seen in cultivation in Europe.

Ochagavia lindleyana (Lem.) Mez, 1853 (Plate 97)

(*lindleyana* after the English botanist Lindley, 1799–1865)
Syn. *Bromelia lindleyana* Lem., 1853; *Rhodostachys carnea* Mez, 1896; *Bromelia carnea* Beer, 1857; *Rhodostachys andina* Phil., 1857–58; *Bromelia longifolia* Lindl., 1851–52; *Ruckia ellemeti* Regel, 1868; *Hechtia elemeti* hort., 1866; *Ruckia allemeetiana* hort., 1874; *Pourretia argentea* hort., 1874

Pl resembling a *Hechtia* or *Fascicularia*
Fo in a dense, flat outspread rosette
LB lineal, to 50 cm long, 2·5 cm wide above the sheath, narrowing to a sharp point; the upper half of the robust spine edged margins are channelled, the underside dense white to grey scaled
Sc very short, raised above the leaves
PB the upper ones similar to the leaves but smaller and with shorter blade, red with a whiter sheath
I glomerate to globose-cylindrical sunk in the rosette, many flowered
FB the outer ones ovate, the margins strongly toothed, longer than the flowers, the inner ones as long as the

sepals
Se to 1·2 cm long, fused to form a tube with short free tips, remaining densely woolly

Pe to 2 cm long, rose coloured
Ha Chile (dry mountain slopes)
CU can be grown in company with cactus, and requiring dry conditions.

Orthophytum Beer, 1854

(Syn. *Prantleia* Mez, 1891; *Sincoraea* Ule, 1908; *Cryptanthopsis* Ule, 1908)
Called the 'straight (ortho) plant' (phytum), this genus is distinguished from the rest of the bromeliads, for the most part, by the long inflorescence which carry normal leaves, reducing in size towards the top. However, there are other species, for example *O. navioides* and *O. saxicola*, which have their inflorescences sitting on the rosette or nestling within it.

The inflorescences are glomerate-spikes of compound flower clusters; at the base of the inflorescence these are widely spaced, crowding towards the tip with leaf-like bracts in the axels.

The few species which make up the genus are found in the dry areas of eastern Brazil; they inhabit rocky ground or are terrestial and are exposed to intense sun and can take extreme dryness. They can be grown with cactus, and in winter should be kept dry and cool.

Orthophytum foliosum L. B. Smith, 1941 (Ill. 234 and 235)
(*foliosum(us)* foliated; because of the be-foliated inflorescence tiers)

Pl with inflorescence 50–60 cm high, forming stolons
Fo loose, reducing in size towards the top
LB narrow-triangular, long pointed, 50–60 cm long, 4–5 cm wide at the base, dull green, underside grey scaled and lengthwise ribbed; margins curled set with brown, hooked, sharp spines
Sc about 30 cm long and 1·5 cm thick, upright, round, green, grey felted scaled
I 10–15 cm long, glomerate-compound flower clusters arranged at the top, those at the base of the inflorescence further apart, becoming closer nearer to the top
SB similar to the leaves with longer blade
PB similar to the inflorescence bracts, the basal ones with longer, leaf like blade and nearly triangular base
Sp short, sessile, 3 to 5 flowered, dense, nearly upright, 2–3 cm long, glomerate
FB triangular, long pointed, 2 cm long, outspread 1·5 cm wide, keeled, green, the margins spined
Se 1·6 cm long, pointed, margins prickled, keeled, green, nerved
Pe white, 1·4 cm long, 3 mm wide, pointed, upright
St and **Pi** enclosed within the flower.

Orthophytum navioides (L. B. Smith) L. B. Smith, 1955 (Plate 118)
(*navioides* because of the similarity with Navia)
Syn. *Cryptanthopsis navioides* L. B. Smith, 1940

Pl stemless, with long stolons

Fo several, in a dense outspread rosette

LS small, not clearly defined, the margins small and densely toothed

LB narrow-lanceolate, to 8 mm wide, weakly lepidote, the margins with delicate 1 mm long spines

Sc absent

I compound glomerate, shaped, few flowered surrounded by the inner rosette leaves

FB narrow-triangular, toothed, the inner ones shorter than the sepals

Fl sessile

Se free, upright, symmetrical, narrow-triangular, pointed, about 3 cm long

Pe white, half ligulate

Ha Brazil (Bahia) on rocky forested slopes, about 500 m high

CU more tender than all other *Orthophytum*; it is an extreme xeròphite.

Orthophytum rubrum L. B. Smith, 1955 (Plate 119)
(*rubrum* because of the red flower bracts of the spikes)

Pl similar in growth to *O. foliosum*, 50–60 cm high

Fo several

LS 2–3 cm long, pale brown scaled, becoming bare and shiny later

LB lineal-triangular, long pointed, to 55 cm long and 2 cm wide, the margins loosely toothed, when young densely scaled with white appressed scales, when older upperside bare and shiny green

Sc lengthened

I tongue-shaped, few compound spikes

SB similar to the foliage leaves, spreading

PB like the foliage leaves, about twice as long as the spikes, spreading

Sp about 4 cm long, 2·5 cm thick, ellipsoid, dense and many flowered

FB spreading, wide-oval, pointed, 2 cm long, toothed, bare, bright rose

Fl white, about 1·5 cm long

Se triangular-pointed, 1·2 cm long, the back ones widely winged, keeled

Pe to 1·5 cm long scaled at the base

St and Pi enclosed

Ha Table mountain near Maracas, in the State of Bahia in southern Brazil

CU because of the outstanding rose colouring of the flower bracts, this is the best of the *Orthophytums*.

Orthophytum saxicola (Ule) L. B. Smith, 1955 (Fig. 67)
(*saxicola* rock inhabiting)
Syn. *Cryptanthopsis saxicola* Ule, 1908

Pl stoloniferous, growing into clumps

Fo several, a small rosette from 10–15 cm in diameter

LB 3–6 cm long, 1·5 cm wide, triangular, pointed, grey scaled at the base, bare towards the tip, green to brownish-green; margins with 2–3 mm long, hooked spines

Sc very short, 1–2 cm long, with the inflorescence concealed in the rosette

I simple, few flowered

SB similar to the leaves

FB similar to the leaves but shorter, about 2–3 cm long, exceeding the flowers

Fl sessile, 1·8–2 cm long

Se 1·4 cm long, with brown spined tip, cutaneous, green with whitish edges

Pe white, 1·3–1·4 cm long, clawed, with, at flowering time, 2 mm wide plate, with 2 ligules at the base

St somewhat extended beyond the flower

Ov 4–5 mm long

Ha Brazil, among the rocks.

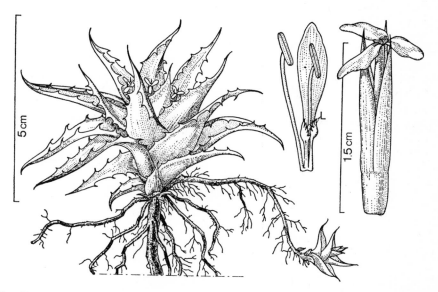

Fig. 67 *Orthophytum saxicola* (Ule) L. B. Smith: 1 habit; 2 flower; 3 petal with scales (L) and stamens.

Portea Brongn., 1856

The genus is named after a French plant collector Dr M. Porte; it contains 5 species, all from Brazil

They are rosette plants with long-lineal, stiff, spined leaves with the undersides scaled.

They are amply branched, dense cylindrical or with very loose inflorescence of which the flowers are generally long stalked, the fruit juiceless berries. Cultivate as for Aechmea.

Portea kermesina C. Koch, 1856 (Plate 120)

(*kermesina(us)* carmine-red; refers to the red inflorescence and primary bracts)

Pl with inflorescence to 70 cm high

Fo several, forming a wide, loose, to 1 m in diameter large rosette

LS wide-oval, 12–16 cm long, 8–11 cm wide, green, upperside lilac blotched

LB tongue shaped—lineal, to 60 cm long and to 6 cm wide, rounded, spine-tipped, green to reddish-lilac, the margins densely set with 2 mm long, lilac coloured spines

Sc upright, much shorter than the leaves, 30–35 cm long, 1 cm thick, round, pale green

I compound, nearly cylindrical, not surpassing the leaves, about 20 cm long; 4 cm in diameter

SB densely imbricated, upright, the scape fully enveloped, with short blade, about 5–6 cm long, red

PB red, upright, densely imbricated (overlaping), spirally arranged, nearly covering the short racemes, the lower about 7 cm long, broadly rounded and short spined tipped

Racemes upright, spirally arranged, loose 4 to 5 flowered, 7–8 cm long, 3 cm wide, with rose-coloured, weakly keeled axel

FB rose, bare to 4 cm long, about as long as the sepal

Fl stalked, 4–5 cm long

Se asymmetric, 1·7 cm long, cutaneous, whitish, fused for up to $\frac{3}{4}$ of their length, with 4 mm long rose-coloured spine tip

Pe 4 cm long, blunt, violet with white tip

Ov about 1 cm long; whitish-rose, staying somewhat white felted

Ha Brazil.

Portea leptantha Harms, 1929 (Plate 121, Ill. 236)

(*leptantha(us)* thin, delicately flowered; on account of the thin flower stalks)

Closely allied to *P. petropolitana*, the difference being the orange coloured flowers and the more ample branching arrangement of the inflorescence

Pl stemless, attaining 1·70 m high with inflorescence

Fo a large rosette with a diameter of 1·60 m

LS broad-oval, about 20 cm long, 14–17 cm wide, both sides brown scaled

LB strap shaped, broadly rounded and with a bent spine at the tip, about 1 m to 1·20 m long, 6–9 cm wide above the sheath, green to reddish-brown, appressed grey scaled; margins brown spined

Sc upright to curving, about 90 cm long, 1·3 cm thick, round, green to pale rose, bare

I 60 cm long, 70 cm wide at the base, pyramidal, loosely compound, racemous-panicled branched, with rounded bare, reddish-brown axels

SB longer than the stem segments (internodes) densely covering the scape, with shorter blade tip, green to pale reddish-brown

PB shorter than the basal sterile section of the branch arrangement of the first order; those on the branch arrangement of the second or higher order scale shaped, insignificant.* Inflorescence branches partially sterile.

FB narrow-triangular, long pointed, green, very small, more or less 5 mm long, bare

Fl 7–9 mm long stalk, to 4·5 cm long

Se 1·4–1·5 cm long, widening out from the base 5 mm wide, asymmetrical, rounded tip with lateral placed green spine tip, bare, smooth, orange, not keeled

Pe 3·2 cm long, 4 mm wide, tongue shaped, rounded at the tip, short spined, with 2 short scoop shaped scales at the base, orange coloured

St and **Pi** enclosed in the flower

Ov 7 mm long, green, bare

Ha Brazil.

Portea petropolitana (Wawra) Mez, 1892 (Ill. 237)

(*petropolitana* named after the Brazilian State of Petropolis)

Syn. *Aechmea petropolitana* Wawra, 1880; *Portea gardneri* Baker, 1889; *Streptocalyx podantha* Baker, 1889.

The length of the branches and the flower stalks are variable. This is a very lovely plant when in flower but it is large; the American plant collector and breeder M. B. Foster introduced from the State of Espirito Santo in Brazil var. *extensa*, which is smaller in all its parts.

Pl stemless, with inflorescence over 1 m high

Fo several, in a dense, tall, to 80 cm in diameter funnel shaped rosette

LS large, oval to round, 15–18 cm long, 10–13 cm broad, both sides dark brown

LB narrow-lineal, over 1 m long, 3–4 cm wide, tip rounded with terminating spine; margins curved and edged closely with brown to 6 mm long spines—especially towards the base

Sc about 70 cm long, round, sturdy, 1 cm thick, upright, yellowish-green, bare

I upright, to 40 cm long, 15 cm wide, loose, multiple compound raceme with light rose coloured axel

SB dense, imbricate (overlapping), lying against the scape, longer than the internodes, upright, green, bare

PB at the base of the inflorescence similar to the scape bracts, to 10 cm long, becoming smaller towards the tip (to 1 cm), lilac-rose, upperside sometimes somewhat white and woolly

FB scale shaped

Fl 1·5–4 cm long stalked

Se to 1·5 cm long, lilac, spine tipped

Pe 3 cm long, 5 mm wide, blue-violet with darker, small tip

Ov 1 cm long, green, at first wax covered

Ha Brazil.

Pseudananas Hassler ex Harms, 1930

Only one species makes up this genus. Like the true *Ananas*, it has a fleshy, berry shaped, edible fruit (syncarpium), but differs from the true *Ananas* in that it does not have the terminal leaf tuft.

* See page 43 for an explanation of orders.

Pseudananas macrodontes (E. Morren) Harms, 1930 (Plate 85, Fig. 68)

(*macrodontes* great resemblance)

Syn. *Pseudananas sagenarius* Camargo, *Ananas macrodontes* E. Morren, 1878; *Ananas microcephalus* Bertoni, 1919; *Bromelia macrodosa* hort., 1878; *Bromelia undulata* hort., 1878

Pl stemless, stoloniferous (in contrast to the true *Ananas*)

Fo several, longer than 1 m, 6–7 cm wide, the margins set with 3·5 cm long spines getting smaller towards the tip; underside grey and scaled

Sc short and thick, white woollen flocked

I dense circular, to 17 cm long and 9 mm thick, without leaf-tuft at the crown of the fruit

SB green, red at the base with reflexed

Fig. 68 *Pseudananas macrodontes* (E. Morren) Harms. Young, fruiting inflorescence.

tip
FB rose coloured, cutaneous, oval-triangular pointed, toothed margins
Fl to 5 cm long, rose-violet; brown coloured after flowering
Fr close to *Ananas*; juicy, sweet smelling, edible, berry shaped

Ha Brazil, Paraguay
CU this plant takes a long time to produce flowers and fruit; this can be lessened by removing the stoloniferous young plants, which are absent with the true *Ananas*.

Quesnelia Gaud., 1842

Comprises about 12 species; they are exclusive to eastern Brazil. The genus is named after a French consul, Quesnel.

They form a more or less large rosette with very stiff, spined edged, pointed or rounded, and spine tipped leaves. The upright or pendant inflorescence are simple-spiked. The petals are free and ligulate; the filaments of the outer stems are free, the inner ones more or less united with the petals for much of their height. Inferior ovary, developing into a juicy berry. Although not very decorative, they make good houseplants.

Quesnelia humilis Mez, 1892 (Plate 122)
(*humilis* low, dwarf)
Syn. *Quesnelia hoehnei* L. B. Smith, 1931

Pl to 20–35 cm high
Fo 10–12, forming a tube shaped, upright rosette
LS 10–13 cm long, 4–5 cm wide, upperside lilac, underside brown-lilac
LB broad-lineal, to 20 cm long, 3 cm wide, dull green, the tip broadly rounded with terminating spine; margins weakly spined
Sc upright to curved over, 4 mm thick, to 30 cm long, white woolly when young
I simple-spiked, to 15 flowered, 7–8 cm long, 5 cm wide, with white woolly axis
SB widely spaced, the basal ones shorter than the internodes; lineal-lanceolate, thin, transitory; red when young
FB narrow-lanceolate at the upper part of the inflorescence, much shorter than the sepals, thin, red, insignificant, hardly scaled
Fl all but upright, sessile, to 6 cm long
Se narrow-lanceolate, blunt, thin, red; bare, weakly white woolly at the tip, to 2·8 cm long and 5–6 mm wide
Pe to 4·3 cm long, lilac-violet, with flared tip
Ov 1·7 cm long, yellowish-red, strongly ribbed
Ha Brazil.

Quesnelia lateralis Wawra, 1880 (Ill. 238)
(*lateralis* lateral; refers to the frequent laterally extending inflorescences)
Syn. *Q. centralis* Wawra, 1880; *Billbergia enderi* Regel, 1886; *Q. enderi*

Gravis et Wittm., 1888

A remarkable bromeliad because the position of the inflorescence differs from the norm in that it arises from a lateral position (Ill. 238 Jl). This however is not its only position; there is also a terminal inflorescence which arises from a permanently short side short without enclosing leaves (Ill. 238 Jt).

Pl with inflorescence 40–50 cm high

Fo 8–12 in a loose, upright rosette from 40–50 cm in diameter

LS 12–14 cm long, 6–7 cm wide; with yellow margins, long-oval, weakly brown scaled

LB strap shape, broadly rounded with terminating spine or short pointed, about 50 cm long, 4–5 cm wide, both sides grey scaled; margins mostly flat and brown spined

Sc to 20 cm long, 4 mm in diameter, upright or rising and curving to pendant, round, green, white woolly, occupying a central (Ill. 235 Jt) or lateral position (Ill. 235 Jl)

I simple, dense, short cylindrical, to 10 cm long, 2–2·5 cm wide

SB always longer than the internodes, upright, covering and lying against the narrow scape, rose to pale rose, grey scaled; margins with few teeth

FB broad-oval, short to long pointed, 2·8 cm long, thin, red, grey scaled at the tip

Fl 3·5 cm long

Se 1·1 cm long, bare, white, rose at the tip

Pe 2·5 cm long, 5 mm wide; blue, with one strongly fringed scale at the base

Ov bare, white, 1 cm long

Ha Brazil.

Quesnelia liboniana (De Jonghe) Mez, 1922 (Ill. 239)
Syn. *Billbergia liboniana* De Jonghe, 1851

Pl stemless, 70 cm high with inflorescence

Fo very few, forming a tube like rosette

LS elliptical, to 10 cm long and to 7 cm broad

LB to 80 cm long and 4 cm wide, lineal, pointed or rounded with terminating spine tip, green, upperside scattered with underside thickly covered with grey scales; margins with widely spaced small spines

Sc upright, 30–40 cm long, 3 mm thick, round, green, bare

I simple, few flowered, upright, to 10 cm long, loose; rachis bright red, bare

SB upright, always longer than the internodes, lineal, to 4·5 cm long, 5 mm wide, thin

FB always near to the position of the inflorescence axis, 0·5–1·5 cm long, long to short-triangular, bare, red, fleeting

Fl arranged spirally, upright to spreading, to 5·5 cm long

Se long-elliptical, blunt, thin, red, bare, 2·3 cm long, 8 mm wide, flared tip

Pe to 5 cm long, blue

Ha Brazil.

Quesnelia marmorata (Lem.) R. W. Read, 1965 (Plate 123)

(*marmorata(us)* marbled; named in respect of the blotched leaves)
Syn. *Billbergia marmorata* Lem., 1855; *B. vittata* sensu Baker non Brongn., 1889; *Quesnelia effusa* Lindm., 1891; *Billbergia speciosa* sensu Wittm., 1891; *Aechmea marmorata* (Lem.) Mez, 1892

Pl stemless, 50–60 cm high with inflorescence

Fo very few, in an upright, tube like rosette

LS not clearly distiguished, 10–13 cm long, 8–9 cm wide, upperside lilac, grey scaled, underside green marbled

LB strap shaped, broadly rounded at the tips with spine, 30–40 cm long, 6–7 cm wide, green, grey scaled, green and lilac marbled, the margins edged with small lilac coloured spines

Sc 30–35 cm long, 7–8 cm thick, upright, bare, round, pale green

I upright, loose compound, more or less 20 cm long, 10 cm wide, red axel, bare, grey 'frosted'

SB cutaneous, shorter than the internodes, 5–6 cm long, about 3 cm wide, rounded at the tip but with a short point, red, grey scaled, nerved

PB as long as the spikes, thin, red, grey scaled

Sp outspread, 3 to 4 flowered, loose, 3–4 cm long, with angular, red, grey waxy rachis

FB absent

Fl sessile, 3 cm long

Se to 1 cm long, short fused at the base, pale greenish blue, asymmetrical, 5 mm wide

Pe violet-blue, 2·2 cm long, rounded tip, 5 mm wide

Ov green, bare, waxy, cylindrical, 5 mm long

Ha Brazil

CU the leaves are very decorative.

Quesnelia quesneliana (Brongn.) L. B. Smith, 1952 (Plate 124, Ill. 240)

(*quesneliana* after Quesnel)
Syn. *Billbergia quesneliana* Brongn., 1841; *Quesnelia rufa* Gaud., 1842; *Q. cayennensis* Baker, 1889; *Q. skinneri* E. Morren, 1930

I with inflorescence about 70–90 cm tall

Fo several, in a dense, broad rosette about 1 m in diameter

LS long-elliptical, to 20 cm long, and to 10 cm broad, upperside lilac, smooth margined

LB tongue-shaped, lanceolate pointed with spiked tip, to 70 cm long, 5–6 cm wide, green, underside much crossbanded and more or less grey scaled; margins densely edged with short brown spines

Sc upright, about 50 cm long, 1·5 cm thick, round, white woolly

I simple, upright, densely strobilate (cone shaped), to 15 cm long, to 5 cm in diameter

SB dense, spirally arranged and completely covering the scape, pointed, prickled, upright, grey scaled, pale-brown-lilac-rose; margins only towards the tip somewhat spined

FB red, dense, imbricate, spirally arranged, upright, strap shaped, blunt, 4·4 cm long, 1·4 cm wide; margins cutaneous, wavy, white scaled, white woolly at the base, completely concealing the flowers

Se 1·5 cm long, 6 mm wide, asymmetrical, cutaneous, partly pale rose; weakly white woolly

Pe 3 cm long, 6 mm wide, tongue-shaped, blunt, white with dark lilac tip, two much cut scales at the base

Ov flattened, white woolly, 1·5 cm long

Ha Brazil

CU a gorgeous plant when in bloom but somewhat fleeting. Closely related to the following species:

Quesnelia rosea-marginata (C. Koch) Carr., 1880

Syn. *Billbergia rosea-marginata* C. Koch, 1864; *B. rubro-marginata* hort. ex Carr., 1880; *Quesnelia rufa* Baker, 1889; *Billbergia skinneri* hort. ex Morr., 1881

This species is distinguished from *Q. quesneliana* by the flat, outspread, flower bracts, the margins of which are not wavy.

From this is also distinguished *Quesnelia testudo* Lindm., 1811, because of its wide blunt flower bracts which in *Q. rosea-marginata* are rounded.

Ronnbergia E. Morren and André, 1874

Short, rosette-shaped plants increasing by stolons. Leaf narrower to a stalk or petiole above the leaf sheath. Margins of leaf blades smooth or toothed. Inflorescence simple-spiked, dense to loose. Flowers blue, sessile. Petals without scoop-shaped scales at base. Inferior ovary

Ronnbergia morreniana Linden and André, 1874 (Fig. 69)

(*morreniana* named after the bromeliad researcher E. Morren)

Pl stoloniferous, to 30 cm high

Fo 20 cm long narrowed stalk or petiole above the leaf sheath

LB oval-lanceolate, short pointed, smooth leaf margins, green, dark green blotched or banded

Sc thin, half upright, short, loosely set with scape bracts

I shorter than the leaves, short cylindrical, to 4 cm long and 5 cm broad

FB scale shaped, very nearly 4 mm long

Fl to 3·8 cm long, upright to angled, blue

Se to 8 mm long, fused to half-way up, broadly-rounded and short tipped

Pe to 3 cm long, the broad-triangular tips flared

Ov inferior, 6 mm long

Ha Colombia.

Streptocalyx Beer, 1857

The family name 'turned-calyx' (*strepto* turned) is derived from the twisted calyx of the first described species *Streptocalyx poeppigii*.

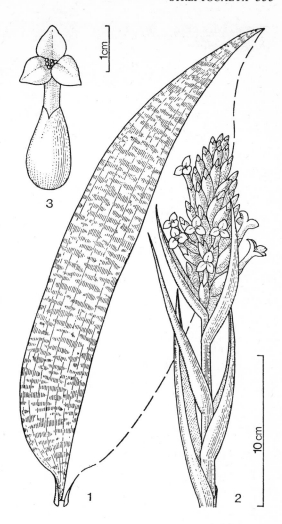

Fig. 69 *Ronnbergia mor-reniana* Linden and André: 1 leaf (left half shown in detail); 2 inflorescence; 3 flower.

The 14 species are epiphytic or terrestial rosette plants with strongly spined leaf margins, imprinted on the leaves, sometimes lying so close to each other as to form a 'pseudo bulb' and frequently infested with ants.

Sessile flowers, blue or white with ligulate petals, in a spike, and the axel large, shell shaped, frequently lively coloured primary bracts.

Long or shortened strobilate inflorescence. Fruit is a juiceless berry. The range of the 14 known species covers Brazil, Guyana to Colombia, to Brazil and eastern Peru.

Streptocalyx angustifolius Mez, 1892 (Ill. 241)

(*angustifolius* narrow leaved)

In Peru, *St. Angustifolius* lives only on old, dead trees and is a typical ant-inhabiting bromeliad. In the large leaf sheaths thousands of biting ants, making the plants quite difficult to collect.

Pl stemless, epiphytic

Fo a great many, narrow, forming a large rosette

LS clearly formed, dark brown; margins not spined, 10–12 cm long, 5–6 cm broad in all forming a pseudobulb

LB lineal, long pointed, green, 30–60 cm long, 2–3 cm wide above the sheath; margins densely spined with 2·5 mm long spines

Sc very short or absent

I nestling, densely strobilate, composed of from 5 rows of compound spikes 9–12 cm long, 5–8 cm broad

PB broadly shell shaped, pointed, enveloping the spikes, delicate rose, brown at the base, as long as the spikes or somewhat shorter, more or less 4 cm long, weakly toothed

Sp nearly upright, 3 to 5 flowered, to 5 cm long, 2·5 cm wide

FB ovate-elliptical; 2·5 cm long; firm, toothed margins, brown, delicate rose towards the tip, grey scaled, sheathing the ovary and the lower part of the sepal

Se 2 cm long, keeled, whitish-rose, long pointed, free, entire

Pe white, 2·5 cm long, 3 mm wide, pointed

St and **Pi** enclosed

Ov 1·5 cm long, white

Ha in the rain-forests from Brazil to north-east Peru

CU an easy plant to cultivate, but requiring plenty of room.

Streptocalyx longifolius (Rudge) Baker, 1899

Syn. *Bromelia longifolia* Rudge, 1805

Very similar to *St. angustifolius* differing from it by the very much longer (to 1·5 m long and hardly 15 mm wide) leaves as well as the serrated margins of the sepals.

Streptocalyx poeppigii Beer, 1857 (Plate 125)

(*poeppigii* named after the German botanist Poeppig, 1798–1868)

Pl stemless, 60–70 cm tall with inflorescence, epiphytic or terrestial

Fo several, forming a flat rosette from 1·50–2 m in diameter

LS large, nearly round, 10–14 cm long, 8–12 cm broad, both sides black-brown

LB lineal, long pointed with robust terminating spine, to 1 m long, 3–4 cm wide, green, becoming reddish-brown in intensive sunlight; margins densely spined; spines of outer leaves 3 mm long, those of the inner leaves only 1 mm long

Sc upright or curving to pendant, robust, round, 15–20 cm long, 2 cm

thick, red, large scaled
I narrow-cylindrical, 30–40 cm long, to 10 cm broad, composed of several spirally arranged compound spikes
SB oval-lanceolate, 9–15 cm long, arranged spirally, standing off the scape, red, grey scaled, toothed margins
PB similar to the scape bracts, oval lanceolate, red, grey scaled, the margin toothed, as long as the spikes and these for the greater part covered
Sp 2 to 8 flowered, sterile near to the tip; rose coloured rachis, large scaled
FB small, kidney shaped (reniform), spine tipped, entire, 5–6 mm long, rose, grey scaled
Fl sessile, loose, bipartite arranged on the rachis, 3–4 cm long
Se 1·9 cm long, free, entire, spine tipped, rose, blue at the tip, scaled to some extent with reniform shaped scales
Pe white with blue tip, 3 cm long, 5 mm wide
Ov cylindrical
Ha rain forests from Brazil to central Peru
CU recommended because of its outstanding decorative inflorescence.

Wittrockia Lindm,m 1891

Named after Wittrock (1839–1914), a Swedish botanist. Stemless rosette plants with lineal, bare, spine margined leaves. Compound inflorescence, glomerate headed and nestling in the rosette. These plants are similar to *Nidularium*, but in *Wittrockia* the coloured 'heart leaves' are absent. There are 6 species, confined to Brazil.

Wittrockia amazonica (Baker) L. B. Smith, 1952

(*amazonica* because of its range—Amazonas)
Syn. *Karatas amazonica* Baker, 1886; *Nidularium amazonicum* Lindm., 1890; *Canistrum amazonicum* Mez, 1891
This species is very similar to *W. superba* (Ill. 242) which differs from *W. amazonica* by having green leaf undersides and pointed, flared petals

Pl stemless, similar in habit to Nidularium
Fo a wide-funnelled rosette from about 60 cm in diameter
LS longish-oval, 11–13 cm long, 7–8 cm wide, upperside green, underside brownish-green
LB to 40 cm long, 4 cm wide above the sheath, at the upper third of the leaf length to 5·4 cm wide, swordshaped, pointed, upperside dull dark green, underside shiny brown-violet, margins finely and densely spined
Sc very short, 7–8 cm long, 1 cm thick, round, white, pale brown felted
I nestling, densely headed made up of few flowered, flat compound spikes, 9–10 cm wide
SB similar to the leaves, green
PB narrow, long, almost triangular, pointed, much longer than the spikes and covering them, giving a star like appearance to the inflorescence, reddish-brown, the margins finely spined
FB roundish, thin, cutaneous, white,

brown scaled, about 2 cm long, longer than the ovary
Se 2 cm long, 6 mm flared, free, bare, green, brownish-red tipped
Pe about 3·3 cm long with hood sha-ped rounded, upright tip, green with narrow, white margin, white at the base, with scales
Ov 7 mm long, white, bare
Ha Brazil.

Wittrockia superba Lindm., 1891 (Ill. 242)
(*superba(us)* splendid, superb)
Syn. *Nidularium karatas* sensu Wawra, 1880 non Lem., 1854; *N. wawreanum* Mez, 1891; *Canistrum cruentum* F. Mueller 1893; *C. superbum* Mez, 1894; *Nidularium superbum* Ule, 1907

Pl stemless, about 40–50 cm high
Fo numerous, forming a large flat rosette from 80–100 cm in diameter
LS broad, roundish, 10–12 cm long, 10–14 cm wide, brown scaled at the base, entire margined
LS long, sword shaped, lanceolate pointed; narrowed at the base, more or less 70 cm long, 4–5 cm wide at the base, at the upper third of the leaf length 5–6 cm wide, green, upperside appressed grey scaled, often red coloured at the tip; margins channelled and small brown spined
Sc very short, 4–5 cm long, covered by the leaves, 1·2 cm thick, round, white, bare
I nestling, glomerate, made up of 5 to 10 dense compound strohlate shaped spikes, 8–12 cm wide
SB dense, upright, robust, triangular, long pointed, prickled, the lower ones smaller and spined, the upper ones larger, not spined and reddish-green
PB similar to the upper inflorescence bracts, reddish-green with prickled tip, about as long as the spikes, weakly grey scaled
Sp densely 15 to 25 flowered, short stalked with white, 1·5 cm long, flattened, brown scaled stalk
FB longer than the sepals, more or less 6 cm long, flared 1·6 cm, the outer ones red, faintly grey scaled, keeled, firm with prickled tip; the inner ones smaller.

9 Sub-family Pitcairnioideae

Key to the Genera (after L. B. Smith)

Identification Stage		Characteristic of Genera	Genus
1	a	Flowers always unisexual (frequently 2 sterile flowers, Fig. 77). Habitat Mexico.	*Hechtia*
	b	Flowers always hermaphrodite (only with *Dyckia maritima* and *Dyckia selloa* (flowers yellow or orange) sometimes unisexual).	see **3**
2	a	Plants large, firm, compact, forming countless single cushion rosettes; inflorescence short, with few long-tubed green flowers; Habitat Argentina.	*Abromeitiella*
	b	Plants of different growth, if cushion-forming inflorescence on an elongated scape or flowers not green.	see **3**
3	a	Flowers in compressed heads (capitate) or in compound panicled heads (paniculate).	*Navia*
	b	Inflorescence differently formed.	see **4**
4	a	Bases of filaments of stamens fused to a tube, united with the corolla petals (Fig. 74, 4); flowers yellow or orange, seldom unisexual.	*Dyckia*
	b	Filaments always free, not united to a tube; flowers always hermaphrodite.	see **5**
5	a	Ovary totally superior.	see **7**
	b	Ovary somewhat partly inferior (Fig. 81, 2)	see **6**
6	a	Flowers conspicuous, large, more or less zygomorphic; petals frequently	*Pitcairnia*

with scoop shaped scales at the base.

b Flowers small, radial; petals without scoop shaped scales at the base.   *Brocchinia*

7   a Petals with scoop shaped scales at the base.   *Deuterocohnia*

b Petals without scoop shaped scales at the base.   see **8**

8   a Seeds narrow, long tailed or pointed at each end, better seen in cross or long section through the ovary by means of a good magnifier.   see **10**

b Seeds more or less broad, at least 3 sides winged (Fig. 87, 2, also best seen through a magnifier). Plants in the main very large.   see **9**

9   a Petals wide, much longer than the calyx at blooming all spirally turned together and changing colour. Generally very large, rosette plants flowering to 10 m high.   *Puya*

b Petals narrow, small, not spirally twisted together after flowering.   *Encholirion*

10   a Seeds or seed positioning (through a magnifier) each end only pointed, not tailed. Filaments at anthesis attached at the back (Fig. 70, 2).   *Cottendorfia*

b Seeds long tailed; plants fairly large.   see **11**

11   a Petals brightly coloured, sticking together after flowering, but not twisted (Fig. 70, 1).   *Conellia*

b Petals white, small, free after flowering; plants small and soft.   *Fosterella*

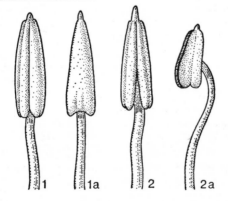

Fig. 70 Stamens of *Fosterella rusbyi* (1–1a) and *Cottendorfia florida* (2–2a), showing different attachments of filaments (front and back views).

Abromeitiella Mez, 1927
Named after the botanist J. Abromeit, the two species forming the genus are found in the high Andes of Argentina and Bolivia. They are interesting terrestial bromeliads, forming very large mats or cushions; very firm, spiny, made up of innumerable small single rosettes (Fig. 7) resulting in a mound or dome (Ill. 371). After flowering, the plant divides at the uppermost leaf axels, and continues to divide thus presenting a very long growth pattern.

Densely arranged leaves, short-triangular, succulent, green or scaled, sharp spiny tip, flared, toothed margins.

Few flowers in a sessile inflorescence, with long, yellowish-green petals. Abromeitiellas are extreme xerophytes, requiring little water in cultivation but a lot of light in order to maintain the compact cushion growth. In winter, the plants can be kept cool. Suitable for growing with cactus.

Abromeitiella brevifolia (Griseb.) Castellanos, 1931 (Ill. 37 and 244)
(*brevifolia(us)* short leaved)
Syn. *Navia brevifolia* Griseb., 1879; *Dyckia grisebachii* Baker, 1889 non *Dyckia brevifolia* Baker, 1871; *Tillandsia chlorantha* Spegazzini, 1899; *Pitcairnia brevifolia* (Griseb.) Fries, 1905; *Lindmania brevifolia* (Griseb.) Hauman, 1917; *Abromeitiella pulvinata* Mez, 1927; *Meziothamnus brevifolius* (Griseb.) Harms, 1929; *Abromeitiella chlorantha* (Spegazzini) Mez, 1935

Pl terrestial and forming large mounds or cushions; densely leaved and much dividing
Fo dense rosette, 2·2 cm long, 7 mm wide, narrow triangular, somewhat succulent about 2·5 mm thick, grey scaled, spined margins
Sc absent
I 1 to 3 flowered terminal from the centre of the rosette
FB oval, to 10 mm long, 6–7 mm wide, cutaneous, pale green, nerved pointed

Fl sessile, to 3·1 cm long, green
Se free, more or less asymmetrical, with small terminating tip, green, bare to 1·3 cm long, 6 mm wide
Pe free, to 3 cm long; 7 mm wide with a 5 mm long, whitish-green scale at the base, intense green
Pi projecting beyond the flower
St as long as the petals
Ov 5 mm long, 3 mm thick
Ha High Andes of south Bolivia and north-west Argentina.

Abromeitiella lorentziana (Mez) Castellanos, 1944 (Ill. 245)
(*lorentziana* named after the German botanist P. Lorentz 1835–1881)
Syn. *Pitcairnia lorentziana* Mez, 1896; *Hepetis lorentziana* (Mez) Mez,

1896; *Abromeitiella abstrusa* Castellanos, 1931
A. lorentziana differs from *A. brevifolia* by being much larger (up to 15 cm long), and with the leaf margins fully spined. *A. brevifolia* is much smaller and the margins are not toothed, except at the base.

Pl terrestial, short stemmed and forming cushions
Fo dense, spiral, flared, triangular with prickled terminating spine, 4–15 cm long, 1–1·5 cm wide at the base, green, grey scaled; margins with and without spines
Sc absent
I 1 to 3 flowered

Fl sessile, to 3·4 cm long, green
Se 1·1–1·6 cm long, 4–6 cm wide, triangular, pointed with spined tip, bare
Pe 2·4–3·2 cm long, 5 mm wide, with a large fringed scale at the base
Pi extending beyond the petals
St as long as the petals, 2·5 cm long
Ha High Andes of north-west Argentina.

Brocchinia Schult. f., 1830

This genus, named after an Italian naturalist G. B. Brocchi (1772–1826) has 5 species in Brazil, Guayana, Colombia and Venezuela.

They are terrestrial rosette plants with large, unspined, flared leaves terminating in a sharp spined tip. The sturdy upright inflorescence scape carries green scape bracts. The inflorescence is loose and branched; flowers are whitish or green, very small, stalked; the petals are shorter than the ovary.

Brocchinia panniculata Schult. f., 1830 (Fig. 71)

(*panniculata* panicled; following the branched inflorescence)
Syn. *Pitcairnia brocchinia* D. Dietr., 1840

Fo numerous, in a dense rosette
LB over 1 m long, 7·5 cm broad, rounded at the tip and terminating with a spine, bare
Sc thick, bare
PB shorter than the sterile section of the branch arrangement of the First Order. See page 43 for explanation of orders.
I loose and much branched (4 to 5 times); inflorescence when young dense brown coarse scaled; lateral branch arrangement of the First Order

(see page 43 for explanation of orders) with long sterile section, the higher with shorter sterile basal leaf arrangement
FB 2·5 cm long, oval, pointed
Fl horizontal to upright, 1·2–1·5 cm long, about 4 mm stalked
Se 2·5 mm long, scale-shaped, underside scarred
Pe longer than the sepals, the edge wavy
Ha Brazil (rain forest).

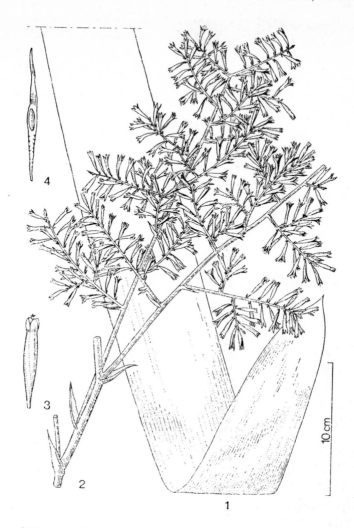

Fig. 71 *Brocchinia panniculata* Schult. f.: 1 leaf; 2 part of the inflorescence; 3 flower; 4 seed.

Conellia N. E. Brown, 1901

The few species which make up the genus are terrestrial and their habitus is similar to that of *Puya*. They are natives of Venezuela and Guayana. Their inflorescence is simple spiked or branched; one or more flowers in the axil of each bract.

Conellia augustae N. E. Br., 1901
(*augustae* after the German Empress Augusta)
Syn. *Encholirion augustae* Schomb., 1846; *Puya augustae* (Schomb.), Mez, 1897; *Caraguata augustae* Benth. et Hook., 1883; *Dyckia augustae* Baker, 1889

Pl short stemmed, to 40 cm high
Fo numerous, in dense rosette, the outer leaves recurved, the inner ones upright
LB to 25 cm long and 1·8 cm broad, up to the middle the margins edged with firm spines
I compound, densely ellipsoid to globose, to 10 cm long and 6 cm wide, indistinctly panicled
PB 2 to 3 flowered in each axil, nearly horizontal, with entire margins; the basal ones broad-oval, spoon-shaped at the base, with short blade, bare
FB lanceolate, to 1·5 cm long, pointed, longer than the leaf stalks
Fl to 2·6 cm long, up to 10 mm long, standing off from the stem, stalks carrying bristles
Se 1·2 cm long, elliptical, rounded at the tip, weakly bristled on the back
Pe 2·4 cm long, broadly rounded, violet
Ha Guayana and Venezuela.

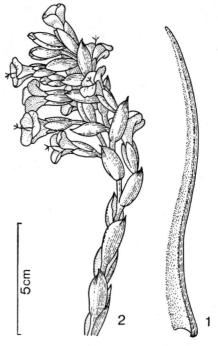

Fig. 72 *Conellia quelchii* (N. E. Brazil) L. B. Smith: 1 leaf; 2 inflorescence.

Conellia quelchii N. E. Br., 1901 (Fig. 72)
Syn. *Puya quelchii* (N. E. Br.) L. B. Smith, 1930
Plant up to 25 cm high; leaves up to 13 cm long; flowers reddish-violet.
Ha Guayana

Conellia carcifolia L. B. Smith, 1951
Ha Venezuela

Cottendorfia Schult. f., 1830
A monotypic Brazilian genus, named after Count Cotta von Cottendorf.
Cottendorfia florida Schult. f., 1830 (Fig. 73).

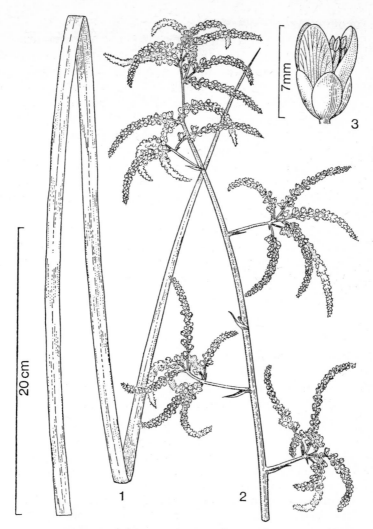

Fig. 73 *Cottendorfia florida* Schult. f.: 1 leaf; 2 inflorescence; 3 flower.

A low growing plant inhabiting the dry region of eastern Brazil. Branched, with leaves to 1 m long, narrow-lineal, grass like, not spined, narrowing to a spine tip, underside white scaled.

Inflorescence loose, branched and panicled, to 40 cm long, with spiked branches to 6·5 cm long.

Flowers small, white with superior ovary. Filaments of the stamens fastened at the back at the anthesis.

Its habitat is on rocks in eastern Brazil.

Deuterocohnia Mez, 1894

The 7 species known so far as forming the genus are terrestrial rosette plants (Ill. 74) with a strongly developed root system and star-shaped rather succulent rosette, leaf margins spined, silver grey scaled or upperside less so. The yellowish or yellowish-green flowers appear on the much branched panicles, those of *D. longipetala* appearing for a year or so with the flowers always being produced afresh.

The species grow on dry, low rainfall western slopes of the Peruvian Andes, where *D. longipetala*, at a height of 600–1,000 m, has formed itself into a stand of pure growth covering the dry steep slopes of rocks in association with cactus. It can also be found in the dry areas of West Chile and Argentina as well as the basin of the Rio Paraguay in Brazil and Paraguay.

The plants are extreme xerophytics; they live in dazzling sun on rocks and with the help of their water storing leaves they can survive the year-long dry periods. In the Mediterranean, the plants can be cultivated outdoors but require some protection in winter.

Deuterocohnia longipetala (Baker) Mez, 1894 (Ill. 246 and 247)
(*longipetala*(*us*) long corolla leaves)
Syn. *Dyckia longipetala* Baker, 1889; *Dyckia decomposita* Baker, 1889; *Puya flava* Willd. ex Baker, 1889; *P. weberi* Schlumn., 1888

Pl group forming, stemless
Fo 10–30 cm long, in flat outspread rosette, terminating in a long pointed tip, the margins set with 3–4 mm long spines, upperside almost bare, with sun-irradiation grey-green or carmine-red, underside silver-grey scaled.
Sc to 5 mm thick, becoming woody to 80 cm long, bare
I ample and loose branched panicled, 50–100 cm long and 30 cm wide, upright or curved over; lateral branch arrangement of the First Order (Ed. refer to page 58 Vol. 1 for explanation of orders), thin, arched, with scale-shaped, brownish leaves at the base, flowering from the same scape axils in the following year

SB the basal ones similar to the leaves, the upper ones blade-less, shorter than the internodes
FB broad-oval, short pointed, scale shaped
Fl to 2·5 cm long, sessile
Se about 8 mm long, oval, obliquely sloping from the tip
Pe lanceolate, yellow, blunt green coloured tip
Ha Western Chile, Argentina, Brazil, Paraguay
CU a very decorative plant at flowering time, which is of very long duration; flowering continues from the same branches so these must not be cut off. Can be grown with cactus.

Deuterocohnia schreiteri Cast. 1935 (Ill. 248 and 249)
(*schreiteri* after the collector Schreiter)

Pl grows into a large clump, 60–80 cm high in flower
Fo numerous, in outspread rosette
LS 3–4 cm broad, white, brownish towards the tip, not sharply defined
LB to 20 cm long, gradually narrowing to a prickled tip, the margins set with firm, brown spines
Sc becoming woody, thin, to 30 cm long, loose set with scape bracts, bare
I loose, 30–40 cm long, with numerous outspread to horizontal compound spikes
SB the basal ones similar to the leaves, margins toothed

PB narrow-lanceolate, long pointed, to 2 cm long, narrowing to the tip; the basal ones with toothed margins
Sp cylindrical-circular, 5–8 cm long, 2 cm thick, many flowered, with sterile flower bracts at the base
FB scale-shaped, long pointed, cutaneous, brownish, to 4 mm long
Fl upright to standing out, to 1·5 cm long, yellow
Se to 6 mm long, oval, greenish at the base
Pe 1·3 cm long, yellow, blunt at the tip, hardly exceeded by the stamens
Ha Argentina

Deuterocohnia strobolifera Mez, 1906
Ha Argentina and Bolivia.

Deuterocohnia chrysantha (Phil.) Mez, 1884
Syn. *Pitcairnia chrysantha* Phil., 1860
Ha Chile (Atacama Desert)
In each of the above species the inflorescence is shortened and more or less strobilate.

Dyckia Schult. f., 1830
Named after the cactus researcher Count Salm-Dyck (1770–1861), this genus contains about 80 terrestrial xerophytic species. They have short, almost tuberous, sometimes creeping stems with strong rooting offshoots, serving the water requirements. The leaves are very stiff, hard, lanceolate, with spine edged margins, terminating in a sharp point, silver white scaled (sometimes only the underside) and form thick, outstretched rosettes — which because of their ability to provide a plentiful supply of young plants, eventually produce a more or less large thicket.

Each species has a very strong inflorescence; raceme, spiked or panicled, with yellow, orange or red coloured flowers.

The frequently clawed petals form plates in the centre region so that the flowers appear triangular from above.

The stamens are, in the main, fused with the petal united tube and are free above this point (Fig. 74, 3–4).

The fruit is a three-compartment, fleshy, brownish, shiny berry with numerous, narrow-winged, keel-shaped seeds (Fig. 74, 5).

The principle ranges of the genus are the dry areas (campus) of Brazil (50 to 60 species), extending to the thorn bush and dry forests of Paraguay, Uruguay, Argentina and Bolivia.

In addition to Dyckia many other terrestrial bromeliads (*Deuterocohnia*, *Hechtia*) are extreme xerophytics, subject to full exposure to sunlight in their habitats and often experiencing year-long

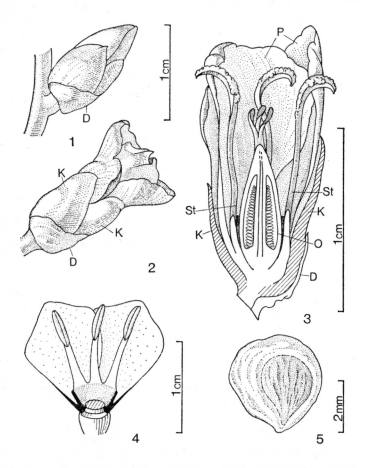

Fig. 74 *Dyckia species* 1 flower bud; 2 open flower, 3 long section through flower; D flower bract; K sepal; P petals; St stamens; O ovary; 4 opened (exploded) flower showing the short filament tube; 5 seed.

drought periods—and with a low temperature of up to 5°C in the winter months. This makes them ideal house-plants and, in general, they remain fairly small (*D. cinera, D. fosteriana* and *D. hebdingii*) are most attractive. They may be grown successfully with many South American cacti.

Many of the species are self-fertile, thus making seed production relatively easy. The remaining plants can be reproduced vegetatively from the numerous offshoots of the young plants.

Dyckia cinerea Mez, 1894 (Ill. 252)
(*cinerea* ash-grey; because of the colour of the leaves)

Pl growing on rocks or in the ground, to 1 m high or higher in flower
Fo dense rosette, broad-triangular, nearly succulent, arched and bent round, about 20–50 cm long, both sides densely ash-grey scaled; margins set with sharp, bent back 2·5 mm long grey spines to the base
Sc upright, 80 cm long or longer, round, 5–7 mm thick
I simple raceme, many flowered, fine grey felted, panicled
SB triangular, long pointed, lying against the scape, shorter than the internodes, thickly grey scaled to the tip
FB the lowest flower bracts shorter than the flowers, but as long as their stalks, recurved at the tip, broad-oval; margins more or less spined, short tipped
Fl 1·7–1·8 cm long
Se to 1·1 cm long, bare or nearly so, short tipped, without spines, margins slightly fringed
Pe to 1·6 cm long clawed, the plate nearly disc-shaped-oval, broad pointed at the tip, hardly keeled
St and **Pi** enclosed in the flower
Ha Brazil
CU this is one of the most charming of the Dyckias.

Dyckia choristaminea Mez, 1919 (Ill. 253)
(*choristaminea(us)* stamens not fused; refers to the free stamens)

Pl growing into a dense clump, 15 cm high with inflorescence
Fo numerous, forming a small rosette, narrow-lineal, to 15 cm long, 5 mm wide above the sheath, pointed tip, prickled, both sides grey scaled, bent towards the tip, set with hooked brown spines
Sc upright, 10 cm long, 2 mm thick, round, green
I simple-spiked, short, few flowered, with white scaled rachis
SB upright, densely set, longer than the internodes, broad-triangular long pointed, 1·7 cm long, 1·1 cm wide, soon drying, thin, nerved, white scaled
FB the upper ones similar to the scape bracts, longer than the sepals and these somewhat covering, about 1·3 cm long, brownish-green, grey scaled, keeled towards the tip
Se 1 cm long, broad-oval to elliptical, pointed, keeled towards the tip, green, soon drying, bare

Pe yellow, clawed with arched out-stretched, 1·7 cm long, somewhat keel-ed plate

St not exceeding the petals, fused only for a short length at the base
Ha dry areas of Brazil.

Dyckia fosteriana L. B. Smith, 1943 (Ill. 254 and 255)
(*fosteriana(us)* named after M. B. Foster)

Pl growing into a dense, cushion forming clump, 50–60 cm high with inflorescence
Fo numerous, in a compact small rosette with a diameter of from 15–20 cm, narrow-lineal, prickled tip, about 9–13 cm long, 8–10 mm wide at the base, both sides dense grey scaled, margins edged with bent prickly spines
Sc upright, about 30 cm long, 5 mm thick, round, green, brownish-green for the upper part, white scaled
I loose, simple-raceme, about 20 cm long, with weak angled dense brown felted scaled axel
SB narrow-triangular, long pointed, upright lying against the scape, longer than the internodes; the margins more or less spined, grey scaled

FB narrow-triangular, the lower ones about 1·5 cm long, the upper 5 mm long, longer or shorter than the sepals, grey scaled with toothed margins
Fl distinct but short stalked, arranged spirally, 1·7 cm long
Se 6 mm long, 4 mm wide, blunt and somewhat hollowed at the base, Orange; margins indistinctly hairy or somewhat spined
Pe 1·2 cm long, clawed, with 7 mm wide, spreading, orange coloured, not keeled plate
St not exceeding the petals; filaments fused at the base, free for 7 mm
Ha Brazil
CU closely related to *D. cinerea* and one of the best of the small *Dyckia.*

Dyckia hebdingii L. B. Smith, 1971 (Ill. 256 and 257)
(*hebdingii* named after R. Hebding, head gardener at the 'Les Cedres' Botanical Garden in France)

Pl stemless, over 1 m high when in bloom
Fo numerous, in a dense rosette, 15 cm high and 20 cm in diameter (Ill. 257)
LS short, to 4 cm broad, 1·5 cm long, shiny, bare, chestnut-brown, white towards the base, sharply set-off from the blade
LB narrow-lanceolate, 15–20 cm long, recurved when old, 1·5 cm long at the base, in a sharp, prickled light-brown flared tip, both sides dense white scaled, margins with thin, hooked-

bent, light brown tipped spines (in intense sunlight the leaves become red-violet coloured)
Sc axillary, to 45 cm long, upright, thin, green, scattered areas of scales otherwise bare
I loose compound, to 40 cm long and to 20 cm wide; axel thin, dark green, scattered with white scales
SB densely set and sub-foliate at the base of the scape, longer about half way, the upper ones shorter than the internodes, narrow-triangular point-

ed, toothed margins, carmine-red, but drying very early

PB narrow-triangular, pointed, to 1·5 cm long, drying early

Sp simple or branched, 15–30 cm long, with 3–5 cm long sterile basal section, open to dense flowered

FB very small, broad-oval, pointed, 3 mm long, brownish-grey-felted hairs, much shorter than the sepals

Fl upright to standing out, sessile to short stalked, to 1 cm long, opening slightly

Se oval-blunt, 4·5 mm long, densely grey felted

Pe to 7 mm long. blunt, yellow, spatula shaped, drying to a grey colour, woolly at the tip

St as long as the petals, fused to a short tube at the base

Pi elongated, at flowering time with its yellow pollen extending about 1 mm beyond the petals

Ha Brazil (Rio Grande do Sul) on rocks.

Dyckia remotiflora Otto and Dietr., 1833 (Fig. 75)

(*remotiflora(us)* remote flowered)

Syn. *D. rariflora* semsu Graham, 1835; *D. rariflora* sensu Lindl., 1836; non Schult. f., 1830; *D. rariflora* var. *remotiflora* Baker, 1889; *D. rariflora* var. *cunnhinghami* Baker, 1889; *D. vaginosa* Mez, 1894

Pl stemless

Fo stiff, outstretched to recurved, in a dense large rosette, 30–40 cm in diameter

LS 1–2 cm long, to 2·5 cm wide, white altogether developing into an onion shape

LB narrow-lineal, long pointed, compact, succulent, to 30 cm long, 1 cm wide at the base, upperside dull green, underside strongly nerved and grey scaled, the margins edged with 1–3 mm long spines

Sc upright, green, bare, somewhat white woolly when young, 50–70 cm long, 4–5 mm thick

I upright, simple-spiked, loose, to 20 cm long with white woolly rachis

SB with long pointed blade, upright, thin, mostly bare, at the upper part of the scape shorter than the internodes and somewhat white woolly

FB broad-oval, spine tipped, 6–7 mm long, straw coloured, white woolly, much shorter than the sepals

Fl 2·2 cm long, almost sessile, spreading

Se to 9 mm long, oval, short tipped, yellowish, weakly keeled

Pe 2·7 cm long, with spreading, 1·2 cm wide, keeled, orange-red plate

St the upper part free

Ha Brazil, Argentina, Uruguay

CU very pretty; a small variable species.

Encholirion Mart., 1830

So called from the Greek *enchos* 'narrow' and *lirion* 'flower', because of the narrow petals.

This genus with some 17 known species occurs in the dry areas of north-

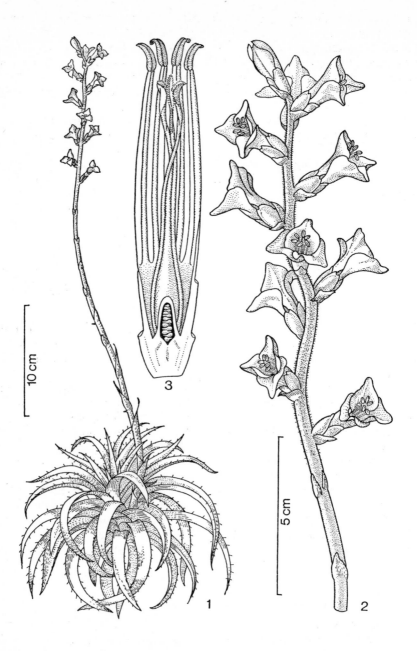

Fig. 75 *Dyckia remotiflora* Otto and Dietr.: 1 flowering plant; 2 inflorescence; 3 longitudinal section through flower.

east Brazil. They are stemless or short stemmed rosette plants. Their lineal leaves, gradually extending to a spine tip, have spined margins, the undersides of the leaves being densely scaled, and forming dense rosettes. The scape is very sturdy. The flowers are yellow or greenish-yellow in a simple raceme, seldom compound. Cultivation in summer is best in full sunlight; in winter, cool and dry.

Encholirion spectabile Mart., 1830 (Fig. 76)
(*spectabilis* conspicuous; because of the conspicuous inflorescence)
Syn. *Dyckia spectabilis* Baker, 1889; *Puya saxatilis* Mart., 1828

Pl stemless, similar in habit to *Dyckia*
Fo a loose very large rosette
LB to 60 cm long, in a long, threadlike terminating tip, the margins set with long to 1 cm—thorns, underside appressed white scaled
Sc very sturdy
I a thick, to 40 cm long, cylindrical raceme with bare axel

FB very small, entire, recurved
Fl 2·3–2·5 cm long, on a 8–10 mm long stalk
Se 7 mm long, oval
Pe 2 cm long, 5 mm wide, yellow
Ha Brazil
CU full sun in summer, cool and dry in winter.

Fosterella L. B. Smith, 1960
Named after M. B. Foster, the American bromeliad grower, this genus of 13 known terrestrial species inhabits the dry forests of South Mexico to Paraguay, Peru and northern Argentina. (Syn. *Lindmania* Mez, 1896.)

Leaves in flat, loose rosette, lying on the ground, with lineal to elliptical, entire or weakly spined, underside scaled, upper side bare blades.

Scape thin, with bracts lying alongside.

Flowers very small, white or greenish, usually pendant facing one way (secund). Filaments of the stamens inserted at the base of the anthesis (Fig. 70, 1).

Of general botanical interest only, though culture is not difficult in half-shade and medium-heavy soil.

Fosterella penduliflora (C. H. Wright) L. B. Smith, 1960 (Ill. 259)
(*penduliflora(us)* pendant flowered)
Syn. *Catopsis penduliflora* C. H. Wright, 1910; *Lindmania penduliflora* (C. H. Wright) Stapf., 1924

Pl stemless, 25–60 cm high with inflorescence

Fo soft, forming a flat outstretched rosette from 30–50 cm in diameter

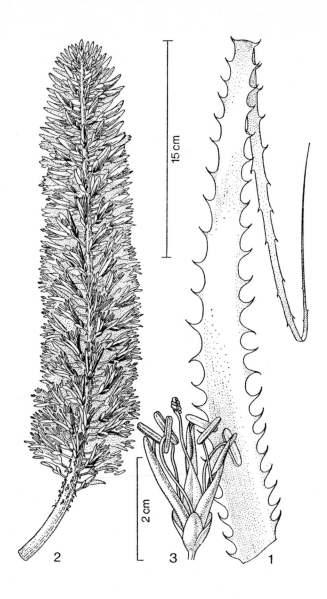

Fig. 76 *Eucholirion spectabile* Mart.: 1 leaf; 2 inflorescence; 3 flower.

LS not clearly identified from the blade
LB lineal-lanceolate, soft, pointed, to 30 cm long and 5 cm wide, narrowing towards the base, dull green, bare, underside sparsely scaled
Sc slender, upright, more or less 25 cm long, 4 mm thick, reddish-brown, round, commencing somewhat white woolly at the base, soon becoming bare
I to 6 to 12 compound loose racemes, to 30 cm long; the basal branches sometimes doubly branched
SB loose, mostly upright, lying against the scape, longer than the internodes, triangular, long pointed, thin, green soon drying
PB small, scale shaped, 7–8 mm long, soon drying
Racemes loose, many flowered, spreading, 15–20 cm long
FB small, insignificant, scale shaped, oval, long pointed, 1–1·5 mm long, green
Fl pendant, secund (facing one way), short stalked, 9–10 mm long
Se lanceolate, blunt, 3·5 mm long, green with white edges
Pe longish-lanceolate, rounded at the tip, 8–9 mm long, 2 mm wide, white
St shorter than the petals
Ha Peru, Bolivia, Argentina.

Fosterella rusbyi (Mez) L. B. Smith, 1960 (Ill. 260)
Syn. *Lindmania rusbyi* Mez, 1901
In South Peru (dry forests of Quillabamba, Province Cuzco) this plant grows at 1,000–1,200 m preferring rock fissures filled with earth where they form large stands. The green cross-banding frequently disappears under cultivation.

Pl short stemmed, 50 cm high when in flower
Fo in loose rosette, mostly upright; hanging down on steep woods
LB about 20 cm long, lineal-lanceolate, threadlike tip, 2·5 cm wide, somewhat narrower above the sheath, but not stalked, pale green, underside dense white scaled, frequently with green cross-banding, margins becoming spined towards the base
Sc upright, slender, bare, green, 20–30 cm long, 2–4 mm thick
I to 30 cm long, loose, double to many times compound racemose, upright, bare
SB entire, upright, longer than the internodes, long threadlike tip, thin, fleeting, white scaled
PB similar to the upper scape bracts, shorter than the lateral branch arrangement of the First Order (see page 43 for explanation of orders) narrow triangular, long pointed, white scaled at the tip, soon drying about 1 cm long
Racemes arching upward, loose 8 to 15 flowered
FB broad-oval, pointed, 1–2 mm long, thin
Fl secund (turned towards the same side), pendant, 3–4 mm long, short stalked
Se oval-elliptical, green, sometimes with brownish tip, 2 mm long
Pe white, 3 mm long, with curled back tips to reveal stamens, pistil and ovary
Ha South Peru and Bolivia

Hechtia Klotsch, 1835

Named after the collector J. G. Hecht, these plants frequently send out from the base short runners which then become plants in their own right.

Leaves are short to long, stiff and succulent in a spiny tapered tip; the margins have stout teeth and both sides or only the undersides are scaled.

The inflorescence scape is very long, set with bracts, the basal ones being similar to the rosette leaves.

Inflorescence is much branched; flowers with short stalks, greenish or greenish-yellow, rarely rose-coloured (*H. rosea*), unisexual or hermaphrodite; one or other, however, is rudimentary.

Infertile stamens exist in the female flowers of *H. rosea* (Fig. 77, 2) or *H. glomerata* (Fig. 78, 3). The ovary is either fully superior—or only half superior (Fig. 77, 3).

The capsule is pointed, deeply furrowed and almost three-divisioned. The seeds are long, narrow and winged on one side (Fig. 78, 4).

The genus contains about 49 species, which are very difficult to categorise.

In contrast to most other bromeliads (where the mother rosette dies after flowering), further inflorescences grow from the rosette even as the side shoots and young plants appear. For that reason the inflorescence appears to come from a lateral position. A consequence of the abundant basal growth of young plants is that a large clump develops and when the old plants in the centre die, there remains what appears to be a 'fairy ring'.

The principle range of the genus extends southwards from Mexico into the southern adjoining states. In the central highlands of Mexico, the plants grow in stands of $4 \, km^2$. These often consist of a single species in the dry, semi-desert-like stony meadows where the only other plants are agaves, cactus of different kinds, tree-like Yuccas, Nolina, Dasylirion and thorn bushes.

Culture is similar to that for *Dyckia*. Hechtias also require long periods of sunshine, dry conditions and medium heavy soil. They are suitable for growing with cactus.

Hechtia argentea Baker, 1881

(*argentea(us)* silvery; refers to the silver-white scales leaves)
Syn. *Dyckia argentea* Nichols, 1886

Pl stemless with short offsets, clump forming
Fo forming a dense rosette, frequently arched and one sided
LB to 30 cm long, 2–3 cm wide, margined with 7 mm long spines, both sides, particularly the underside, thickly white woolly
Sc sturdy, white woolly
I loose branched panicle with white

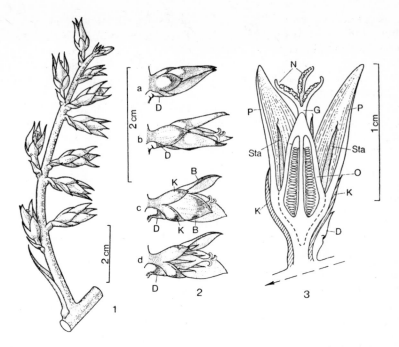

Fig. 77 Branch of inflorescence of a female flower of *Hechtia desmetiana* (Baker) Mez: 2a-d development of a female flower of *Hechtia rosea* Morr.; 3 longitudinal section through female flower; D flower bract; K sepal; P petal; Sta the sterile stamens; O ovary; G pistil; N stigma.

woolly axil
FB broad-oval, pointed
Fl in dense, cylindrical, to 5 cm, long spikes, sessile, about 7 mm long
Se 5 mm long, elliptical, blunt tip,

white woolly
Pe free, elliptical, white, rounded and
Ov densely felted
Ha Central Mexico on stony rock-slopes (near to Tehuacan).

Hechtia epigyna (Plate 126)
(*epigyna*(*us*) with superior ovary)

Pl stemless, 80–90 cm high with inflorescence
Fo a dense rosette from 40–50 cm in diameter
LS indistinct, white, smooth, 3–4 cm long, 4–5 cm wide, finely spined margins

LB 34–45 cm long, 3 cm wide above the sheath, triangular, long pointed, fresh green, both sides grey scaled; margins weakly channelled, thickly set with about 2 mm long, arched, white spines
Sc about 50 cm long, upright, 4 mm

thick, round, bare, green
I 30 cm long, 5 cm wide, 15 to 20 loosely compound racemous inflorescence branches
SB at the base of the scape longer, at the upper regions shorter than the internodes, upright, narrow, 4–5 cm long, the margins of the lower ones spined
PB much shorter than the racemes, 1–1·5 cm long, narrow, soon drying, almost triangular, long pointed, bare
Racemes spreading, 3–4 cm long, 2 cm wide, 15 to 20 flowered
FB 2–3 mm long, much shorter than the flower stalk, lilac, pointed, weakly toothed margins
Fl with stalk 5 mm long
Sc free, 3 mm long, lilac, rounded tip
Pe lilac, 4 mm long, with rounded tip
Ha Mexico.

Hechtia glomerata Zucc., 1840 (Ill. 261, Fig. 78)

(*glomerata*(*us*) gathered together; refers to the clustered flower elements of the inflorescence

Pl stemless, thickly set with offsets at the base
Fo numerous, in a dense outspread rosette
LB to 60 cm long, lineal-lanceolate, narrowing to a sharp spine tip, the margins set with 7 mm long spines, underside silvery scaled
Sc sturdy, laterally placed, the bracts lying closely against the scape
I elongated, with brown scaled axis interrupted by clustered flower elements, these vey dense, to 2 cm in diameter (Fig. 78, 1)
FB oval, narrow pointed, smaller than the sepals
Fl sessile, small, to 7 mm long
Se 4 mm long, broad oval-scale shaped, underside brown scaled
Pe free, oval, rounded tip, white
Ov almost superior, white haired (Fig. 78, 3)
Ha Mexico.

Hechtia marnieri-lapostollei L. B. Smith, 1961 (Ill. 262 and 263)

(named after Julien Marnier-Lapostolle)

Pl small, forming loose mounds
Fo few in a loose rosette
LS about 2 cm long, bright, chestnut-brown towards the tip
LB narrow-triangular, recurved, tapering to a prickled tip, the margins rigidly spined, both sides dense grey scaled, to 13 cm long and 2 cm wide, succulent
Sc bare at the top, flattened, 4 mm thick
I loose-compound with horizontal, 4–5 cm long, short stalked spikes
SB broad-oval, with long-lineal blade, exceeding the internodes
PB similar to the upper inflorescence bracts
Fl the female ones much larger than the male ones, white

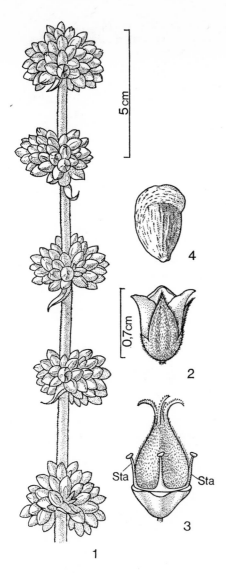

Fig. 78 *Hechtia glomerata* Zucc.: 1 part of the inflorescence of a female plant; 2–3 flower with stamens; 4 seed.

Hechtia montana Brandegee, 1899
(*montana(us)* growing on mountains)

Pl forming large clumps which when old form into rings or lengths of plants **Fo** numerous, in a dense upright or outspread rosette

LB 25–30 cm long, 2–3 cm wide, tapering to a prickled tip, margins spined, upperside green and striped, underside white scaled

Sc 80 cm long, compact
I 40–50 cm long, compound to many flowered, to 8 cm long, racemose inflorescence branches; axil white scaled, bare at fruiting time
SB the basal ones leaf like, 10–15 cm long, the margins toothed
FB 3 cm long, 8 mm wide, long toothed getting smaller towards the tip

Fl 6–8 mm long stalked
Se 3 mm long, triangular pointed
Pe to 6 mm long, cream coloured
Fr short pointed
Ha the mountains at San Jose del Cabo in Baja California, coastal mountains of San Carlos Bay, Guyamas, north west Mexico.

Hechtia rosea Morren, 1889 (Plate 127, Fig. 77, 2)
(*rosea*(*us*) rose coloured)
Syn. *Hechtia roezlii* hort. ex Baker, 1889

Pl stemless
Fo a flat, outstretched rosette
LB 50–60 cm long, 2 cm wide at the base, thick, succulent, lineal, long pointed, upper side greenish brown, both sides grey scaled; margins strongly spined; spines to 4 mm long, bright
Sc about 60 cm long, upright, bare, to 1 cm thick
I to 70 cm long, loose, biprimate to many flowered, spirally arranged, outspread to recurved spikes
SB set apart from each other, shorter than the internodes, upright, prickled, grey scaled
PB similar to the inflorescence bracts,

much shorter than the sterile sections of the spikes, these to 30 cm long, 3 cm wide, many flowered
FB 4 mm long, much shorter than the sepals, green, broad-oval, short pointed
Fl almost sessile, to 1·5 cm long, loose, spirally arranged
Se 5 mm long, pointed, rose coloured, bare
Pe 1·3 cm long, almost triangular, pointed, red
Ov almost superior, 5 mm long
Ha Mexico
CU one of the few rose to red coloured Hechtias.

Hechtia tillandsioides (André) L. B. Smith, 1951 (Fig. 79)
(*tillandsioides* like a tillandsia)
Syn. *Bakerantha tillandsioides* (André) L. B. Smith, 1934; *Bakeria tillandsioides* André, 1889; *Tillandsia glaucophylla* hort. non Baker; *Vriesea glaucophylla* hort. non Hook.; *Hechtia purpusii* Brand., 1920

Pl stemless
Fo numerous, in a dense outstretched rosette
LB to 20 cm long, and 2 cm wide, long, tapering to a threadlike tip, recurving, the margins not spined, both sides densely silver grey scaled
Sc thin, curved, double as long as the

rosette leaves
I to 1 m long, loose-double-racemose, to 15 cm in diameter
SB few, narrow-lanceolate, lying against the scape
PB small, scale shaped, about as long as the sterile basal sections of the raceme, these arched, to 12 cm long,

loose at the base, densely flowered to the tip

FB small, insignificant, almost 2 mm long

Fl to 6 mm long, 5 mm long bare stalk, thin hermaphrodite

Se about 4 mm long, flared, elliptical, red

Ha unknown (Colombia? L. B. Smith gives the locality as being Mexico).

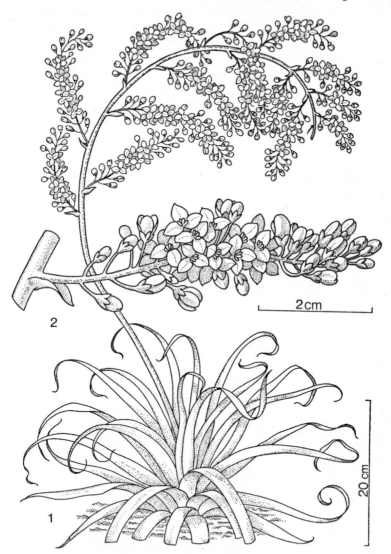

2cm

20 cm

Fig. 79 *Hechtia tillandsioides* (André) L. B. Smith: 1 flowering plant; 2 part of inflorescence.

Navia Schult. f., 1836

Stemless plants lying flat on the ground (Fig. 80, 1), often in cushions or with elongated, thick, bare stems where the leaves have dried and fallen off (Fig. 80, 2).

Leaf blades are lineal-lanceolate, spine tipped, the margins smooth or finely toothed, bare or with the undersides covered with star-shaped hairs. Sessile inflorescence (Fig. 80, 1) or elongated, simple or compound (Fig. 80, 2). Flowers are sessile or stalked, in a simple head (Fig. 80, 1) or in a series of clusters spaced along the scape (Fig. 80, 2). Sepals pointed (Fig. 80, sa, K), the two back ones covering the first.

Petals are unified into a slender tube, with spreading plates, hooded at the tip (Fig. 80, 2a). Capsule, with seeds naked or without wing (Fig. 80, 1a).

The species are found in Colombia, Venezuela, Brazil and Guyana.

Navia acaulis Mart. ex Schult. f., 1830 (Fig. 80, 1–1a)

(*acaulis* stemless)
Syn. *Dyckia acaulis* Baker, 1889

Pl stemless or with very short stem
Fo numerous, in a dense rosette lying on the ground
LB to 12 cm long, 1 cm wide, pointed, margins finely toothed, light green
Sc absent
I globose, nestling, few to many flowered
FB lanceolate-pointed, 5 mm long
Se equal (4–6 mm long) the front ones keeled or winged
Pe longer than the sepals, white
Ha Brazil, Colombia.

Navia caulescens Mart. ex Schult. f., 1830 (Fig. 80, 2–2a)

(*caulescens* forming a stem)
Syn. *Dyckia caulescens* Baker, 1889

Pl with several clearly developed stems or branched, flowering to 3 m high
Fo the young upper leaves of the rosette erect, the old ones recurved
LS not clearly defined, margins smooth
LB 9–12 cm long, 5–12 mm wide, when young with undersides covered with star-shaped scales, the margins densely and finely toothed, pointed tip
Sc clearly developed, but shorter than the leaves
I elongated, like a string of pearls to a series of cluster headed flowers spaced along the scape
SB lanceolate, pointed, mostly as long as the internodes
PB lanceolate, pointed, the lower ones exceeding the flower clusters, these about 1·5 cm in diameter and semi-globose or globose, many flowered
FB broad-oval, to 4 mm long, entire
Fl white, 7 mm long, sessile
Se about 5 mm long
Ha Colombia.

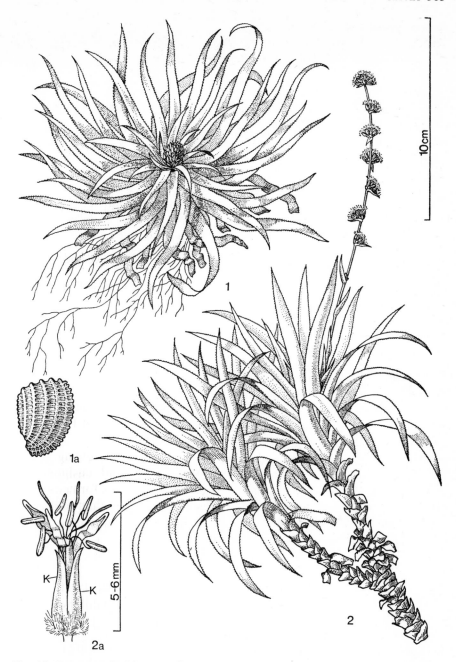

Fig. 80 *Navia acaulis* Mart.: 1 flowering plant; 1a seed; 2 *N. caulescens* Mart. ex Schult; f part of flowering plant; 2a flower.

VARIETY
var. *minor* Schult., 1830
The leaves are very narrow, hardly 5 mm wide.

Navia heliophila L. B. Smith (Ill. 243)
(*heliophila*(*us*) sun-and light-loving)

Pl forming a short stem, the older part of the stem completely set with dead leaf bases
Fo numerous in a dense, outspread, terminal rosette
LS completely covered, not visible
LB lineal, 2 cm wide, terminating in a sharp prickled tip, to 25 cm long, flat, the margins densely saw edged, almost bare
Sc absent
I compound, very dense, globose, 25 mm in diameter
PB lanceolate-triangular, saw-edged, shorter than the globose, short stalked spikes
FB oval, pointed, clearly shorter than the sepals, brown woolly at the base
Fl very short and insignificantly stalked, white
Se lanceolate, pointed, 6 mm long, the back ones sharply keeled and fused for 2 mm of their height
Pe narrow-elliptical with spreading plates
Ha Colombia, on the edge of the Savanna and Catinga bush at 200–300 m.

Pitcairnia L'Hérit., 1788
Named after a London doctor Pitcairn (1711–1791), this genus is second only to *Tillandsia* in having the most species. There are about 180 and, in addition, a very varied selection of growth forms.

There are species with short, almost tuber-like axes, owing to the abundant branching forming a mass of clumps and cushions (*P. heterophylla*); others with 1 m-long underground runners formed by the old leaf sheaths (*P. melanopoda*); species with more than 1 m long, simple or branched stems and climbing epiphytics, ascending the tree by the sympodial (superimposed branches appearing as a single stem) rooting rhizome (*P. scandens*, *P. riparia*).

There is also a multiplicity of leaf shapes. One group of species is heterphyllic (heteromorphic, dimorphic—having leaves of different forms on the one plant). In *P. heterophylla* (Plate 129, Fig. 83), the catophyllis (rudimentary scale leaves) are chlorophyll-less but the leaves are green. In its habitat, the latter frequently drop off at the beginning of the dry season and commence new growth at the start of the rainy season. Flowering occurs without leaves during the dry season.

With many other species, only one sort of leaf is grown and these are superficially alike (homomorphic). These leaves are narrow-lineal,

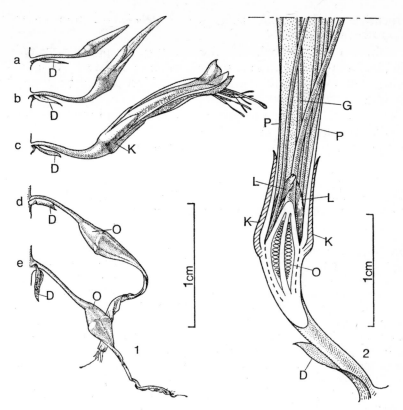

Fig. 81 *Pitcairnia xanthocalyx* Mart.: 1a-e development of flower; 2 longitudinal section through a developed flower; D flower bract; K sepal; P petal; O ovary; G pistil; L ligule.

lanceolate or strap shaped, frequently on a narrower part of the leaf resembling a stalk; the blade margins are spined or smooth.

The inflorescence scapes are either short or long, thin to stout, set with leaf like or membraneous scape bracts.

The very large, but fleeting, white, yellow, greenish or red-flowers comprise a simple raceme, with spike or compound inflorescence; they are slightly zygomorphic (yolk shaped) (Fig. 81, 1c), and then sometimes curled back with petals frequently having ligulae at their bases (Fig. 81, 2 L).

From a half-inferior ovary (Fig. 81, 2) the capsule develops to disperse numerous, small spindle- or keel-shaped seeds.

The genus ranges from southern Mexico, the West Indies to northern

Argentina and Peru; but one species (*P. feliciana*) is found in tropical West Africa.

Culture is only difficult with the leaf droping species; a definite 'rest period' is required, with only the minimum of moisture. As soon as the first leaf shows, plently of water and fertilizer should be given. These plants need a lot of light.

The leaf retaining species require half shade and moderate temperature. In summer plenty of air is beneficial.

Propogation is by division of the rhyzome or from seed.

Pitcairnia andreana Linden, 1873 (Ill. 264)
(named after André, the bromeliad researcher)
Syn. *P. lepidota* Regel, 1873

Pl 20 cm high with inflorescence
Fo homomorphic, few in a loose rosette (leaves all alike)
LB lineal-lanceolate, long pointed, to 35 cm long, 3 cm wide, underside densely appressed grey scaled, upperside green with less scaling
Sc short, 4–6 cm long, often arched ascending, round, green, grey scaled
I upright, simple loose raceme, few flowered, to 10 cm long; rachis grey scaled
SB similar to the leaves, much longer than the internodes
FB narrow-oval, long pointed, the basal ones longer than the 1 cm long flower stalks, green, grey scaled
Fl upright, stalked, arranged around the rachis, with stalk about 6·5 cm long
Se lanceolate, pointed, not keeled, green, weakly scaled
Pe 5·5 cm long, orange at the base, yellow at the tip
St about as long as the petals; anthers 7 mm long
Ha Colombia.

Pitcairnia atrorubens (Beer), Baker, 1881 (Plate 128, Ill. 265)
(*atrorubens* black-reddish; the colour of the flower bracts)
Syn. *Phlomostachys atrorubens* Beer, 1857; *Puya warszewiczii* H. Wendl., 1861; *Pitcairnia lamarcheana* E. Morren, 1889; *Neumannia atrorubens* K. Koch, 1889; *Lamproconus warszewiczii* Lem., 1889; *Pitcairnia atrorubens* var. *lamarcheana* Mez, 1896

Pl 60–90 cm high with inflorescence
Fo heterophyll, the outer ones much reduced, the inner ones forming the foliage
LS triangular-oval, underside brown scaled, upperside bare, white
LB above the sheath, narrowed stalk 20–30 cm long, grey-brown scaled, black spined, channelled, widening to lanceolate, 50–80 cm long, 5–10 cm wide, upperside weakly grey scaled, underside bare
Sc sturdy, upright, 20–30 cm long, 1 cm thick, grey brown woolly

I simple, upright, densely-spiked, 20–30 cm long, 3–5 cm thick, blunt
SB upright, densely overlapping, longer than the internodes, enclosing the scape, the lower ones with longer, greenish-red tip to the blade, the upper ones with reduced blade, dark-red, brown felted
FB broad-oval with triangular, flared blade, 5–7 cm long, fully covering the sepals, dark to light red, seldom yellowish bare, grey to brown felted when young
Fl almost sessile, 7 cm long
Se 2·5 cm long, pale yellow, red edged at the tip, grey felted
Pe lineal, broadly rounded and pointed, of unequal length, to 6·5 cm long, pale yellow, with fringed scales at the base
St shorter than the petals
Ha Costa Rica, Panama, Mexico.

Pitcairnia feliciana (A. Chev.) Harms and Mildbr., 1938 (Fig. 82)
(*feliciana* after Jacques-Felix)
Syn. *Willrusellia feliciana* A. Chev., 1932 (Liliaceae)
This is the only known bromeliad to be found outside the Americas. (N.B. *Rhipsalis cassutha*, another American plant, is also found in Africa and Madagascar.)

Pl with tuberous rhizome, forming large clumps, 50 cm high when in flower
Fo several, heterophyll (dimorphic); *Catophylls* with larger sheath and shorter narrower blade, margins with firm black spines. *Foliage* leaves longish-lanceolate, to 50 cm long, 1·5–1·8 cm wide, long pointed, the margins smooth with isolated teeth, perennial
Sc upright, thin, to 30 cm long
I a simple, 15–20 cm long raceme
SB the basal ones similar to the foliage leaves the upper ones smaller
FB small, scale shaped, lanceolate-pointed, as long or smaller than the flower stalks, 5–7 mm long
Fl upright when young, then horizontal at the anthesis, pendant after flowering 4–4·5 cm long, lively orange-yellow, weakly zygomorphic, 6–10 mm long stalked
Se longish-lanceolate, firm, 15–22 mm long
Pe longish-lanceolate, with rounded tip
Ha Kindia in Guinea, Africa on rocks.

Pitcairnia heterophylla (Lindl.) Beer, 1857 (Plate 129, Fig. 83)
(*heterophylla(us)* different leaves; because of the leaves of different forms on the one plant)
Syn. *Puya heterophylla* Lindl., 1840; *Puya longifolia* C. Morren, 1840; *Hepetis heterophylla* Mez, 1896

Pl with short-tuberous, many-branched rhizome and therefore forming a large group, about 10 cm, occasionally 20 cm high when in flower
Fo heterophyllic, the outer ones, the catophylls, strongly spined, the inner ones (Fig. 83, 1), foliage leaves; catophylls with large semi-circular to oval

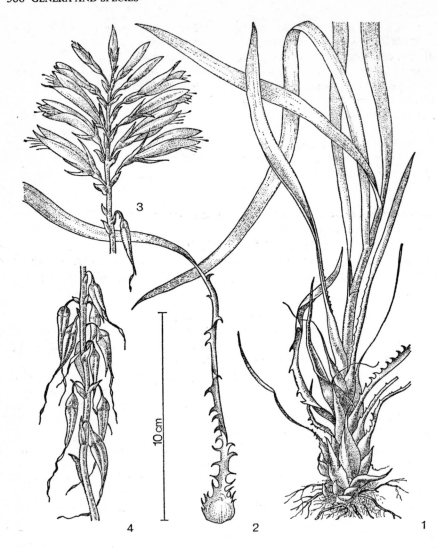

Fig. 82 *Pitcairnia feliciana* (A. Chev.) Harms and Mildbr.: 1 habit; 2 lower part of leaf; 3 part of the inflorescence; 4 fruit formation.

shaped dark chestnut brown sheath and narrow-lanceolate, strongly spined blade (Fig. 83, 2); foliage leaves long-lanceolate, to 70 cm long and 1·3 cm wide, upper side green, underside at first white woolly, later bare, the blade dropping off before flowering (Fig. 83, 1 N); sheath remains with the catophylls

Sc very short or absent

I a loose, very much shortened, 3 to 12 flowered raceme

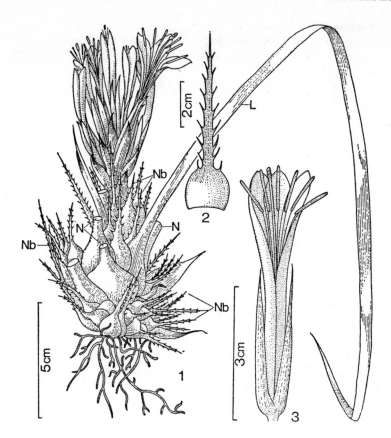

Fig. 83 *Pitcairnia heterophylla* (Lindl.) Beer: 1 flowering plant, foliage leaf (L) is ready to fall, leaving leaf sheath at scar N; NB basal leaves, one shown enlarged (2); 3 flower.

SB where evident, oval, pointed, the basal ones black spined, thin, white woolly

FB similar to the upper scape bracts, entire, shorter than the sepals

Fl upright, 3 mm long stalked, to 6 cm long

Se narrow-lanceolate, pointed, 3 cm long, thin, woolly

Pe lineal, to 5·5 cm long, bright red (seldom white) with a ligule at the base

Ov half-superior

Ha from South Mexico to Panama, Venezuela to North Peru. In Peru, it grows at a height of 1,500–2,500 m on rocky summits or in earthy crevices; it makes a wonderful picture to see the plants in bloom against the cloudless blue winter sky.

CU requires a strict resting period after leaf fall and during flowering and a very bright and sunny position. When the first leaf begins to grow water may be given

Pitcairnia pungens H. B. K., 1816

(*pingens* prickled; in respect of the prickled catophylls)
Syn. *P. laevis* Willd. ex Schult., 1830; *P. concolor* Baker, 1881; *Hepetis pungens* Mez, 1896
Distinguished from *P. heterophylla* by its greater size (up to 30 cm long), its thinner inflorescence and the sessile yellow red flowers.

Pitcairnia integrifolia Gawl., 1812 (Ill. 266)

(*integrifolia*(*us*) entire leaved, entire margined)
Syn. *P. graminifolia* hort., 1827; *P. decora* A. Dietr., 1847; *P. alta* Hassk., 1856; *P. graminea* Beer, 1857; *P. tenuis* Mez, 1896; *P. hartmannii* Mez, 1919

Pl over 1 m high with inflorescence
Fo several, dimorphic, the outer ones strongly spined, brown, the inner ones, foliage leaves
LS 1–2 cm long, 3–4 cm wide, dark brown
LB the foliage leaf to 70 cm long narrowed at the base but not stalked, about 1–1·5 cm wide and with the margins brown spined, the upper part 2–3 cm wide, tapering to a threadlike long point, smooth margins, upperside dull green, underside densely grey scaled
Sc round, green to brown, sparsely white woolly
I pyramidal, loosely compound, upright, 40–50 cm long, 20–30 cm wide. Rachis dull-red, part somewhat white woolly
SB longer than the internodes, the basal ones leaflike, the upper ones narrow-triangular, tapering to a thread-like point, weak white woolly, smooth margins
PB green, triangular, long pointed, much shorter than the basal ones, stalked, the sterile portion of the *raceme* 20–30 cm long, the basal ones branched more than once, spreading
FB small, insignificant, all somewhat shorter than the flower stalks, cutaneous, red, weak white woolly
Fl about 1 cm long stalked, 4 cm long without stalk, loosely supported, spirally arranged, red, part weak white woolly
Se lineal, pointed, 1·5 cm long, 4 mm wide, not keeled, bare to weak white woolly, bright red
Pe to 3·7 cm long, 8 mm wide, bare, bright red, with a ligula at the base
Ha Venezuela, Trinidad.

Dimorphic (heterophyllic) species

P. caldasiana Baker, 1889, Brazil; *P. camptocalyx* André, 1888, Colombia; *P. corallina* Linden and André, 1873, Colombia; *P. cuzcoensis* L. B. Smith, 1932, Peru; *P. echinata* Hook., 1853, Colombia; *P. ensifolia* Mez, 1894, Brazil; *P. exima* Mez, 1906, Peru; *P. fuertesii* Mez, 1913, Antilles; *P. kniphofioides*, L. B. Smith, 1935, Colombia; *P. longifolia*

(Morr.) Beer, 1857, Mexico; *P. megasepala* Baker, 1881, Colombia; *P. sprucei* Baker, 1881, Brazil, Colombia, Peru; *P. subpetiolata* Baker, 1881, Peru; *P. theae* Mez, 1896, Costa Rica; *P. tuerckheimii* Donn-Smith, 1888, Brazil, Colombia, Peru.

Trimorphic species

Pitcairnia trimorpha L. B. Smith, 1955
The specific name is given because of the 3 different leaf forms. The outer basal leaves are strongly spined, brown catophylls (*P. heterophylla*); the next are entire margined, elliptical, 5–8 cm long, brown when dry; the innermost are normal green leaves which fall during the dry period. Inflorescence to 1 m high; flowers in a simple seldom branched raceme. Found in Colombia.

Homomorphic species

Most *Pitcairnia* species are homomorphic leaved and produce only green leaves all alike, which either persist or drop at the dry season. Some examples are given below

Pitcairnia maidifolia (C. Morren) Descne., 1853–1854 (Fig. 84)
(*maidifolia*(*us*) maize leaved)
Syn. *Puya maidifolia* C. Morren, 1849; *P. maydifolia* Descne, 1851; *P. funkiana* Linden, 1850; *Pitcairnia funkiana* A. Dietr., 1851; *Neumannia maidifolia* C. Koch, 1856; *Phlomostachys densiflora* Beer, 1857; *Pitcairnia maisaifolia* hort. ex Beer, 1857; *P. zeifolia* C. Koch, 1857; *P. oerstediana* Mez, 1896; *P. macrocalyx* Hook., 1853; *Hepetis funkiana* Mez, 1896; *H. maidifolia* Mez, 1896; *H. oerstediana* Mez, 1896

Pl with short rhizome and loose leaved rosette, flowering to 1·3 m
LS narrow-oval, blackish, underside brown
LB at the base in a 5–10 cm long narrowed stalk, then long lanceolate-pointed, 50–100 cm long, 6–10 cm wide, like maize leaves, when old both sides bare, green
Sc upright, sturdy, large scaled
I a simple at first dense, later loose, 10–45 cm long raceme
SB the basal ones leaf like and exceeding the internodes, the upper ones scale shaped, oval, pointed
FB broad-oval, pointed, 3–3·5 cm long, about as long as the sepals, green or yellow, frequently green tipped
Fl at first upright, then spreading at the anthesis, 5–6 cm long, short stalked
Se asymmetrical, broad-oval to elliptical, blunt, 2·6–3 cm long, green or yellow bare
Pe broad-lineal, pointed, 5–6 cm long, white or greenish-white
Ha Costa Rica, Honduras, Venezuela, Colombia, Guayana, Surinam.

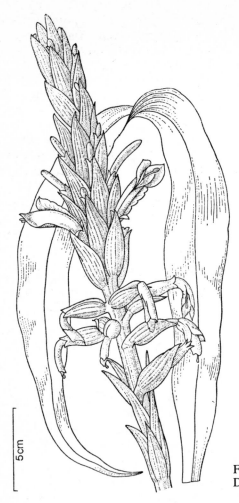

5cm

Fig. 84 *Pitcairnia   maidifolia* (Morr.) Descne; inflorescence with leaf.

Pitcairnia melanopoda L. B. Smith, 1963

(*melanopoda(us)* black footed; refers to the black rhizome)
One of the smallest and a most interesting species because of its grass-like leaves and black rhizome. In its natural habitat, it grows in moss and lichen cloud forests in pure sand; it is a very difficult plant to cultivate

Pl 1 m long, underground black rhizome formed from old, covering black leaf blades; to 30 cm high with inflorescence

Fo homomorphic, perishable, the sheaths remaining
LS narrow-oval, almost black, 4–6 cm long, underside grey scaled

LB narrow-lineal narrowing to a long threadlike tip, to 17 cm long, 5 mm wide at the base, at first white woolly, later becoming bare, margins smooth and channelled
Sc 15–17 cm long, upright, round, white woolly, especially at the nodes, about 2 mm thick
I simple, moderately loose at least at the base, about 7 cm long, with white woolly rachis
SB the basal ones longer than the internodes, the upper ones shorter, green, narrow-oval, long pointed, bare

FB the basal ones almost as long as the sepals, the upper ones shorter, narrow-triangular, 8–10 mm long, 3 mm wide at the base, green, bare
Fl upright, on short 2 mm long white woollen stalks
Se longish, rounded at the tip, 8 mm long, 2 mm wide, green, bare
Pe 1·4 cm long, the upper part 7 mm wide, pale yellow, without scales at the base
Ha North Peru, Pomacochas, 2,100 m (Province of Chachapayas).

Pitcairnia nigra (Carr.) André, 1888 (Plate 130, Ill. 267)
(*nigra(us)* black; because of the black-purple petals)
Syn. *Neumannia nigra* Carr., 1881; *Pitcairnia gravisiana* Wittm., 1889; *Hepetis nigra* Mez, 1896

Pl forming a long stem, 80–100 cm high
Fo a few, loosely arranged, trimorphic, the outer ones with short triangular sheaths (4 cm long, 3 cm wide) and narrow, spined blades (to 8 cm long), the inner leaves with long stalks about 30 cm long and elliptical-oval blades, entire, pointed (30–37 cm long, 10–14 cm wide), green, at first grey scaled, but soon becoming bare
Sc short, upright, 10–20 cm long
I simple-spiked, upright, 10–50 cm long, 6–7 cm wide; rachis dense brown woolly
SB densely overlapping, covering the scape, upright, elliptical, pointed, spined, red with green tip

Fb densely overlapping, the rachis fully covered, spirally arranged, elliptical, pointed, arched and flared, about 6·5 cm long, bare, bright red
Fl sessile, more or less 10 cm long
Se thin, nerved, 3·6 cm long, 8 mm wide, lanceolate, pointed, cutaneous, white, red at the tip
Pe 10 cm long, 1·5 cm wide, pointed, the lower part yellowish-white, the upper part dark purple-black, with a single grey scale, bare, with ligule at the base
St not exceeding the petals
Ov wholly superior
Ha Colombia, on trees (Ricaurte, 1,200 m) and Ecuador, on grass slopes (Tondapi).

Pitcairnia riparia Mez, 1913 (Ill. 268)
(*riparia(us)* growing by rivers or streams; the typical growth situation of the species)
The type plant was originally found by A. Weberbauer, a botanist, in the

valley of the River Tabaconas in the Province Jaën in northern Peru. The plant was again found, this time by Professor Rauh in 1970 near the Hacienda Udima at 2,000 m in the valley of the River Sãna in southern Peru

Pl with elongated, rooting, climbing stolons, ending in a few leaved rosette which finished growing when the inflorescence commences to form, and then new stolons grow, so that sympodial chain of shoots results; epiphytic
St 30–50 cm long, to 1 cm thick, densely set with firm, piercing, dark brown spine margined scale leaves which gradually become 5–10 cm long, spine tipped and smooth margined towards the rosette base
LB narrowed but not stalked above the sheath, 4–6 cm long, 1–1·5 cm wide, margins spined, upperside brown, underside grey scaled; the blade-like portion 30–40 cm long, 4–5 cm wide, long pointed, underside lightly grey scaled, upperside bare
Sc upright, about 30–40 cm long, round, brown-red, at first grey felted, later bare
I simple, loose raceme, 10–15 cm long
SB loosely spiral, always longer than the internodes, the basal ones with longer blade-tips, toothed, the upper ones smooth, grey felted, later bare
FB oval-lanceolate, long pointed, about as long as the sepals or somewhat shorter, not spined
Fl with the 1 cm long stalk to 6·5 cm long, nearly upright
Se triangular, long pointed, keeled, about 2 cm long, dark-red
Pe 6 cm long, shiny red, with scales
Ov perfectly superior
Ha northern Peru.

Pitcairnia scandens Ule, 1906

Very similar to *P. riparia*, the difference between the two being that the inflorescence bracts of *P. scadens* are shorter than the stem internodes, whereas the inflorescence bracts of *P. riparia* are longer than the stem internodes.

These two species are interesting bromeliads, because by being able to 'climb' trees by means of their rooting stolons, they may be classed as climbing plants.

Similar types are: *P. adscendens* L. B. Smith, 1943, Colombia (at Buenaventura, near the coast); *P. volubilis* L. B. Smith, 1954, Colombia (Narino: Las Mesas); *P. macranthera* Andrè, 1888; Colombia; *P. elongata* L. B. Smith, 1949, Colombia, Ecuador; *P. bakeri* (Andrè) Andrè ex Mez, 1896, Colombia, Ecuador; *P. oblanceolata* L. B. Smith, 1937, Costa Rica, Panama, Colombia.

Pitcairnia tabuliformis Linden, 1862 (Fig. 85)

(*tabuliformis* table, table shaped; because of the flat, outspread, circular table-like rosette of leaves)

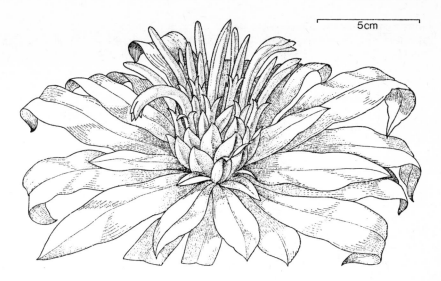

Fig. 85 *Pitcairnia tabuliformis* Linden; flowering rosette.

Pl stemless or with short stem
Fo homomorphic (all alike) perennial, in a flat, outspread small rosette
LB narrowed above the sheath but not stalked, longish-lanceolate, pointed tip, to 1·6 cm long and 3 cm wide, bare, entire margined
Sc absent
I nestling, many flowered, globose to capituliform, surrounded by small inner rosette leaves, to 6·5 cm in diameter
FB lanceolate, pointed tip, bare
Fl short with thick stalk, to 3·5 cm long, yellow, slightly zigomorphous
Se to 1·9 cm long, triangular-pointed, clearly keeled
Pe tongue-shaped, rounded at the tip
Ha southern Mexico (Chiapas)
CU this plant requires a winter rest period.

Pitcairnia xanthocalyx Mart., 1848 (Ill. 269, Fig. 81)
(*xanthocalyx* yellow calyx; because of the bright yellow coloured sepals)
Syn. *OP. flavescens* Bak. non hort., 1877; *P. sulphurea* C. Koch non André, 1857

Pl stemless or forming a short stem, multiplying strongly into a large clump
Fo homomorphic, perennial, narrowing above the sheath but not stalked, sword-shaped, to 1 m long and 2·5 cm wide, the margins entire or sparsely small spined, upperside bare, under- side scaled
Sc upright, shorter than the leaves
I for the most part simple, racemous, upright 50–70 cm long, many flowered, loose, cylindrical
SB upright, longer than the inter- nodes, triangular-oval, terminating in

a long thread-like tip, underside scaled
FB lineal-lanceolate, to 13 mm long, outspread or recurved
Fl long-stalked, upright in the bud stage, horizontally standing off at flowering time, downward curved after flowering
Se to 15 mm long, triangular-pointed, not keeled, frequently barbate at the tip, bright yellow
Pe pale yellow, 40–50 mm long, wide pointed, the stamens hardly surpassed, ligulate
Ov half inferior
Ha Antilles (Santa Domindo)
CU an easy free flowering plant to grow.

Puya Molina, 1782

The name *Puya* is derived from the native Chilean. The numerous species making up this genus vary from grass-like to cushion-building dwarf forms, with inflorescences to 10 cm high, through to the giants of the bromeliads up to 10 m high *Puya raimondii* from the Peruvian high Andes (Ill. 25).

Some Puyas are left with only the living branching stem after the flowers die. These branches unite forming very thick decumbent or upstanding stems, which root and give rise to new stems. In the course of years, these stems form massive groups of large rosettes, covering wide areas; the original old stem and branches gradually die off.

In the high Andes, Puyas are seldom found as isolated single plants. They usually occur en masse, often extending to areas of four square kilometres on great expanses of stony, grass covered slopes. For example, *P. roezlii*, a plant of steep rock walls, forms in central Peru (the Rimac Valley) at 2,000–3,000 m, a single plant community as a result of its great profusion.

The general characteristics of the genus are as follows.

Leaf sizes are very variable, in dense, outspreading or upright rosettes. Sheaths are clearly formed, white, stem clasping. The young blades are upright, with mature blades outspread to folded back, stiff, tapering to a spine tip, margined with very robust and long hooked spines at the tip and towards the base. The upperside is mostly bare with the underside more or less densely scaled.

Inflorescence is very variable in size (10 cm to 5 m), simple-racemed or spiked, frequently compound, often cylindrical (*P. fastuosa*, Ill. 271). With the compound inflorescence, the branches are either sterile at the top (Fig. 86) or fertile towards the tip (Fig. 87).

Primary bracts, flower bracts and sepals are usually flocculose, easily flaking off.

Flowers are often conspicuous, colours ranging through white, pale

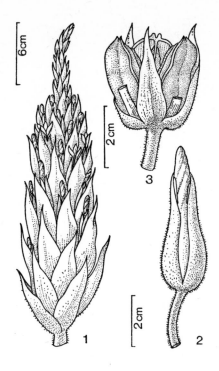

Fig. 86 *Puya raimondii* Harms: 1 part of the inflorescence; 2 flower bud; 3 outsprung open capule.

greenish-yellow, a remarkable sea-green, ice-blue or violet and are pollinated by humming birds and starlings. After flowering, the petals roll up into a spiral (Fig. 87)—a character of all the Puyas.

The ovary (Fig. 86, 3) has numerous small, narrow or wide winged seeds (Fig. 17 and 87, 2).

The genus comprises about 100 terrestrial species, their range being throughout the Andes of Colombia, Equador, Peru, Bolivia and northern Chile. In Peru, the genus has a wide vertical range from 800 m (*P. lanata* in northern Peru) to 4,800 m, to the beginning of the perpetual ice (Ill. 274; *P. rauhii, P. raimondii*). Indeed, they are the highest growing bromeliads and it is noteworthy that the largest bromeliad, *P. raimondii*, grows at 4,000–4,300 m, and that the 2-m tall *P. nivalis* occurs at the snow line (4,800 m) in the Sierra Nevada de Santa Martha in Colombia.

The upper habitats of the genus are the perpetually humid, cloudy and wind-swept, often with snow storms (the Paramos of Colombia and North Ecuador). The lower habitats are the periodically dry grass meadows of the Punas of Peru, Bolivia and northern Chile. Here are many Puyas wintering in frost at −20°C; often they are snow covered and

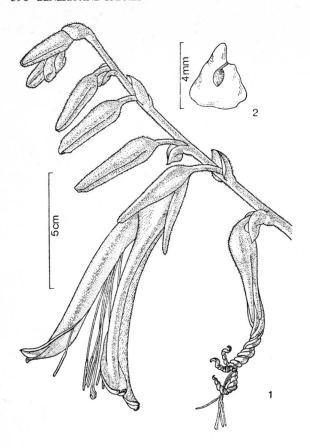

Fig. 87 *Puya ferraginea* (Ruiz and Pav.) L. B. Smith: 1 part of an inflorescence branch; 2 seed.

at other times are exposed to intense solar radiation and month long droughts. Many Puya species are found in Peru in the extremely dry inter-Andean valleys (the Mantaro and Apurimac Valleys of southern Peru) (Ill. 275). All the species are able to endure the extremes of climate of their habitats by reason of their strong xeromorphic construction. The continued existence of *P. raimondii* is very seriously threatened because of its destruction by Indian shepherds who, since there is no woody material available in the grass lands (or Punas) use the Puya for their camp fires — its stem contains inflammable resins.

Cultivation of Puyas is successfully carried out by seed raising; germination and growth are relatively quick, but they do not flower for years.

Puya alpestris (Poepp. and Endl.) Gay, 1853 (Fig. 88)
(*alpestris* belong to the Alps; meaning the Chilean Andes)
Syn. *Pitcairnia alpestris* Bailey, 1916; *Puya whitei* Hook f., 1868;
Pourretia caerulea Miero, 1826; *Pitcairnia caerulea* Baker, 1889

Pl forming a stem or trunk, becoming very large
Fo in a dense rosette, longer than 1 m, 2–2·5 cm wide, the margins with about 5 mm long spines, underside densely white scaled
I very large, loose-panicled, from numerous dense compound panicles, those at the tip sterile and carrying bright rose bracts
FB oval, short pointed, bare or hairy, thin, delicately cutaneous, shorter than the sepals
Fl to 5 cm long with 7 mm stalk, deep dark blue with a green-metallic shimmer, purple-red after flowering

Fig. 88 *Puya alpestris* (Poepp. and Endl.) Gay; habit and part of the inflorescence.

Se lineal, rounded to pointed, about 2·3 cm long, greenish, bare or hairy
Pe to 4·5 cm long, longish lanceolate
with blunt tip
St with bright orange-yellow anthers
Ha on dry hills of central Chile.

Puya caerulea Lindl., 1840 (Ill. 250)
(*caerulea(us)* blue; because of the blue colour)
Syn. *Pitcairnia caerulea* Benth, 1883; *Pourretia rubricaulis* Miero, 1826; *Puya rubricaulis* Steud., 1840

Pl large, robust
Fo numerous in a dense rosette, to 60 cm long, the margins with stout hooked spines, underside farinaceous
Sc sturdy, 1–1·5 cm long
I many flowered, densely panicled, long pyramidal to bunched, to 40 cm long and 15 cm thick, with dense woolly axil when young; branches to 15 cm long, upright fertile to the tip
FB lanceolate, long pointed, underside bare or hairy, shorter than the sepals, however, longer than the flower stalks
Fl stalk to 6 mm long, 5·3 cm long, deep dark blue
Se to 3·4 cm long, lineal, rounded but short spine tipped, underside densely white haired
Pe to 5 cm long, lanceolate
Ha Chile.

Puya chilensis Molina, 1782 (Plate 131, Fig. 89)
(*chilensis* endemic to Chile)
Syn. *Pitcairnia chilensis* Lodd., 1830; *Puya suberosa* Molina, 1810; *Pitcairnia coarctata* Pers., 1805; *Pourretia coarctata* Ruiz et Pav., 1802; *Puya coarctata* Fisch., 1852; *Renealmia ramosa-lutea* Tuille, 1725
The plant shown (Plate 68) is distinguished from the type plant by its strong, woolly haired flower bracts

Pl more than 1 m tall, with distinct candelabra-like branched stem
Fo numerous, in a dense rosette, 80–100 cm long, to 5 cm wide, the margins with very stout, 1 cm long hooked spines
I compound, loosely panicled, 1–1·5 m long; inflorescence branches racemose, whith sterile bracts at the tip;
axis red brown scaled
FB elliptical to pointed, to 3·5 cm long, bare, shorter than the sepals
Fl yellow or green, to 7·5 cm long, sessile or 1·5 cm long stalked
Se to 3·5 cm long, lineal, greenish, bare or hairy, rounded at the tip
Pe to 6·5 cm long
Ha Mountains of central Chile.

Puya fastuosa Mez, 1906 (Ill. 270 and 271)
(*fastuosa(us)* very beautiful; because of the conspicuous inflorescence)

Pl very sturdy, blooming 1·5–2 m high, forming a stem, much branched
and forming groups
Fo numerous, in a dense rosette with

20 cm

Fig. 89 *Puya chilensis* Mol; habit and part of the inflorescence.

large, white, stem clasping sheaths, the inner ones, when young rigid upright, the outer ones, horizontal to bent back when old, 40–60 cm long, 4 cm wide at the base, densely set with 6 mm long, hooked bent spines

Sc robust, very thick, 50–70 cm long with yellowish-white easily removed wool

I compound, densely cylindrical to club-shaped, 30–60 cm long and to 15 cm thick; all the parts are enclosed in a thick, yellowish woollen mantle, out of which only the flowers and the brown, parchment-like tips of the primary bracts show

PB very large, oval at the base, narrowing to a long spined tip, margins toothed, dense woolly at the base becoming bare towards the tip, much longer than the flower racemes; these few to 10 flowered, to 8 cm long, completely covered with wool, fertile to the tip, 4 cm long

FB to 3 cm long, firm, broad elliptical, leather brown, densely woolly haired, exceeding the sepals

Fl 3·7 cm long on thick, loose haired

stalks about 7 mm long

Se to 2 cm long, elliptical with rounded tip, thick woolly haired on the outer side

Pe light green, to 1·7 cm long, wide tongue shaped with bent back blunt tip, hardly showing through the wool

of the inflorescence

Ha Ecuador to Peru, though a wide ranging Puya; in northern Peru it grows in bogs; in southern Peru as a typical riverside plant at altitudes of 2,800–3,500 m.

Puya ferruginea (Ruiz and Pav.) L. B. Smith (Fig. 87)

(*ferruginea(us)* rusty, red brown; because of the rust coloured hair of the flower axil and sepals)

Syn. *Pitcairnia ferruginea* Ruiz et Pav., 1802; *Pourretia ferruginea* Spreng., 1825; *Pitcairnia asterotricha* Poepp. et Endl., 1838; *Puya grandiflora* Hook., 1861; *Pitcairnia imperialis* Harms, 1929; *Pitcairnia echinotricha* Bak. et André, 1889

Has a wide range in Peru and is not variable. It forms both stemless, relatively small rosettes, however also developing a sturdy trunk; the inflorescence is either simple-racemed or much branched, the flowers yellowish-green or violet. It is distinguished from the similar *P. mirabilis* mainly by the branching and the thick brown woolly inflorescence

Pl stemless or forming a stem

Fo humerous, in a dense outspread rosette, to 50 cm long and 5·5 cm wide, margined with 15 mm long black spines, underside black scaled

Sc upright, sturdy, densely rust brown scaled, later becoming bare

I many flowered, loosely branched, in outline broad-pyramidal, more than 1 m long and to 50 cm in diameter; axils dense red-brown haired; flower racemes 10 to 30 flowered, upright to spreading

FB upright, broad-oval, pointed or oval, 6 cm long, shorter than the sepals

Fl stalked, to 14 cm long

Se to 4·7 cm long, thick, firm, rounded, underside densely rust red haired

Pe white to pale violet, long tongue-shaped rounded at the tip, spirally rolled up after flowering

Ha Peru and Bolivia, from 1,000–3,000 m, clinging to rocks

CU it flowers freely in cultivation but frequently dies afterwards.

Puya laxa L. B. Smith, 1958 (Ill. 276)

(*laxa(us)* loose; because of the loosely arranged leaves)

An interesting species, differing from all other Andean Puyas, because of its succulent, very narrow, almost bristly scaled leaves forming not too dense rosettes, arranged along the thin little stems

Pl with inflorescence to 1 m high, forming a stem 25 cm long branching

from the base and group forming

Fo loose, spirally arranged round the

thin stem, the basal ones outspread
LS broad-triangular, 3–4 cm long, to 6 cm wide at the base, shiny dark-brown on the outside, shiny light brown on the innerside, bare; upper half of the margins spined and without spines at the base
LB nearly succulent, narrow, ending in a long threadlike tip, to 60 cm long, 2 cm wide above the sheath, underside densely covered with white, spreading, to 2 mm long scaly hairs, upperside thick white felted; margins bent up with 5 mm long, upward bent brown spines; tip not spined
Sc upright, about 30 cm long, 8 mm thick, round, green, white scaled, soon becoming bare
I loose-compound with 4 to 6 spreading racemes, to 70 cm long, narrow-pyramidal, to 40 cm wide
SB few, spaced apart, upright with leaflike blade, longer than the internodes, white scaled
PB oval-triangular, long pointed, to

3·5 cm long, soon drying, grey scaled; margins entire
Racemes 30–35 cm long, many flowered; rachis white scaled, later becoming bare
FB loose, spirally arranged, 2 cm long, 7 mm wide, white scaled, thin, soon drying and then nerved
Fl 4 cm long with stalk, secund on one side; stalk 8–9 mm long, white woolly
Se 1·7 cm long, 7 mm wide, lanceolate, green to pale red, white scaled
Pe 3 cm long, 5 mm wide at the base widening to 1 cm towards the tip, dark violet with, on the outside a green centre stripe; tip rounded and flared, scattered with white scales
St locked in the flower
Ov 6 mm long
Ha Argentina
CU extraordinarily tough, resistant to a dry atmosphere and because of its small size especially suited to being grown along with cacti. It flowers readily in cultivation.

Puya lanata (H.B.K.) Schult. f., 1830 (Plate 132)
(*lanata(us)* woolly; because of the woolly haired inflorescence)
Syn. *Pourretia lanata* H.B.K., 1815; *Pitcairnia lanata* Diedr., 1820

Pl forming a stem and grouping, flowering to over 2 m high
Fo to 80 cm long, 4 cm wide at the base, silver grey to reddish, both sides densely scaled, underside grooved, margined with hard, hooked spines
Sc 1–1·5 m long, with 2–5 cm thick woolly haried axis
I simple, dense cylindrical, 70–100 cm long and 10 cm thick with strong woolly haired axis
SB densely set, overlapping each other, longer than the internodes, to 12 cm long and 4 cm wide, long pointed, with finely toothed margins, dense

woolly haired when young, becoming bare later, grey or green
FB long-triangular, pointed, about 6 cm long at the base, 1·5–1·8 cm wide, terminating in a sharp, long, recurved tip, underside dense white woolly, upperside light to leather brown and nearly bare, as long or longer than the sepals
Fl very short, stalked, to 8 cm long
Se 4·5 cm long, the back keeled, pale green, dense white woolly, free to the base
Pe to 8 cm long, pale greenish to yellow, 2 cm wide, recurved at the tip

St with yellow filaments and orange coloured anthers
Ov 1 cm long
Fr about 3 cm long, shorter than the flower bracts
Ha only in North Peru and not (as incorrectly stated by Metz) in Colombia. It was found in 1970 in the Valley of the River Jequetepeque, in massive stands, between 1,300 and 2,300 m high.

Puya medica L. B. Smith, 1953 (Plate 133, Ill. 251)
(*medica*(*us*) curative; the plant is used medicinally by the natives for treating pneumonia)

Pl short stemmed, to 20 cm high with inflorescence. Rosettes forming loose cushions
Fo numerous, forming a small low rosette from about 15 cm in diameter
LS to 2·5 cm long, 5 cm wide, bare, the inner and outer faces white at the base, brown above
LB to 12 cm long, 1·3 cm wide above the sheath, narrow-triangular with terminating spine, both sides densely grey scaled; margins set with brown spines about 2 mm long
Sc very short, to 4 cm long, upright, 7 mm thick
I simple-spike, dense, cylindrical, about 10 cm long, 2 cm thick; rachis green, grey scaled, covered by the flower bracts
SB upright, longer than the internodes, similar to the leaves
FB spirally arranged, triangular-oval, spine tipped, 2·5 cm long, 2 cm wide at the base, rose coloured, grey scaled, margins irregularly spined
Fl almost sessile, upright, 2·2 cm long
Se oval, pointed, 1·2 cm long, 5 mm wide, greenish-white, grey scaled, shorter than the flower bracts
Pe dark blue, 2 cm long, 7 cm wide, tip widely rounded and with small notch
St not exceeding the flower
Ov 5 mm long
Ha Northern Peru (Department of Cajamarca), on rocks.

Puya mirabilis (Mez) L. B. Smith, 1968 (Ill. 277 and 278)
(*mirabilis* wonderful; because of the large flowers)
Syn. *Pitcairnia mirabilis* Mez, 1906; *Pitcairnia mirabilis* var. *tucumana* Castelanos, 1929
Vegetatively similar to *P. ferruginea* and distinguished from it by the absent rust-red hair and the simple raceme

Pl stemless, with inflorescence exceeding 1 m high
Fo numerous forming a loose rosette
LS 3–4 cm long, 3–4 cm wide, white to brownish, with finely toothed margins
LB 60–70 cm long, narrow lineal, long pointed, 1–1·5 cm wide at the base, underside appressed grey scaled, upper side grey scaled only on the lower half; spined margins
Sc sturdy, upright, round, green, to 90 cm long, 1 cm thick
I upright, simple-racemed, loose, about 50 cm long, 16 cm thick

SB similar to the leaves, the basal ones with longer recurved blades, the upper ones with shorter ones, upright, longer than the internodes, green, soon drying; spined margins

FB loose spirally arranged, outspread, soon drying, shorter than the sepals, 3–4 cm long, triangular-long pointed; spined margins

Fl 10–11 cm long, 1 cm long stalked

Se 5 cm long, 1 cm wide, green, grey scaled

Pe 10 cm long, 2 cm wide, greenish to white, spreading at the tip

Ha Argentine and Bolivia

CU a very fine, small free flowering plant.

Puya nana Wittm., 1916 (Plate 134)
(*nana*(*us*) small or dwarf; because of the small growths)

Pl stemless or short stemmed

Fo in a dense, flat outspread rosette, narrow-lanceolate, to 60 cm long and 2·5 cm wide, the margins set with sharp, firm brown spines, upper side bare, underside densely scaled

Sc short or absent; the inflorescence is formed into a compound head and nestles in the centre of the rosette as in *Nidularium*.

I a short, compound spike bent one way

SB similar to the leaves but shorter

PB similar to the inflorescence bracts but narrower, densely brownish haired

FB nearly as long as the sepals, densely brown scaled

Fl short stalked, to 4 cm long, bluish green hardly opening

Se longish-lanceolate, to 4 cm long, 6 mm wide at the base, keeled, terminating in a sharp tip, nearly as long as the petals, underside densely haired

Pe lanceolate, rounded, 4 cm long, 9 mm wide

St a little longer than the petals

Ha Bolivia, near Samaipata.

Puya raimondii Harms, 1928 (Ill. 25, 272 and 273, Fig. 86)
(*raimondii* after Raimondi, an Italian botanist, who studied the Peruvian vegetation)

Syn. *Pourretia gigantea* Raimondi 1874

This gigantic plant dies without vegetative offspring after the ripening of its seeds. The age of a plant when blooming cannot be estimated for certain, the growth is extremely slow, but mature plants have been estimated to be at least 80–100 years old. Although millions of small, winged seeds are released (Fig. 17), and seed raising is easy in cultivation, this does not happen in its habitat of stoney grassy slopes. It appears that only few suitable germinating conditions are found. Rauh covered all the area and found only a few young plants. It would appear that local

Indians with their herds of animals are at least partly responsible for this state of affairs, breaking down the plants before they have seeded and burning them for warmth. In former times *P. raimondii* had a larger range within Peru than it has today. It is now found only in small groups.

Pl covered with long old leaf sheaths, to 50 cm thick, 2–5 m high, unbranched stem, blooming to 10–12 m high
Fo at the head of the stem in a dense, globular rosette, stiff and inflexible, standing out from all sides, 1–2 m long, 6 cm wide at the base, terminating in a sharp point, upperside shiny green or reddish (under the sun's influence), when old, changing to straw coloured, underside densely appressed scaled, margined with very stiff, long, brown, hooked 1·5 cm long spines after flowering, dying and dropping off
SC very thick, 0·50–2·50 m long, densely set with inflorescence bracts, the basal ones similar to the leaves, with drooping blades, the upper ones scale shaped, bladeless, bent back, the shaft completely covered
I club-shaped, 3–5 m long, 50 cm in diameter at the base, tapering towards the top, compound with many racemed component inflorescences; these to 30 cm long, downwards curved, densely flowered with long, sterile ends and long white woolly axils
FB to 6 cm long, elliptical-oval, long pointed, entire, much longer than the sepals, haired at the base, bare towards the tip
Fl to 5·5 cm long with 7 mm long, thick, hairy stalk
Se to 4 cm long, delicately cutaneous, elliptical-pointed, hairy at the base, clearly nerved
Pe about 5 cm long, yellowish-green, tongue shaped (ligulate)
Ha Central and southern Peru (Cordilla Negra, Blanca and near Lampa at Lake Titicaca) at 3,800–4,300 m.

Puya gigas André, 1881
This very large puya—6–10 m high is found in the Paramo peat-moss bogs of Colombia (Cordilla Pasto).

In contrast to *P. raimondii*, it is persistent in its growth, and the great rosettes form loose cushions.

Besides these giant bromeliads, there is also the following range of dwarf plants without doubt of interest to collectors and growers which can also be grown with cactus.

Puya humilis Mez, 1896
This blooms in the high Andes of Bolivia when only 15 cm high and forms loose-to-dense cushions. Leaves are up to 20 cm long and 7 mm wide, with spined margins, underside grey scaled; inflorescence upright, cone shaped, few flowered, to 6 cm long with blue or violet flowers.

Fig. 90 *Puya volcanica* Cast; flowering plant.

Puya volcanica Cast., 1921 (Fig. 90)
An outstanding decorative plant growing between 2,000 and 4,000 m in
Argentina. It reaches 30 cm in height when in bloom. The numerous
20–30 cm long, 1·5–2 cm wide leaves build up into a dense, flat outspread
rosette, from the centre of which the 15 cm long stalked, cone shaped,
simple-spiked inflorescence emerges. Flowers are violet, on very short
stalks.

Further small species are: *Puya tuberosa* Mez, 1896, flowering to 30 cm
high in Peru; *Puya paupera* Mez, 1906, flowering to 22 cm high in Bolivia;
Puya brachystachya (Bak.) Mez, 1896, Syn. *Pitcairnia brachystachya*
Baker, 1889, flowering to 40 cm high in Colombia; *Puya lineata* Mez,
1896, flowering to 40 cm high in Colombia.

Bibliography

André Eduard *Bromeliaceae Andreanae; Discription et Historie des Broméliacées recoltées dans la Colombie, L'Ecuador et la Venezuela* Paris, 1889.

Baker, J. G. *Handbook of the Bromeliaceae* London, 1889.

Beer, J. G. *Die Familie der Bromeliaceen* Vienna, 1857.

Bromeliad Society Bulletin. Vols. I–XXVIII. *Journal of the American Bromeliad Society*, Los Angeles, California.

Castellanos, A. *Bromeliaceae.* In *Genera et species plantarum Argentinarum*, Vol. III. 1945.

Chodat, R., et Vischer, W. *Les Broméliacées.* In *La Végétation de Paraguay. Bull. de la Soc. Botan. de Genève* Vol. VIII. 1916.

Foster Mulford B. *Brazil, Orchid of the Tropics*, 1945.

Gilmartin, A. J. *The Bromeliaceae of Ecuador* 1972.

Gilmartin, A. J. *Las Bromeliacias de Honduras*, Ceiba 11(2), 1965.

Harms, H. *Bromeliaceae.* In (Engler, A., and Panttl, K., eds) *Die natürlichen Pflanzenfamilien* Vol. 15a, Leipzig, 2, Edition 1930.

Mez, C. *Bromeliaceae.* In (Engler, A., ed) *Das Pflanzenreich*, Vol. IV, 32. 1956 *Physiologische Bromeliaceen-Studien* I: *Die Wasserökonomie der extrem atmosphärischen Tillandsien. Pringheims Jahrbücher f. wiss Botanik* Vol. 40, 1904.

Morren, E. Various articles and first time descriptions of bromeliads in *La Belgique horticole* Vol. XX (1870) to Vol. XXXV (1885).

Picado, C. *Les Bromeliacées épiphytes, considérées comme milieu biologique. Bull. scientifique de France et Belgique* Series 7, Vol. 47. 1913.

Rauh, W. *Bromelienstudien* Parts I–VI, Heidelberg.

Richter, W. *Anzucht und Kultur der Bromelien mit besonderer Berücksichtigung den für den Handel wichtigsten Arten. Grundlagen und Fortschritte im Gartenund Weinbau*, Part 76. Stuttgart, 1950 (out of print).

Richter, W. *Zimmerpflanzen von heute und morgen: Bromeliaceen* Dresden 1962.

Rohweder, O. *Die Farinosae in der Vegetation von El Salvador* Hamburg 1956.

Schimper, A. F. W. *Botanische Mitteilungen aus den Tropen*, Part 1 *Die Wechselbeziehungen zwischen Pflanzen und Ameisen* Jena, 1888: Part 2 *Die epiphytische Vegetation Amerikas* Jena, 1888.

Smith, L. B. *The Bromeliaceae of Brazil.* In *Smithsonian Miscellaneous Collections* Vol. 126. 1955. (Reprinted as *Contribution from Reed Herbarium* No. XXVI, 1977).

Smith, L. B. *The Bromeliaceae of Bolivia.* In *Rhodora* Vol. 71. 1969.

Smith, L. B. *The Bromeliacea of Columbia.* In *Contributions from the US-National Herbarium* Vol. 33. 1957. (Reprinted in *Contribution from Reed Herbarium* No. XXVI, 1977).

Smith, L. B. *The Bromeliaceae of the Guayana Highland.* In *Memoirs of the New-York Botanical Garden,* 1967.

Smith, L. B. *The Bromeliaceae of Panama.* In *The Flora of Guatemala, Fil ediana* Vol. 24, Part 1, 1958.

Smith, L. B. *The Bromeliaceae of Peru.* In (F. Macbride, ed) *Flora of Peru,* Part 1, No. 3. Field Museum of Natural History, Chicago, 1936.

Smith, L. B. *Notes on the Bromeliaceae,* No. I–XXXIII in *Phytologia* (1953–1971) (reprinted).

Smith, L. B. *Studies in the Bromeliaceae* Nos. I–XVII in *Contributions from the Gray Herbarium of Harvard University,* in the Vols. 1930–1946 (Parts I–XIV) and *Contributions from the US-National Herbarium* Vol. 29, 1949–54 (Parts XV–XVIII) (reprinted).

Smith, L. B. *Flora Neotropica.* Monograph No. 14 (Pitcairnioideae) (Bromeliaceae) Vol. 1, 1974.

Smith, L. B. *Flora Neotropica.* Monograph No. 14 (Tillandsioideae) (Bromeliaceae) Vol. 2. 1977. Vol. 3 in course of preparation.

Smith, L. B. *Flora de Venezuela* (Bromeliaceae). 1971.

Smith, L. B. and Pittendrigh, C. S. *Flora of Trinidad and Tobago* Chicago, 1967.

Standley, P. C. *Flora of Costa Rica, Botanical Series* Vol. 28, Part 1. Field Museum of Natural History, Chicago, 1937.

Tietze, M. *Physiologische Bromeliaceen-Studien* III *Die Entwicklung der wasseraufnehmenden Bromeliaceen-Trichome. Zeitschr. f. Naturw. Halle* Vol. 78, 1906.

Index

As well as reference to page numbers, this index indicates line illustrations (Fig.), black and white photographs (Ill.) and the colour plates (Pl.). Species names in parentheses are synonyms. In addition, entries of a more general nature and which refer principally to Part I are also included.